THE AUTOMOBILE AND THE ENVIRONMENT

MIT Press Series in Transportation Studies

Marvin L. Manheim, editor

Center for Transportation Studies, MIT

1. *The Automobile and the Environment: An International Perspective*, edited by Ralph Gakenheimer, 1978

THE AUTOMOBILE AND THE ENVIRONMENT:
An International Perspective

Prepared by the Organisation for Economic
Co-operation and Development

Edited by
Ralph Gakenheimer

The MIT Press
Cambridge, Massachusetts, and London, England

This book was set in IBM Univers by The Blue Ridge Group, Ltd. and printed and bound by Halliday Lithograph Corporation in the United States of America.

Library of Congress Cataloging in Publication Data

Organization for Economic Cooperation and Development.
The automobile and the environment.

(MIT Press series in transportation studies ; 1)
Includes bibliographies.
1. Urban transportation—Environmental aspects—Addresses, essays, lectures. 2. Automobiles—Environmental aspects—Addresses, essays, lectures.
I. Gakenheimer, Ralph A. II. Title. III. Series.
HE305.073 1978 388.4 78-2493
ISBN 0-262-07070-7

CONTENTS

VI

SERIES FOREWORD

The field of transportation has emerged as a recognized profession only in the last ten years, although transportation issues have been important throughout history. Today, more and more government agencies, universities, researchers, consultants, and private industry groups are becoming truly multimodal in their orientations, and specialists of many different disciplines and professions are working on multidisciplinary approaches to complex transportation issues.

The central role of transportation in our world today and its recent professional status have led The MIT Press and the MIT Center for Transportation Studies to establish The MIT Press Series in Transportation Studies. The series will present works representing the broad spectrum of transportation concerns. Some volumes will report significant new research, while others will give analyses of specific policy, planning, or management issues. Still others will show the interaction between research and policy. Contributions will be drawn from the worldwide transportation community.

This book, the first in the series, focuses on a pressing policy issue of concern in many countries—the automobile in the urban environment—and draws upon several bodies of research. Thus, it exemplifies one of the styles of work that will be presented in this series.

Marvin L. Manheim

PREFACE

Here is a set of papers that organize the data on transport-related environmental problems and the options for policy response better than any we have seen. They do so at the international level, with attention to circumstances in the developing as well as in the postindustrialized countries. Their flavor is one of hope; they implicitly share the view that the problems are solvable if governments address them firmly. The approaches examined are highly practical, most of them being relatively undemanding of institutional adjustments or changes in social values. Showing their independence from the hardware-oriented approaches prominent in the 1950s and 1960s, the authors share limited faith in technological innovation for the solution of transportation's environmental problems.

First versions of these papers were written in 1972–1974 for the Environment Directorate of the Organisation for Economic Co-operation and Development (OECD). OECD has since begun to tackle the problems of transportation and the environment in several productive ways. An important conference, which displayed case evidence on urban traffic policies, was held in 1975, and the proceedings were published as *Better Towns with Less Traffic* (Paris, 1975). The most promising cases were then evaluated by the Environment Directorate, and a first report was published as *Evaluation of Traffic Policies for the Improvement of the Urban Environment* (Paris, 1976). But the papers in this volume have not been superseded. They are the source papers that have formed the basis for further surveys and analyses of particular cases.

By giving equal weight to data from several nations, these papers help to remove the nationalist myopia from transportation studies. While the institutions and values particular to a country set some limits on the options that can be taken, the worldwide awakening of environmental concern has laid the way for exchanging experiences. Environment as a topic turns the American reader in particular outward toward the experiences of other countries, especially those of Western

Europe, where observations of the problems and attempts at their solution are the most advanced in the world. Therefore this book serves several important purposes by providing both structured data on the problems and access to foreign experiences.

The paper on "Urban Public Transport" is an excellent review of initiatives, primarily in Western Europe and the United States, in operations, planning, and technology. Originally prepared in 1972, the monograph was revised for this volume by Robert Burco, now Director of Transportation in the Department of Transportation of the U.S. State of Oregon. He is the first to note that major new technological innovation is unlikely and, in so doing, sets the mood for the other papers. His more promising observations are on reorganization of existing technologies and advances in operations management. Burco's final recommendations are a difficult charge for governments that would facilitate substantial innovation. He points out that it is necessary for governments courageously to recognize the need for technological change, rather than stay on comfortable ground with policies that do not require it. He points out the need to relate government to industry in order to stimulate innovations that government will be committed to use by prior agreement. He stresses the need to coordinate comprehensive studies with expectations of technological change so that secondary and tertiary consequences may be anticipated.

J. Michael Thomson's "Methods of Traffic Limitation in Urban Areas" makes an important contribution just in terms of the typology it creates to organize the many actions under the rubric "traffic limitation." The basic categories are "restraint" (by which trips are obstructed), "restriction" (by which they are discouraged), and "avoidance" (by which travel desires are kept from arising in the first place). In continuing with subcategories, physical and legal, direct and indirect, the author creates groupings of similar techniques that form bases for choice. His discussion is a complete review of experiments with these methods up to 1972. Since that time there have been many more trials but few generically new experiences. The recent experiment with an auto-restricted zone in Singapore is an exception. The OECD is planning to hold a seminar on all new experiments in 1978.

There is a minimum of overlap between these first two papers because Thomson deals mostly with private automobiles, whereas Burco deals primarily with public transit. In addition, Thomson deals with the limitation of traffic, whereas Burco deals primarily with raising the level of service throughout the system, though at the same time inducing environmental advantages.

It should be noted as an observation on the two papers that no country has the lead in testing possible applicable solutions. Only the pool of experience from all of them produces a useful experience. Thomson concludes that traffic limitation methods are available and proven for any normal set of objectives that might be pursued.

"Implications for the Planning Process," by Marvin L. Manheim and John H. Suhrbier, is built on years of intensive research on incorporating environmental considerations in project decisions in the United States. More than the other papers, this one is built on the contributions of a single country, but the special concern these matters have received in the United States over the last decade justifies this concentration. The methods proposed in this paper may require considerable adaptation for use in other countries, where concepts such as compensation and social equity are viewed differently.

The Manheim-Suhrbier piece emphasizes project evaluation and the relationship between short-range actions and longer-range concerns. The latter is an important contribution to the book, since the focus of the other parts is very much on the short-range environmental problem. It may be very important to ask what kind of cities these urgent, short-run policies are leading to.

Wilfred Owen's paper is the only one that deals extensively with the developing countries. It is based on rich arrays of data resulting from the painstaking data gathering for which his work is notable. These countries present different problems than the others in that they are less responsive, in most cases, to environmental problems but, at the same time, are suffering much more rapid disruptions because of increasing population and automobile use. In developing countries the problems of environmental amenity and development are to some extent in conflict in seeking the attention of public policy. Because these countries are in phases of rapid growth, their choices of options are important. And the variety of available options is wide; after all, as Owen points out, countries can be affluent with few automobiles and poor with many automobiles. Nonetheless, connections between this section and the rest of the book are evident. Calling upon his broad background, Owen moves easily between the developing and postindustrialized countries, seeking suggestions for solutions in one from the other.

The volume ends where the investigation of the Environment Committee began, "Automotive Air Pollution and Noise." This report of the Ad Hoc Group includes a wealth of data on the impingement of automobiles on air quality and noise levels and on the effectiveness of particular actions to reduce them. These analytic data provide a valuable resource for planning that has not previously been available to the American reader. The categories of actions are quite different from those of Thomson, because they are oriented to environmental issues rather than to the broader objectives his work addresses.

These papers are all of vital significance today. The variety of dates of first publication, however, requires some explanation of the editorial procedures used. Robert Burco's paper, written in 1972, has been extensively updated by the author, so that the version included here is current. We are most grateful to him for taking time from his present duties to undertake this work. The Manheim-Suhrbier paper (1974) required virtually no updating since the process issues it addresses have suffered no obsolescence over the intervening time. Wilfred Owen did some editing on his paper (1973), and we further revised it together with the papers of Thomson (1972) and the Ad Hoc Group (1972). Our way of managing these papers was as follows. We did not update the individual cases reported, since the fact that all of them could not be updated would leave different time cut-offs for reporting in the same paper. At the same time, the text was adjusted to read logically for 1977. That is to say, references to processes in transition at the time of writing were deleted and data mentioned as current at the time of writing were marked with the date in the text. When such data were meant to suggest a current condition that has changed in the interim, the data were deleted. As a result, the papers are no longer confusing to today's reader who is aware that they were written at the dates mentioned or whose recollection of the dates has lapsed. There have been additional cases of interest since the completion of these reports, but very few events that have measurably changed the perspective conveyed by these papers.

Ralph Gakenheimer

The editor wishes to acknowledge the assistance of David Damm and Ann Rappaport in the preparation of these papers.

THE AUTOMOBILE AND THE ENVIRONMENT

ENVIRONMENTAL IMPLICATIONS OF OPTIONS IN URBAN MOBILITY: Observations of the Environment Committee of OECD

INTRODUCTION

Between 1960 and 1970, while the number of people in OECD countries increased less than 11 percent, the number of motor vehicles almost doubled. In 1972 there were 230 million automobiles and trucks registered in the 23 OECD countries—one for every 2.3 people in North America and Australia, and one for every 10 people in Europe and Japan. In the single decade of the sixties their number increased by 100 million. Almost 46 percent of the gain was in Europe, 37 percent in North America and 17 percent in Japan and Australia.

The outlook is for still more motor vehicles, and particularly private cars. One projection suggests that the 1972 automobile total of 200 million units in OECD countries may grow by another 120 million, reaching 320 million cars by 1985. This would be more than three times the number that was in operation ten years ago. Naturally, such projections are beset by uncertainties, depending as they do on assumptions about economic growth, popular demand and the various factors that will cause "saturation" in different countries. What cannot be denied is that problems related to the use, numbers and design of automobiles are likely to become increasingly acute in many areas unless action is taken.

The impact of these trends is being magnified by the fact that automobiles are heavily concentrated in metropolitan areas. While throughout the OECD territory as a whole there are roughly five cars per square kilometer, in densely settled cities like Paris, Tokyo or Chicago the number of cars per square kilometer is 200 to 300 times higher. Almost two-thirds of all cars are registered in urban agglomerations of 100,000 people or more. If present urbanisation trends continue, the outlook is for further concentration of automobiles and for an increasing number of heavily congested towns.

This is the background against which the Environment Committee launched in March 1971 its inquiry into the public policy implications of the impact of motor vehicles. Underlying its action was a desire to answer certain questions that pose a major predicament for all OECD governments: How is the conflict between the advantages of motorisation and its detrimental side effects to be resolved? What is the balance to be struck between the desire for private transportation and other societal needs? Above all, what innovations in policy are needed to create an environment in which man and the automobile can live in harmony?

1 THE WIDER CONTEXT OF AUTOMOBILE POLICIES

Pollution control, of course, is an important aspect of a much wider problem. But even if all motor vehicles were to be made virtually pollution-free and noiseless, their sheer numbers would still generate a tremendous impact on the environment and on the quality of our lives. Concern with eliminating automotive emissions should not be allowed to divert attention from other, equally urgent problems.

THE RISING DEMAND FOR FUEL

One such problem is the rising demand for motor fuel. Road transport in 1970 was consuming 21 percent of world oil production; passenger cars alone accounted for 12 percent. Based on projections of vehicle registrations, and assuming present fuel consumption patterns, the fuel requirements of the world automobile fleet may double by 1985. Certain emission control devices may increase these requirements still further.

Such trends have serious implications. Rising world demand is likely to bid up oil prices and the combination of greater demand at higher prices may severely affect the balance of payments of many nations. According to one forecast, the cost of oil imports will increase by 1980 from $23 billion to $31 billion in Europe; from $18 to $24 billion in the United States; and from $12 to $16 billion in Japan. How to minimise this drain on foreign exchange will become a major challenge to the governments of many countries.

These matters form only a part of the wider problem of world energy supply and the economics of the international oil industry, and are highly complex. To discuss automobile fuel demands in isolation from other energy demands is to oversimplify many aspects. But the concepts in the following paragraphs will be important in the continuing discussion of how best to use the world supply of fuels of all kinds.

One approach would be to make the available reserves of relatively cheap fuel go further. Often overlooked in policy consideration is the influence of motor vehicle weight on fuel consumption. Weight is the single most important parameter affecting fuel economy: a 2000 kilogram automobile consumes twice as much fuel as its 1000 kilogram counterpart. The scope for conserving fuel by reducing vehicle weight is greatest in the United States. In Europe a smaller average size of cars leaves less scope for this approach to fuel conservation, though some saving in consumption would flow from counteracting trends toward heavier passenger vehicles.

Other determinants of fuel consumption include the manner in which vehicles are driven, traffic conditions (stop-and-go driving is particularly wasteful of fuel), engine design, engine maintenance, and such power consuming accessories as automatic transmission and air conditioning.

More fuel-economy oriented automobile designs could be promoted in a variety of ways. Differential taxes on cylinder displacement have helped to give rise to a whole family of small-engined, fuel-efficient European cars. A differential tax on vehicle weight or unit fuel consumption might achieve similar results while giving manufacturers greater design freedom. In countries where the price of fuel is still relatively low, heavier taxes on motor fuel might constitute another method of encouraging higher fuel performance cars by making purchasers and manufacturers more fuel-economy conscious. Of course, any rise in the price of fuel irrespective of tax increases would have a similar effect. Taxes on power-consuming accessories, especially air conditioning, might also achieve some overall fuel savings.

To the extent that such policies increase the number of small cars, they could come into conflict with the desire for safer cars. But if oil shortages and balance of payments make it imperative to conserve fuel, current design philosophies may have to be re-examined and other possibilities explored. Is adding weight the only way to build safety into cars, or are there alternative approaches to protecting car occupants in collisions? Would size still be a relevant factor if all cars were smaller? Are there other effective methods of reducing highway accidents which do not require changes in the design of vehicles?

Competition for fuel resources can also be expected to prompt questions about the relative economy of different forms of transport. Comparison of the modes currently in use makes it clear that conventional cars, regardless of their size, are among the least efficient consumers of energy. As a first step toward a more rational use of energy for transport, fuel economy should join safety, noise abatement and air pollution as a factor in the design of motor vehicles. As a second and more far-reaching step, it will be necessary in the shaping of urban transport policy to consider explicitly the energy consumption implications of alternative modes of transport.[1]

Current trends toward ever-rising transport demands for fuel could be deflected in various ways. Improvements in communications and changes in the structure and density of urban development could reduce the need for travel. Heavier emphasis on investment in public transit could, if coupled with measures to restrain the use of cars, bring about changes in commuting habits. Public automobile service, car pools, and other schemes that increase vehicle occupancy could improve the fuel efficiency of automobile travel. Increasing the opportunities to walk and bicycle could also diminish the use of cars. The combined effect on energy requirements of these and similar developments could be far-reaching and merit detailed analysis.

MOBILITY THROUGH PUBLIC TRANSPORT

Energy consumption, while important, is not the only problem raised by over-dependence on automobiles. Growing congestion in urban areas is providing an independent stimulus for a rethinking of current transport policies. Cities that have tried to accommodate cars through ambitious highway building programmes have found that streets remain congested, commuting becomes increasingly difficult and that the quality of the urban environment continues to be eroded. Traditional remedies aimed at adapting the city to the car have clearly shown themselves to be ineffectual.

In an attempt to find an alternative approach, an increasing number of cities are re-examining the potential of public transport. Additional impetus has been lent to such reappraisals by the growing opposition of residents to the construction of new urban highways and by the necessity to meet the needs of people who have no cars, or cannot drive, and who depend entirely upon public transport.

Intensive efforts are currently underway to develop new concepts of public transport service that could offer viable alternatives to the automobile. No single approach yet envisaged is likely to offer a level of service comparable to that provided by the car. But used in combination, as elements of a comprehensive strategy, they extend a hope to urban dwellers of obtaining a degree of mobility and convenience that has traditionally been the exclusive preserve of motorists. Local conditions may influence the mix of policies adopted by individual cities, but they are likely to include the same four basic themes everywhere: improvements in bus operations, the development of innovative rail transit systems, the sharing of cars and their use as a form of public transport, and the introduction of selective limitations on the use of private automobiles.

IMPROVING BUS SERVICE: A FIRST PRIORITY

Buses are today a principal means of public transport for the majority of urban residents. Yet in most cities buses have become ineffective for lack of determination to untangle them from traffic congestion created by other vehicles. A first priority for many cities is therefore to overcome this deficiency, and it can be done in a number of ways that are by now well known. Buses may be accorded priority at traffic lights; they may be alloted exclusive use of entire city streets or individual lanes; busways may be created on arterial highways leading towards city centres; and short stretches of busways may be tunnelled under the most congested parts of the inner city.

An example of a busway is the Shirley Highway in the United States, a heavily travelled commuter artery leading to Washington, D.C. Twelve miles of the center section of the freeway are devoted exclusively to buses. One year after its opening bus ridership increased by 51 percent, and car traffic in rush hours had been reduced 33 percent. For those who continue to use cars there is less congestion; for those who use the bus there is a 30 minute saving in time each way.

In Bologna and in Reading extensive networks of one-way streets have been introduced that are expressly designed to permit the establishment of against-the-flow reserved bus lanes. Such lanes become, in effect, "guideways" for the buses and enable them to provide a service with the regularity and dependability of a busway.

In many cases, however, the creation of exclusive bus lanes or bus-only streets may be difficult from a traffic management standpoint. That is why a growing number of cities are laying emphasis on providing an unobstructed but not necessarily exclusive right of way for public transit vehicles. In Southampton, in the United Kingdom, a vehicle metering system allows buses unrestricted entry onto a main arterial highway leading to the city centre while private vehicles are admitted onto the same highway by means of computer-controlled traffic lights in such numbers as to ensure a free flow of traffic. Travel time for bus commuters over three miles has been reduced by ten minutes. This example demonstrates that improvements in bus operation can be had at a relatively low cost and at no sacrifice in road space, and that seemingly simple innovations can produce significant benefits in terms of improved transport service.

RAIL TRANSIT SOLUTIONS

The last ten years have seen extensive development of rail rapid transit in many countries. One reason for this surge of activity has been the recognition that the capacity of streets in high-density urban centres is too limited for exclusive reliance on buses. In London, for instance, about 69 percent of all commuters to the central area arrived by rail in 1972. In Tokyo 76 percent of commuters used mass transit and in Paris the figure was over 80 percent. The subway was the typical means of transport of all these travelers, at least for part of the trip.

However, underground mass transit cannot be expected automatically to produce relief from street congestion. This is borne out by the experience of Toronto, Montreal and San Francisco. These cities have discovered that decisions to build high-capacity underground lines have promoted additional development, particularly around stations, leading to the crowding of more employment into the city centre and inevitably to more traffic. Construction of rail rapid transit may thus, in the absence of land use controls, exacerbate rather than relieve traffic congestion.

Another fruit of the experience built up over the past decade is that the high-cost, high-capacity underground transit systems are increasingly unsuited to the land use and travel patterns evolving in most cities. According to one view, the conditions for an economically viable investment in such underground systems include a population of two million, a density of 14,000 persons per square mile, 100,000 jobs in the central district and the same number of commuters in peak hours. On these criteria, few cities in countries that do not already possess or have under construction a mass rapid transit system could justify building one.

A solution increasingly recognised to be better suited to the density characteristics of modern cities is the light railway or semi-metro. In essence, a semi-metro consists of light rail cars operating on exclusive tracks, generally on the surface but descending into tunnels in the style of conventional underground mass transit in city centres and at other places where surface space is at a premium or requirements for amenity are high. Semi-metros are in operation in Brussels, Cologne, Rotterdam, Frankfurt, Stuttgart, Essen and Vienna.

The semi-metro is a direct descendant of the old tram which fell into neglect in the 1950's until the rapidly rising costs of underground rapid transit construction once again revived belief in their economic attractiveness for medium-sized cities.

Cities that have maintained the tramways of an earlier generation more or less intact have the strongest incentive for adopting the semi-metro solution. Since investment in securing rights of way in outlying areas has already been made, only short stretches of tunnel in the heart of the city are needed to obtain a rapid transit system.

The semi-metro fills a gap between the high-capacity, high-cost underground rapid transit and the low-capacity and slower buses and trams operating on city streets. Combining, as it does, extensive coverage in the outlying areas with fast, unobstructed access into the city centre, and providing capacities able to service passenger flows in all but the densest urban corridors, the light rail transit is among the most promising service innovations in urban transport that have emerged in recent years. Along with the bus rapid transit it represents a major breakthrough in the search for less expensive, environmentally compatible transport systems that could serve economically the relatively dispersed travel patterns of today's urban areas.

MAKING BETTER USE OF CARS: PUBLIC AUTOMOBILE SERVICE

The bus and the railway, however, cannot meet all of the transportation requirements of a city, and in particular they cannot provide the flexibility, the convenience, and the door-to-door service of the automobile. It may be possible, however, to incorporate passenger vehicles as part of the public transportation system.

At its most elementary level the idea is illustrated by the "car pool", which combines the group-riding characteristics of a public transport vehicle with the flexibility of a private automobile. The "bus pool", involving the chartering of bus service by groups of commuters living in the same suburban location, is in use in Hamburg, Ottawa, Washington, D.C. and is gaining currency elsewhere. Both approaches can be effective in reducing congestion and the demand for parking space in the central city, while preserving some of the convenience of personal transportation.

More institutionalised forms of group riding, known in different countries as "jitneys", "shared taxis", "peseros", "dolmus", or "sheruts", are rapidly expanding in popularity in such cities as Istanbul, Teheran, Hong-Kong, Singapore, Tel Aviv, Caracas, Bogota and Mexico City. Indeed, in all these cities the bus has already lost considerable ground to the more convenient public auto service.

In Teheran, for example, 2,000 jitney taxis running on fixed routes carry about 100 million passengers per year—or 50,000 passengers per taxi. If a fleet of jitneys were to be substituted for the bus fleet which now carries 700 million passengers a year, it would take only some 14,000 jitney taxis to assume the entire burden of the Teheran bus system. Manila is already moving toward this kind of solution. The six to ten passenger "jeepneys" share responsibility for public transportation with the bus and at the same fares. Manila buses still number 3,800, but the superior service offered by the jeepney is already displacing the bus as the preferable, low-cost method of getting around.

While providing convenient and efficient transport in denser urban areas, such a fleet of cars intensifies the competition for street space with other motor vehicles. A public automobile network, therefore, could be more effective if accompanied by a restriction on the use of private cars, at least during rush hours, or by giving group-riding vehicles preferential treatment, for example by allowing them to share exclusive lanes and exclusive transit streets with the bus.

The concept of shared use of passenger automobiles can take other forms, too. In Montpellier in France, a fleet of small cars was put at the disposal of individuals who joined the service on a subscription basis. Special parking facilities were set aside throughout the central city for the distinctively marked cars which could only be picked up and returned at designated locations.

In each case the underlying objective is to provide personalized transport service with the help of fewer but more intensively used cars. The service is designed for both central city residents and for commuters from outlying areas who require the use of a car in the central city during working hours.

Yet another version of group riding has sprung up in the form of demand-responsive systems, colloquially known as the "dial-a-ride" service. The vehicles do not have fixed routes and schedules; instead, they respond to individual travel requests as they are received. This concept employs small-sized vehicles such as minibuses to provide door-to-door transportation over relatively dispersed areas to people who telephone for service. Demand-responsive systems are undergoing trials in a number of communities around the world: Regina, Toronto and Bay Ridges in Canada; Emmen in the Netherlands; Nose near Osaka in Japan; Haddonfield (New Jersey), Rochester (N.Y.) and Ann Arbor (Michigan) in the United States; and Maidstone, Abingdon, Harrogate and Harlow in the United Kingdom. The territories best served by such systems have low to moderate population densities, scarcely threatened by congestion. But the service fulfills an important social function by attempting to meet the needs of those people in the automobile-dominated suburbs who are either too young, too infirm or too old to drive or who, for a variety of other reasons, are deprived of the use of a car.

The experience of Teheran, Manila, Monpellier, Regina and other cities suggests that passenger cars can be shared in a variety of ways. So far no effort has been made to combine the different modes of operation, but sufficient evidence exists to suggest that extensive and well-organised fleets of public automobiles could provide convenient transportation service for a significant fraction of urban travellers. Social benefits would include reduced traffic congestion and reduced demand for parking space in central areas. More significantly, such services offer the best promise of narrowing the gap that today separates car owners from the carless.

THE PROMISE OF NEW TECHNOLOGY

Thus far the strategies considered for improving urban mobility have been evolutionary changes in the existing modes. However, there are also intensive efforts to develop entirely new systems that might offer a level of service approximating that provided by the private car.

There are two such families of technologies under development today. One is Personal Rapid Transit (PRT) consisting of small cabins operating on guideways over a network covering an extended urban area. The other is "people movers", systems designed to carry large numbers of passengers at moderate speeds over relatively short distances in high activity centres such as the central business districts of large and medium-sized cities, airports and other densely developed places characterised by heavy flows of people and an intensive use of land.

Personal Rapid Transit, under development in Germany, France, the United States and Japan, relies on networks of light-weight guideways that might be constructed at grade, underground or elevated, perhaps over streets and through buildings. Small cabins would provide non-stop service from boarding stations to destinations at the push of a button. The technology of such systems that could operate in a truly area-wide service at the required degree of safety and reliability is still in its infancy, nor are their capital and operating costs well established. The environmental impact and aesthetics of an extensive network of elevated guideways traversing an historic city centre are likewise in doubt. Nevertheless, a fully automated system would carry an economic advantage over conventional rail and bus systems that is hard to ignore. Continued research and development of PRT systems, therefore, appears warranted, at least to the point where their technological feasibility, costs, environmental impact and user acceptability could be determined in a full-scale public demonstration.

The technology of "people movers" for high activity centres is in a more advanced stage of development, with a number of prototypes in existence in Germany, France, Japan and the United States. These systems take the form of either automated vehicles moving on exclusive guideways; or of continuous conveyors, moving at speeds lower than those associated with vehicular systems but higher than those of the conventional moving belts or escalators.

Several such systems have or will soon reach the stage of demonstration. In the United States, shuttle systems employing either individual automated cars or small trains are being installed at several airports to overcome the growing distances between parking lots, main terminal and airport boarding areas.

It would be premature to envisage the widespread introduction of such new urban transport systems within the next few years, but steady progress in their development can be expected. By the 1980's, when the economics of driverless operation is likely to become an important influence on the design and choice of transport systems, "people movers" could begin to play a major part in serving the transportation needs of metropolitan areas.

LIMITING THE USE OF CARS

The improvement of public transportation must go hand in hand with a deliberate policy to curtail the use of cars within the congested parts of urban areas. For there is a growing realisation that city centres cannot accommodate large numbers of automobiles, that cars are not necessary and even counterproductive in such a setting, and that improvements in public transport service often depend on freeing streets from traffic congestion.

By now there is firm evidence that limiting the use of automobiles in urban areas is politically, commercially and technically feasible. Public opinion surveys in France, Sweden, Germany and the United Kingdom have shown that people strongly favour a car-free environment. Merchants are increasingly coming to the conclusion that streets freed of traffic bring concrete benefits in terms of bigger crowds, more customers and increased sales. Perhaps the most convincing proof of the practicality and public acceptance of traffic restrictions is their growing use. An investigation carried out in 1972 by the OECD Sector Group on the Urban Environment disclosed over 100 cities which have banned traffic from portions of their central districts and dedicated them to pedestrian use.[2] An impressive number of them—Vienna, Munich, Leeds, The Hague, Genoa, Turin, Essen, Cologne, Bonn, Düsseldorf, Rome and Copenhagen, to cite a few outstanding examples—have networks of pedestrian streets extending across a considerable part of the city centre.

Limiting the use of cars over wider areas, though in a more selective manner, is also proving to be feasible. London, Zurich, Stockholm, Amsterdam and Minneapolis have declared certain streets off-limits to private cars while continuing to permit the operation of public transit vehicles—thus aiming to improve the conditions for pedestrians and public transport alike. The cities of Bremen, Göteborg, Bologna, Liverpool and the inner suburbs of Stockholm have had good results in reducing the volume of traffic entering the central area by dividing the area into separate sectors and prohibiting the direct movement of cars from one sector to another, while diverting through traffic to a perimeter road.

The most successful ventures to curtail traffic have occurred in cities which have adopted balanced and comprehensive policies for facilitating movement and improving access in the city. One conclusion to be drawn from these successes is that plans to restrict or restrain the use of cars must be accompanied by a concurrent investment in public transport, in highway improvements (especially those designed to divert traffic), in traffic management and in pedestrian schemes. Proper integration of and balance between these various elements should be the cornerstone of any strategy for urban mobility.

REDUCING TRANSPORT REQUIREMENTS

In the long run, policies that concentrate merely on accommodating travel demand will not be enough. Lasting solutions will also require some degree of control over the factors that generate demand for transportation. These factors include the location of housing and activities, their densities, their physical arrangement and the interrelation among work places, shopping and community services.

In newly urbanising areas it is possible to balance transport supply and demand by building a total urban environment in which housing, employment, community services and transportation are physically integrated and designed to work together. It is more difficult to control these conditions in existing cities. The land use arrangements have already established the patterns of movement, and the demands on the transport system in the short run are fixed.

But this is not to say that the problems of the established cities are beyond solution. Even an existing urban area is in a constant state of change and a start can be made to guide the process of growth in ways that will reduce transport problems rather than aggravate them. This can be done by providing housing and a good environment close to places of employment so as to reduce the necessity for long commuting trips, and by clustering related activities in order to create self-contained neighbourhoods in which distances between homes, schools, shops, playgrounds and community facilities are small enough to be covered on foot or bicycle. On a larger scale, planning can begin on restructuring urban areas into multi-centered regional cities linked by efficient high-speed public transport as well as expressways. Patently, these are relatively long-term goals whose achievement will require careful planning, public consultations and in many countries new powers to intervene in the workings of the private land market.

A STRATEGY FOR URBAN MOBILITY

The issues raised in this section concern the future role of the automobile in our urban society. The problem is not an easy one to resolve, for decisions about transportation involve choices between competing interests and may affect some people beneficially and others adversely. Neither the premise that the automobile can be dispensed with nor the belief that it can serve as the exclusive instrument of urban mobility are likely to be politically feasible or operationally useful: individual and mass transportation satisfy different travel requirements and therefore both hold a legitimate place in an urban transportation system.

A wiser view, therefore, might be that the automobile and public transport should complement one another rather than compete with each other. Both should serve as elements of one integrated urban transportation system whose purpose is not just to move people but also to enhance the quality of the city and to contribute to the broader community objectives.

This larger perception of transportation has forced a reappraisal of the conventional wisdom of transportation planning. It has become increasingly apparent that qualitative origin-and-destination studies and linear travel demand projections have become self-fulfilling prophesies, and that the most promising solutions are likely to be found neither in massive investment in more urban highways nor in expensive high-capacity underground systems but rather in innovative low-cost improvements in existing systems of the type described above.

Such improvements can be implemented on a short time scale. Valuable progress can therefore be achieved during the next few years, giving cities much needed respite while new technology and innovative land use designs are being developed to provide more lasting solutions.

Substantial improvements in the quality of public transport, however, may often prove impossible to achieve without a reduction in the volume of traffic that crowds into the city centres. There is a strong case, therefore, to be made also for deliberately curtailing the use of private cars in central areas—not in the negative sense of denying people access to personal transport but as a positive action to improve the operation of public transport. The restraint of auto traffic to benefit public transport can be accompanied by schemes to totally ban vehicles from smaller areas in the interest of environmental improvement.

In the final analysis, of course, the kind of transportation strategy which each nation requires will depend on the goals which are set for metropolitan areas and the quality of their environment. If the goal is to preserve a strong central city with a high population density and a viable centre, then efforts should be focussed on traffic management and transportation services which would increase freedom of movement and enhance the pedestrian environment within the central area and improve the radial flows to and from the outlying suburbs.

If, on the other hand, a decentralised, low-density urban form is the preferred urban strategy, primary attention should be devoted to the design of public transportation services which could adequately complement the private car in serving a dispersed travel pattern and provide adequate accessibility to all urban residents regardless of whether or not they own a car.

For most countries, of course, the urban goals and the resulting needs for transportation improvements cannot be stated as a blunt dichotomy. Each country and each metropolitan area will require a mix of transportation and pedestrian improvements that will benefit a wide range of residents—those living in the central cities as well as those in the suburbs, those who are captives of public transportation as well as drivers, bicycle users and pedestrians. But whatever mix of transportation services the goals of a given urban area may dictate, one thing is clear: to bring about the kind of improvements which will assure urban mobility while preserving environmental quality, a vigorous and sustained level of innovation will be required.

Strong initiative and positive action by central governments will encourage such innovation. Even the largest and wealthiest municipalities and transit companies are reluctant to take the risk of introducing major changes in service for fear that these will meet with public indifference or hostility and thus turn out to be a bad investment and a political liability. Central governments could help to reduce these risks by promoting a sustained programme of real world experimentation. Full-scale experiments are often the only way to test public response to a change in service and to evaluate its wider impact on the urban area. They are thus a powerful device for reducing uncertainty and lessening the constraints that inhibit successful innovation.

A concerted programme of demonstration projects, covering a wide spectrum of low-cost improvements in transportation, including methods to limit the use of cars in central areas, would provide decision makers with a broader range of alternatives from which to choose the elements of a transportation strategy calculated to enhance mobility while contributing to larger community goals.

NOTES TO CHAPTER 1

1. For a fuller discussion of the impact of the motor car on energy resources, see "The Motor Car and Natural Resources" by Gerald Leach, Inquiry into the Impact of the Motor Vehicle on the Environment, Working Paper No. 4, OECD, Paris, October 1972.

2. "Policy towards the Creation of Vehicle Free Areas in Cities", Sector Group on the Urban Environment, OECD, Paris, October 1972. See also "Traffic-Free Zones in German Cities" by J. Kühneman and R. Witherspoon, OECD, Paris, August 1972.

II

URBAN PUBLIC TRANSPORT: Service Innovations in Operations, Planning, and Technology

Robert A. Burco

ACKNOWLEDGMENTS

This study of emerging public transport alternatives to private automobile travel in urban areas was undertaken over a period of fourteen months to develop background material for consideration in policy discussions of urban mobility and environmental quality. It sought to report, as currently as possible, the present state of development and planning for a variety of recent urban transport initiatives in O.E.C.D. countries of Western Europe, North America and Japan as they stood in the period from 1971 to 1972.

It was intended as an initial overview of the breadth of activity in a newly revitalized and important field of urban endeavor. As such, it was intended to open discussion of future urban transport alternatives within O.E.C.D. member countries and to elicit comment and review as a working document. Because of the extensive scope and relatively limited resources devoted to the study and the accelerating pace of developments in the urban transport field, it is almost inevitable that some activities may have been overlooked, that subsequent developments not receive the degree of coverage afforded to earlier ones, and that some factual errors may be contained in the work as it now stands.

Material for the original report was gathered in three extensive field visits: in Europe from July 19 to October 6, 1971; in North America from January 3 to February 24, 1972; and in Japan from June 25 to July 7, 1972.

Without the generous cooperation of many individuals and organizations in local and national governments, industry, universities, and private laboratories in numerous countries, this effort would not have been possible. A large debt of gratitude is owed to these individuals and organizations for the frankness and completeness of the information discussed during the field visits. (A complete list of the organizational contacts made in the conduct of the study is included in Appendix B.) A particular acknowledgement is owed to the Institute of International Studies of the University of California Berkeley which provided supporting staff and office facilities for the portion of the study carried out in the United States.

Only a representative sampling of the many written materials available in this broad field has been included in a selective bibliography, but addresses of contacts with additional information on specific systems, demonstrations, or research have been given where possible in descriptive footnotes throughout the body of the report. Two particularly important documents that have appeared since this paper was first published should be cited for overall reference. One, an OECD Conference report on "Better Towns With Less Traffic" published in 1975 treats a variety of service innovations in public transit in joint application with traffic and pedestrian improvements in selected cities current to early 1975.[1] The second, a report by the Office of Technology Assessment of the United States Congress takes technological developments of automated guideway systems up through their development in a comparable period.[2]

There has been no attempt to fully revise this document to bring the whole field up to date some five years later, however, important observations have been attempted in retrospect and status reports on prior demonstrations identified where possible.

A final caveat: this report is a *working document*. Since much of what will be said is one man's interpretation of the views and actions of others, I must personally take responsibility for the subjective elements introduced, and for any errors of fact that may be reported.

SUMMARY

This study of emerging public transport alternatives to private automobile travel in urban areas was undertaken to develop background material for consideration in policy discussions of urban mobility and environmental quality. It seeks to report, as currently as possible, the present state of development and planning for a variety of recent urban transport initiatives in O.E.C.D. countries of Western Europe, North America and Japan. It is intended as an initial overview of the breadth of activity in a newly revitalized and important field of urban endeavor. As such it is intended to open discussion of future urban transport alternatives within O.E.C.D. member countries and to elicit comment and review as a working document during the coming months.

The policy framework of this report deals with the scope of future needs for public transport, the effects of mobility on the urban environment, the planning process itself, and particularly the roles of government—most notably central government—in fostering the various kinds of innovations needed to open up the range of choice available to policy makers (and indeed to travelers) in dealing with issues of urban mobility and urban environment.

The focus on innovation that the report takes is one which seeks to recognize that at least two kinds of technical progress are possible: (1) incremental advances in the day to day operations of existing transit systems and (2) the development of entirely new modes of urban travel. In addition, much of the potential for improved urban transport can be found to lie, not in the technology of the transport systems available today or tomorrow, but in the administrative and planning innovations that will influence how such systems are implemented.

Vast improvements in urban mobility and urban environment are possible through incremental advance of existing transport technologies, as will be pointed out in detail in chapters on bus operations and a variety of rail services. But such improvements in existing public modes fall far short of yielding the technical capabilities that will be needed to bring about a variety of entirely new transport services that may soon be in demand.

The potential demand for such new services arises in part from the trend toward physical restructuring of urban areas. Among the new urban forms that are emerging are a multiplicity of widely dispersed activity centers; the extensive growth and spread of urban regions of low to moderate densities; and the relative isolation of important, but environmentally noxious, regional facilities such as major airports. The demand for new transit services arises as well from personal characteristics of the population, such as increasing affluence, flexibility and variety of life styles, desires for privacy and personal freedom, and feelings of environmental and social responsibility.

The creation of only a few new functionally innovative systems of urban public transport is perhaps needed, or even possible. But one or two such systems, capable of reponding far more flexibly to individual demands for service and environmentally less noxious than existing modes, might be quite revolutionary, perhaps as revolutionary as the private auto has become, in dealing with the aspirations of urban dwellers and the quality of their environment.

But technology alone—whether in improving the operations of existing modes or in inventing entirely new ones—is not enough to ameliorate the transportation difficulties of the major cities of the world. Institutional innovations (in the form of planning, administrative, and financial initiatives governing the operations and interrelationships of transport modes, urban development, and even social and economic activity) may contribute as much or more to the resolution of both present and future urban transport difficulties.

Despite the many and varied attempts at innovation observed in the course of this report, the prospects for realistically implementing much of the promise of such innovations, particularly those of substantial technological novelty, are at present quite marginal. Too little government and public concern has thus far been expressed in defining the nature of the problems that these new technologies should help ameliorate, and the processes by which they might be tested and selectively introduced to gain acceptance as an alternative to existing transport modes. Thus the possibilities for undue delay and embarrassing failures in applications are significant, given the present array of organizational and political factors bearing on the whole structure of institutions involved in urban transit development.

These dim prospects do not lie in technological problems concerning new systems, nor in the transportation characteristics of existing ones—as a number of creative engineering organizations, whose work is detailed in the body of this report, are making significant progress in many technical areas at this time. They lie, rather, in the gaps of communication and responsibility between those institutions with specific transportation missions and those with differing tasks, such as enhancing the quality of the urban environment, or promoting the application of science and technology to human affairs.

Since the original version of this report appeared in 1972, exactly such problems of implementation have plagued experiments like the PRT installation at the University of West Virginia in Morgantown, the Krauss-Maffei Transurban demonstration originally scheduled with Ontario Ministry of Transportation and Communications backing in Toronto and abruptly cancelled with the failure of continuing West German governmental research and development support to Munich-based manufacturers, or installation of prototype CVS and ARAMIS automated systems demonstrated in test tracks in Japan and France during the mid-1970s. Following critical Congressional review of the entire United States and international automated guideway transit field in 1974-75, a U.S. program of reduced scope and expectations has been reinstated through the Urban Mass Transportation Administration, with extensions to the Morgantown, West Virginia system, as well as limited downtown peoplemover experiments in five U.S. cities currently programmed for funding (1977). The path ahead for such innovations as these and others is likely to remain rocky indeed, yet they remain, nevertheless promising beacons to the future.

The final chapter addresses some of the roles that government, largely national government, can play in accelerating the pace and insuring the quality or relevance of innovation in urban public transport. There are possibilities as well, however, for both local governments and international cooperative bodies to contribute to this effort. The principal focus at each level of potential government action is in overcoming the gaps in present institutional arrangements that limit the understanding, the resources, or the responsibilities necessary to deal with urban transportation policy in the context of overall environmental quality and innovative transport service possibilities.

Together, a wide array of service innovations in operations, planning, and technology may be called for in any comprehensive and successful attack on the complexity and variety of urban transport difficulties that plague major cities. A different mix of technological and administrative measures may be called for too, in the diversity of national urban cultures and geographies present within the member countries of O.E.C.D.

2 INTRODUCTION

Public transport once thrived because, short of walking, it was the only means of urban transportation available to large numbers of people. Whether or not it was comfortable, convenient, reliable or fast was incidental to the main issues of whether it was there at all, was cheap enough to be afforded by those needing transportation, and, at least in its early years of private ownership, was able to turn a profit.

Such days are far removed from the present where it now seems that almost no amount of comfort, convenience, reliability, and even speed provided by public transit is sufficient to compete with the private automobile for patronage when riders have a real choice—that is, when they (1) are physically able to drive; (2) can afford the purchase and upkeep of a personal vehicle; and (3) have an adequate network of streets, highways, traffic controls, and parking facilities for uncongested operation of their vehicle when and where they choose.

Competitive policies pitting autos and transit against each other in terms of comfort, convenience, speed, and apparent cost, do not seem to have worked well in the past, and show no particular merits as a policy for the future. Yet this has been the policy framework that has dominated most urban transport planning—or, more accurately, nonplanning—ever since the automobile began making major inroads on transit patronage in the late 1940's in North America and in the 1950's and 1960's in Western Europe, Japan, and other industrialized and urbanized countries around the world.

To the degree that competition for the "free" choice of the discretionary traveler is allowed to guide both future transit and highway investment in urban areas, then wasteful and conflicting policies will result. It is such a framework, particularly in the United States, that has led to the effort to push freeways into dense urban centers where many feel that they do not belong, and to propose high speed and high capacity rail transit systems in spread out urban areas where highway networks are already extensive and existing densities or urban activity do not warrant such huge transit investments.

For if everyone with a choice concerning mobility were allowed to choose, wouldn't everyone drive everywhere, except for a few of the elderly, the young, the handicapped, and the decreasing number of those too poor to afford a car? And that, in effect has been the actual policy: one of encouraging the fullest possible motorization for each city (compatible with some standards of preservation of historic core areas, parklands and amenity) along with, of course, a minimal background of transit service for those left out by this policy, the so-called "captive riders." There seemed nothing gravely wrong with this policy—at least there didn't appear to be until recently, particularly if mobility were the only question examined in urban policies towards automobiles and transit.

But as we all are finding out, it may well not be possible any longer, because of environmental as well as economic factors, to construct and operate all the networks of streets, highways, traffic controls, and storage facilities required to give every person the possibility of driving their own vehicle anywhere and at any time. Instead, it has now become an urgent necessity to look for other systems of mobility that can provide alternatives to the auto, both because many people will never have access to a car and because the network of adequate facilities will never be built that can accommodate uncongested operation of motor vehicles at all times and places, at least with the growth rates of automobile ownership that are expected over the next several decades.

Thus the policy framework of this report must deal with the scope of future needs for public transport, the effects of mobility on the urban environment, the planning process itself, and particularly the roles of government—most notably central government—in fostering the various kinds of innovations needed to open up the range of choice available to policy makers (and indeed to travelers) in dealing with issues of urban mobility and urban environment.

The focus on innovation that the report takes is one which seeks to recognize that at least two kinds of technical progress are possible: (1) incremental advances in the day to day operations of existing transit systems and (2) the development of entirely new modes of urban travel. In addition, much of the potential for improved urban transport can be found to lie, not in the technology of the transport systems available today or tomorrow, but in the administrative and planning innovations that will influence how such systems are implemented.

And since the aim of this report is to share the differing experiences of O.E.C.D. member nations, some activities that may now be considered merely good standard operating practice in one place may be viewed by another as an innovation, worthy of trying in a slightly different geographic, governmental, or social context.

Cities and their surrounding urban regions are constantly changing: rebuilding, expanding, decaying, altering their geography, economy and demography to meet dynamic social, economic and technological forces. But at any one time, the amount of present investment in the way things are is enormous.

In addition to a great, observable physical plant—homes, industries, social service facilities, communication and transportation networks, and basic supply systems—making up an urbanized area, there are less visible, but very real, social, economic and political institutions which govern the allocation of resources within this physical plant and the processes by which change occurs. Thus there is, in speaking of urban innovation, not only a great inertia in what is already being done, but in the processes of continuing to do more of the same.

The perspective of the present weighs on the future both in terms of what may be done and how it may be done. A lack of appreciation by all parties involved of this perspective and what it means in terms of the pace and process by which useful change can occur is the single largest barrier to successful innovation in urban public transport today.

Substantial investments—particularly in the largest cities of Europe, North America, and Japan—already have been made in existing modes of urban public transport, as well as in highways, streets, and motor vehicles. It is certainly a reasonable priority, as it was in every country visited, to make these systems perform as well and as efficiently as possible, using a significant portion of the available urban transport funds to do so.

But just to keep them operating much as they have in the past—and often that means a quite distant past—is not sufficient in a world of changing life styles, rising affluence and new demands for comfort, convenience and flexibility. The functions performed by existing rail rapid transit systems, buses, trams and suburban railroads are not static, nor is the competition that they face from the private auto and the highway network, nor even the physical structure of the urban areas they were designed to serve, and in part created.

A recognition of these changing circumstances surrounding mature transport technologies can lead to innovation in their use and in their planning. But for such innovations to have major influences on ameliorating future transport problems a sensitivity is needed to the complexity of the urban world external to the transportation function itself. The additional consideration in this report of planning and administrative reforms is aimed at demonstrating the means by which such incremental innovations in existing technologies can be focused on major, comprehensive improvement of urban mobility and environment.

Moreover, the level and quality of *new investment* in existing modes—both in places that already have them, such as London, Chicago or Paris, and in entirely new application such as Munich, Lyon and San Francisco—is crucial. If admittedly the quality of service of existing transit modes has failed often to compete efectively with the private automobile and its public street and highway infrastructure, it is partly the disproportionate allocation of resources to transit and highways that has been at fault, at least in a number of OECD countries.

New money for capital investment in greatly different proportions between transit and highways than has been the case in recent decades, particularly in North America, will be necessary. The discussions throughout this report of a variety of exciting public transport innovations take this fact as a premise.

On the other hand, so much misinformation and wishful thinking about the technological end of the innovation spectrum has been found in the press and elsewhere that a hard worldwide look at the reality of present research, development and demonstration progress is also sorely needed.

The new technologies that receive so much play in the Sunday supplements can be seen in several perspectives—most of them self-defeating. When portrayed as almost here today they lull the public and government alike into a complacency about a future already taken care of; if only we can deal for a few more years with the present.

But these technological innovations are often far from almost here; major public investment will be required to move them beyond the relatively inexpensive stage of laboratory models and even test track prototypes. And still—if and when they do arrive—they may well not represent a solution to any particularly relevant problem.

When portrayed for long enough as almost here, yet never delivered in practical terms, such "innovations" take on aspects of crying wolf too often or prematurely, so that those with "practical" decision-making responsibilities can dismiss new technologies all together.

And when the future of urban transportation is portrayed in glamorous technological terms, it is even easier to view the entire problem, present and future, as a technical one simply requiring engineering and money for solution. Therefore, holding out great technological promise for the future of urban transportation has significant dangers if handled solely as the preserve of industrial firms and science fiction writers, whether in public or private employ.

There is, to be sure, much promise in technological innovation for urban public transport, and a number of creative engineering minds are working in this area. But too little government and public concern has thus far been expressed in defining the nature of the problems that these new technologies should help ameliorate, and the processes by which they might be tested and selectively introduced to gain acceptance as an alternative to existing transport modes and environmental constraints. They have thus, for the most part, been left hanging in an artificial world of technical sophistication and social naiveté.

To break out of this unworldly isolation will be difficult and represents a major challenge at this time. To harness useful new technological innovations in urban transport, as well as in many other fields, is going to call for much more sophisticated government and public participation in defining goals and priorities for technologies, as well as much more sophisticated awareness of what real and imaginary problems are that need to be addressed. This is another critical instance of the perspective of the present needing to guide the future, rather than letting a perspective of the future misguide the present.

For new technologies to really contribute new systems of urban public transportation in a relevant, timely, and comprehensive way, they must be viewed as relating to the same, messy institutional, physical, and social complexity that the present systems have to cope with. Homes dislocated by a sophisticated new mode will cause much the same problems as homes displaced by highway construction today, perhaps even more so. Huge investments in elite services like high speed suburban trains may cause more social resentment than investments in roads that many could use. And the list goes on.

In focusing on these aspects of new technologies and systems, care has been taken to make this report not just one more shopping list of innovative concepts. That may have its place—or did have several years ago—but for policy considerations concerning what to do now and in the future about motor vehicles and the urban environment it is necessary to speak critically about where we are today in the way of commitments toward innovative developments and about timetables for demonstration, evaluation and even installation of at least some new systems, whether conceptually the best or not. Conceptual refinement will be necessary as we go along, as it has been largely ignored to date by any but the technological innovators.

Not only must concepts be distinguished from commitment, but functional innovation of new urban transport services must be distinguished from mere technological innovations in how one goes about doing any given thing. By functional innovation is meant the creation of a new function or public service, to be served by the system that did not exist before.

The creation of only a few new functionally innovative systems is perhaps needed, or even possible. But one or two entirely new, in functional or service terms, systems of urban transportation might be quite revolutionary, perhaps as revolutionary as the private auto has become, in dealing with the aspirations of urban dwellers and the quality of their environment.

On the other hand ten different suspension systems or propulsion systems for a tram or a train might make almost no real difference. Most of what appears to be new is new only in technological terms. And that is the tragedy of an unsophisticated bridging of the perceptions of cities and the public and industrial producers and their government sponsors.

There is a strong tendency today to get a lot of technological innovation because it is relatively easy and showy (and possibly also because it is expensive in the right ways) without functional innovation. This report thus expects to call a train a train and a tram a tram, whether they are on rubber tires or a cushion of air.

It also hopes to look for something more in all of the technical activity around the world, and to point out the nature of the few functional innovations in urban public transport that have a revolutionary potential for the users and for the cities themselves. They are few, but vastly exciting. It would be a pity if they were overlooked by mistaking glamour for innovation.

3 NEW POTENTIALS IN PUBLIC TRANSPORT SERVICE

Only a limited number of public transport modes now exist, and these have existed basically in their present form for roughly a hundred years, some of them even more.[1]

Certainly technological changes have occurred that have altered the means of propulsion, suspension, even guidance of such modes. For example, the tram moved from horse traction to cable operation or self-contained steam or gas power, and then to electricity. But when talking about progress it is clearly necessary as mentioned earlier to distinguish between such technological change and change of a more fundamental kind that affects *what* services are rendered, and not just how they are rendered.

That this distinction is often confused is evident in just one example of many that could be cited—that of rediscovering and modernizing the tram. This seems the current technological preoccupation of public transport programs in those countries that discontinued tram usage some time ago to make way for the private auto and the bus. Where the tram was not discontinued but modified in its application it seemingly has continued to thrive without major need for technological wizardry.

This point will be a central one of this chapter: that incremental advances in public transport operations—those which reflect a capability to provide a *new or improved service* with existing or evolutionary technology—are to be distinguished from mere technological changes which render the same service more efficiently from the viewpoint of the supplier of the service (city or transit authority or company). These latter changes are important too of course, but they have been treated extensively in the technical operations and hardware literature of the transit industry, and will be dealt with only briefly here.

It is partly because transit services have been viewed as somewhat static that the majority of innovative thought in this field has been directed at improvements in efficiency, in carrying out the given task better. But it is precisely because the society (and its physical embodiment in urban form) has itself changed so overwhelmingly that a static view of transit services has now become so anachronistic.

Where the service is new, even if the technology is old, we have the possibility for substantial improvement. But where the technology is new, yet the service is old, we have some cause for doubt in appraising how much real improvement has been made.

In spite of these cautionary words, the irrational aspects of how technologies and services are perceived can admittedly influence the acceptance and use of a product or service, as all good advertisers have known for years. An old tram service, rediscovered by a forgetful government and public and then called something new and packaged in shiny new technology, may have the potential for revolutionizing peoples' perceptions of public transport and thereby broadening usage and support for *real* innovations in service—through changes in operations, planning, and even new technologies that promise to do something different than before. In this way the confusion in people's minds between which innovations are real and which are perceived sometimes have been of value to policy making for urban transport.

Here, however, we will be taking the perspective that the confusion is *never* a blessing when it is in the minds of the policy makers themselves. And we we will seek to clearly distinguish such kinds of innovation as we progress in our discussion of both incremental advances in existing public transport services and service innovations in new transit technologies.

INCREMENTAL ADVANCES IN TRANSIT OPERATIONS

What, then, is really new in the operations of existing public transport? There are two basic kinds of progress possible: (1) new or improved services which can be carried out by innovative uses of existing technology, and (2) more efficient—or less environmentally objectionable—performance of existing transit services made possible by incremental innovations in transit technology or in the ways in which it is applied.

Together, these advances offer considerable potential for improved urban public transport services, for complementary uses of traveled rights of way by public and private transport, and for environmental improvement of urban areas, particularly the most densely developed of these.

At least five characteristics of urban transport operations differentiate the services provided by the several generic modes now available.[2] For the primary differentiation of these modes one may helpfully specify: (1) the scale of urban geography served; (2) the size of the vehicle; (3) the degree of exclusiveness of the traveled right-of-way; (4) the flexibility of the routes and schedules, including transfers; and (5) the manpower requirements for operating personnel (see appendix IIA).

While successive waves of economic, technological and social changes have altered the quality and acceptance of the various services rendered by these public modes over their long histories, the types of service actually provided have altered little in all this time. The basic service required of a transport system is to get a passenger from an origin to a destination, making up a single trip. But the kinds of persons comprising an urban population may be vastly heterogeneous, as may be the nature of their origins, destinations and times of travel.

The potential densities of travel between various general locations are influenced by the distribution of land use activities (commerce, industry, services, residences, recreation, etc.) and the income and other individual characteristics of the inhabitants. At the same time the presence or anticipated presence of at least a minimally adequate transport service influences the locational choices as well.

Where travel densities are high and incomes low, one kind of service becomes operationally viable; where travel densities are low but incomes high, another—sometimes quite different—service becomes viable, perhaps both in economic and operational terms. At such extremes the public transport options are well defined; they set out the limits of the range of services that need to be provided in a heterogeneous physical and social urban environment.

Relatively cheap, high or moderate density line-haul transport service along a fixed route and with a fixed schedule has been the dominant style of mass public transit for as long as most of us can remember. On the other extreme has been the elite public transit option of the taxi—an individualized transit device with a driver for hire to go almost anywhere at one's bidding for a price relatively few have been willing or able to afford, at least on any regular basis.

Within these extremes much is possible, although until recently almost nothing was done to bridge the very broad gap between them. Emerging efforts to do this—to provide (1) higher quality mass transit and (2) lower cost, nearly individualized transit—are the most exciting aspects of the incremental advances in public transport operations currently afoot. Lower residential densities, rising incomes, growing environmental awareness and a number of other factors are causing the distinctions between mass and individualized transit expectations to blur considerably.

SERVICE INNOVATIONS THROUGH NEW TECHNOLOGY SYSTEMS

Vast improvements in urban mobility and urban environment are possible through incremental advance of existing transport technologies, as will be pointed out in detail in chapters 4 and 5. But such improvements in existing public modes still fall far short of what would be needed to bring about a variety of entirely new transport services that may soon be in demand.

The potential demand for such new services arises in part from the trend toward physical restructuring of urban areas. Among the new urban forms that are emerging are a multiplicity of widely dispersed activity centers; the extensive growth and spread of urban regions of low to moderate densities; and the relative isolation of important, but environmentally noxious, regional facilities such as major airports.

The demand for new transit services arises as well from personal characteristics of the population, such as increasing affluence, flexibility and variety of life styles, desires for privacy and personal freedom, and feelings of environmental and social responsibility. The combination of these physical and personal changes suggests the need for several transport services, at present virtually unmet by any existing mode of travel. And each of these would seem to have great potential in the near future if they could be provided at reasonable costs and with minimal environmental intrusion.

The activity centers that are developing at high densities within existing central cities—and within their suburbs as well—are a product of many social, economic, and technological forces. Certainly one such force is the economics and technology of high-rise building construction, related in part to land costs, construction methods, and the vertical mobility of the elevator.

Just as the elevator technology has been a contributing factor to the possibility of high-rise building construction, so a functionally equivalent *horizontal* elevator technology would seem to many observers to be an innovation that could overcome some of the limitations in mobility at these high-density activity centers. At present, economical parking facilities for personal autos, surface street congestion, discomforts of over-long walking distances, and access congestion all place limits on the possible size and amenity of such complexes.

No commonly available transport mode has yet been planned to overcome these limits in a totally satisfactory manner, although minibuses, special bus and pedestrian malls, and the denser parts of subway networks have been used somewhat for this purpose in existing downtowns undergoing revitalization. And special shuttle trains, low-speed conveyors, buses, and individual cars have begun to fill this need in large airports and transport terminals.

But to do the job of circulation well enough for these centers to realize their full potential would seem to require a technology or technologies capable of higher speeds, larger capacities, and greater convenience. A range of such new system initiatives to fill the transport requirements of these centers will be described in chapter 7 on activity center circulation.

The spread-out nature of much current urbanization generates only moderate densities of traffic in many corridors of travel. In addition the distribution and quantity of activity centers throughout this general "background" level of development insures that both origins and destinations of most trips in urban areas will be widely scattered. Together these factors signal a strong need growing for genuinely area-wide transit systems, rather than continued reliance on line-haul systems interconnected merely through transfers.

The growing public desire and ability to pay for comfort, convenience, privacy, and reduced travel time also indicate a number of characteristics that such an area-wide system should provide. Among them are responsiveness to individual origin and destination desires; private or small group occupancy of travel compartments; minimum transfers and delays, and ease of access to originating and terminating locations.

A generic name for such a functional innovation in transit service that has come into widespread, even international usage, is Personal Rapid Transit (PRT). Its general characteristics are small vehicles, automatically controlled, operating on exclusive guideways, with off-line stations and individual choice of origins and destinations throughout an extensive, area-wide network.

INSTITUTIONAL FACTORS IN PUBLIC TRANSPORT INNOVATION

But technology alone—whether in improving the operations of existing modes or in inventing entirely new ones—is not enough to ameliorate the transportation difficulties of the major cities of the world. Institutional innovations (in the form of planning, administrative, and financial initiatives governing the operations and interrelationships of transport modes, urban development, and even social and economic activity) may contribute as much or more to the resolution of both present and future urban transport difficulties.

While this brief report cannot hope to do justice to the wide variety of such institutional innovations that are being applied within various O.E.C.D. member countries, some treatment of this subject matter is necessary here, if only to avoid the pitfall of appearing to offer purely technological solutions to complex urban problems involving social, economic, and political dimensions of great importance and difficulty.

Chapter 6 is included in this report in an effort to give technical readers some feel for the range of non-technological possibilities being advanced as part of the world-wide assault on urban transportation difficulties. An entire book could and should be written on this subject alone. Rich experience in the innovative planning and administrative aspects of urban public transport will doubless be developed within the next few years to complement the technical literature now appearing.

Together, a wide array of service innovations in operations, planning, and technology may be called for in any comprehensive and successful attack on the complexity and variety of urban transport difficulties that plague major cities. A different mix of technological and administrative measures may be called for too, in the diversity of national urban cultures and geographics present within the member countries of O.E.C.D.

4 BUS OPERATIONS AND THE MANAGEMENT OF STREET AND HIGHWAY FACILITIES

Buses, long the mainstay of most urban transit operations around the world, have not changed much in basic design or operation for many years. That is, they hadn't until very recently, when the realization that transit must be substantially upgraded in a hurry placed transportation planners on the spot to innovate improved services. Moreover, planners were asked to develop these services by flexible and relatively inexpensive means—largely within the limits of existing investments in streets and highways, the traffic controls governing the use of those roadways, and the public vehicles (namely, buses) that use, or share, those facilities with private autos.

Just as the tram in certain applications is considered an intermediate substitute technology for rail rapid transit, so the bus in some of the new applications that resulted may be viewed as an interim technology until there is the demand, land use, or technology for more specialized and advanced systems. Nevertheless, many of the emerging concepts for employing buses in inventive, new ways offer capabilities that will be difficult to exceed with any other mode of transit, either existing or envisioned.

The huge public and private investment already existing in the urban street networks of most large cities and their surrounding regions does not bid casually to be duplicated, if such can be possibly avoided. Thus any system of public transport that can use existing rights-of-way all the gamut from local streets to limited-access, high-speed freeways has much going for it, not only in the short run, but in the long run as well.

But this does not include the same old slow, often long and circuitous, area-wide coverage routes that have been the general characteristic of many urban bus operations. These operations, even when equipped with the newest of rolling stock and operating on moderate to short headways, seem to hold and serve only those people unable to travel in automobiles. That, of course, is a valuable service, but not one with great potential for attracting riders with an automobile available to them, who are thus able to avoid the delays, transfers, inclement weather and other drawbacks such a bus service offers by its very operational nature. Rather, what does offer the potential of substantial shifts from automobile to transit usage in particular kinds of trips, and improved services to "captive" transit riders for other types of travel already well served by autos, is a variety of highly differentiated new services that are possible in addition to the basic, already existing "background" service.

These new services make use of a range of vehicle types from quite small mini-buses, capable of carrying perhaps a dozen or so passengers, to large, articulated bus "trains" with double or triple the capacities of standard present vehicles. Some of the services are highly dependent on schemes giving priority for the use of existing street and highway capacity rather than on any special new characteristics of the vehicles themselves. And others take advantage of communications methods and computers to add flexibility in their ability to serve dispersed origins and destinations in low density areas.

Advances are of course also being made in the more standard operations of urban bus fleets to yield greater efficiencies, higher comfort, lower air and noise pollution, and higher reliability of existing services. But these do not constitute the type of *service* innovations discussed in this report. A wide variety of articles and reports already exist in the trade and professional literature of transit operations on these matters, and they will not be summarized here.

Innovations in bus operations that can result in significant innovations of service will be discussed instead, under the following four headings, each categorizing a different transport function performed: bus rapid transit; demand-responsive buses; activity center circulation; and central city operations.

BUS RAPID TRANSIT

Bus rapid transit is basically the concept of operating a more or less conventional bus on its own exclusive right-of-way, independent of the vagaries and congestion of regular street or highway traffic. As such it operates much like a rail rapid transit system, but with lower hourly capacity and at lower capital cost.

Quite often the large capacities of 40,000–50,000 passengers per hour possible with a rail rapid transit link are vastly in excess of the possible transit demand for a given corridor which could nevertheless still benefit from high-speed, essentially line-haul transit service. Providing an exclusive lane of an existing, already congested highway or constructing a separate busway for the uncongested operation of buses at moderately high speeds yields a medium-capacity rapid transit operation with buses rather than rail cars.

Since the average capacity of a freeway lane is roughly 1,500 automobiles per hour, and the occupancy figure for individual automobiles is about 1.6, the person capacity of such a lane is roughly 2,400 persons per hour, a number often contrasted with the 40,000-50,000 figure of a single rail line by proponents of rail rapid transit on efficiency grounds. But the same freeway lane, if given over to exclusive bus operations, or a busway similar to it, could handle at capacity from 120 to 180 vehicles per hour, or more,[1] depending somewhat on speeds and vehicle characteristics, and with relatively large numbers of persons per vehicle, say 50 for example, could therefore carry on the order of 6,500-9,500 passengers per hour.

Where such an allocation of lanes takes away from the already limited and congested space available to individual motorists, the time delays to the remaining motorists may initially exceed the time advantages to the transit passengers overall, particularly where the exclusive lane is operating far short of capacity. But it is partly the inducement of avoiding congestion and time delay that is at the heart of this means of encouraging people to shift from auto to transit ridership for increased efficiency in the use of existing highway facilities. Were there not a long-range potential for improved utilization of such facilities, such a lane allocation, with its interim inefficiencies while the changeover is being made, would probably not make for good transport practice.

In a number of situations where this concept might be applied there is a third option to that of either taking an existing lane of rush hour traffic out of service or constructing an entirely new facility. That is to take a lane of existing highway out of service in the non-rush hour direction, where it is only lightly traveled and thus essentially surplus, and mark it off temporarily for rush-hour exclusive bus operation in the opposite direction, thereby adding a new lane of capacity at low cost and without penalizing existing auto traffic over its present level of congestion.

None of these possibilities are merely concepts at this point; all are being tried, and in most instances, being judged as

4.1 Exclusive bus lanes on Shirley Highway for commuter bus rapid transit service between Virginia suburbs and Washington, D.C.

successful. Three examples from the United States can be cited.

The first example is the Shirley Highway Express-Bus-On-Freeway Demonstration Project using a special lane of an interstate highway connecting the Virginia suburbs with downtown Washington, D.C. (figure 4.1). There the reconstruction of what was originally a four-lane freeway with a wide median, into an eight-lane freeway with three lanes each way and two reversible lanes in the median, provided the opportunity for express bus service using the reversible lanes.

In September 1969, exclusive bus lanes began operating during the morning peak period over the 4.9 miles of the Shirley Highway (I-95) where construction of the reversible lanes had been completed. Even over this relatively short stretch of highway, time-savings of 12 to 18 minutes per trip were already being realized.

By April of 1971 four more miles of temporary busway had been completed and connected with a new bridge leading directly into Washington. The result at this point was a combined time-savings of 30 minutes for buses over automobiles for the same portion of the Shirley Highway trip.

As of September 1971, the busway was carrying 41 inbound morning trips with 1,752 passengers. And planning was underway for the introduction of 90 new buses in 1972, as well as for the continued operation of the busway as a demonstration for at least the three remaining years that reconstruction of I-95 would require.[2]

By November 1974, over 400 bus trips were being made during the 6:30 A.M.–9 A.M. with 50 of them being private carriers and an additional 2,600 carpools, making a total of 35,000 people using the facility during this peak morning period. The busway is no longer considered a demonstration project and is a permanent part of the Washington metropolitan transportation system. It has served not only as a valuable transportation corridor for express service but has relieved other arterials in the area so that local traffic has been able to operate in a less congested manner.

Another example of the application of this concept is the construction of an exclusive busway 11 miles long in a corridor between Los Angeles and El Monte, California, which is built partially within the center of, and partially along the side of, the existing San Bernardino freeway. The most important aspect of this project is the cooperation between the Federal Highway Administration in planning, funding, and evaluating this type of permanent highway transit facility. The estimated passenger volume will be 4,000 persons per hour in the peak-hour direction with an average bus speed of 40 miles per hour and a travel time of 18 minutes.[3]

By late 1976, 17,000 daily riders were using the facility, with 75 buses operating in the peak hour. Carpools have been added to the facility and, by May 1, 1977, full mix use of carpools and buses is planned at an average operating speed of 50 miles per hour.

Finally, the third example is the reverse lane operations of two and a half miles of a heavily congested route feeding cars and buses for entry into New York City. An exclusive bus lane has been provided since December 1970 on a section of the I-495 (New Jersey Turnpike) approach to the Lincoln Tunnel. The bus-only lane is an additional eastbound lane which ordinarily carries westbound traffic. Thus there are now four lanes eastbound and two lanes westbound daily during morning commute hours.

During the first quarter of 1971, the normal morning volumes recorded were 794 buses carrying 33,400 passengers. The peak-hour (8:00-9:00 A.M.) volumes were 471 buses carrying 21,200 passengers. Based on preliminary estimates, the average transit rider is saving 15 minutes per inbound trip. Due to the special geometry of the facilities, however, no similar reversal of lanes is possible in the evenings for the outbound commuter trip.[4]

In the years since 1971, a number of additional busways have been introduced. One is functioning in the Banfield Freeway in the city or Portland, Oregon, taking advantage of a limited widening opportunity in a freeway at a time when resurfacing was necessary.

A second busway recently introduced amidst great political clamor involved the Santa Monica Freeway in Los Angeles where no new capacity was added for automobiles while an existing lane was taken away for use of high-occupancy vehicles and buses. This Los Angeles experiment was terminated after a judicial finding that an environmental impact statement was needed to be provided for the bus and high-occupancy vehicle facility.

The key difference in the Portland and Los Angeles cases seems to be the issue of whether or not the busway is added as new capacity to the facility or is obtained by taking away existing auto capacity.

While bus rapid transit differs little from other fixed or exclusive guideway systems during the line-haul portion of its travel through a heavily traveled urban corridor, there can be certain quite noticeable differences at both the lower-density, suburban end of the trip and at the higher-density, center city end as well. The same bus that operates at moderately high speeds in its exclusive freeway lane can also operate at lower speeds in mixed traffic on local streets to pick up and deliver passengers relatively close to their residences before starting the high speed run, or at its conclusion.

Likewise, the same bus can operate on downtown city streets once it leaves its special line-haul facility, to distribute its passengers close to job locations throughout the urban center. In the latter case, however, mixing with normal street traffic at rush hours downtown can create substantial delays, thereby reducing the time-savings door-to-door of bus travelers in comparison to auto travel.

To overcome such street delays on the downtown portion of both morning and evening trips a variety of special traffic priorities for buses can be instituted. These include bus-only lanes at certain times on surface streets, special turning privileges at intersections, and traffic signal priorities weighting timing sequences in favor of buses. The variety of such aids to central city bus operations on surface streets is discussed more fully in a section later in this chapter.

The potential for buses to operate both on their own fixed rights-of-way exclusive of other traffic, and on surface streets mixed with other traffic, albeit often with priority treatment, can be thought of as a form of dual-mode operation. While the dual-mode concept often carries with it trappings of automation and rail guidance on the line-haul portion of the trip, such is not central to the successful operation of a dual-mode bus rapid transit service. The key is a combination of local street distribution coupled with exclusive right-of-way operation through a high traffic density corridor.

Automation of operations on the busway does have certain limited advantages in reducing the width of the busway needed and in terms of increasing the theoretical capacity of a single lane of right-of-way in a number of vehicles. While such improvements were the aim of a British research and development effort called the Throughways Bus, they were found to be insufficient to warrant costly development efforts.

The ultimate advantage of experiments with automated bus operation on separate rights-of-way is not so much in terms of large size bus vehicles, since few corridors not already served by rail rapid transit would require the added capacity, but of smaller sized public vehicles as the forerunners of the first automated highways to demonstrate dual-mode auto operations as well. (This subject will be further treated in a section on dual-mode systems in Chapter 9.)

Thus bus rapid transit is already here today, albeit in a limited way, and it has performed sufficiently well to have American authorities highly enthusiastic about its prospects. It is also here in the beginnings of a dual-mode operation with priority treatment on congested city streets in Washington, D.C., and it shows signs of growing rapidly in application over the next few years.

DEMAND-RESPONSIVE BUSES

Demand-responsive bus operations are at almost the opposite extreme of the efficiency/capacity spectrum of bus services from that of bus rapid transit described above. The key concept in demand-responsive operations is to make the local service (as opposed to line-haul) of buses as flexible and personalized as possible, resembling taxi operations in origin-to-destination service over relatively small service areas.

Benefits do not accrue from reducing traffic congestion, shifting much mileage from automobile to transit, or gaining great efficiencies or reduced costs in transit operations. The areas best served by such systems are already low to moderate density areas, scarcely bothered by congestion at all; the trips involved are mostly short, and the inefficiencies and costs of such small bus/collective taxi service may be high when compared to more conventional bus operations.[5]

But there is a basic social inequity to be overcome by bringing such transit service to carless inhabitants of the preponderantly moderate density portion of present urban areas, an inequity which is created by the dominance of private transport in these areas to the almost complete exclusion of economically viable conventional local bus transportation.

The choice of the auto for local travel by the vast majority of those inhabitants in low density areas with a choice available to them has thus left those without autos without public transportation as well. Since little opportunity for economies of scale exists in such local areas, the justification for providing a relatively expensive transit service may well have to come on social equity grounds, if it comes at all. But a concept and its supporting technology exist now in a wide variety of demonstrations, and even a few fully functioning systems, to rectify this inequity at costs somewhat less than subsidized taxi operations fulfilling the same function might cost.

The operating demand-responsive systems, whether fully functioning systems (as in the Canadian cities of Toronto, Ontario and Regina, Saskatchewan), or small-scale local experiments (such as those in the American cities of Batavia, New York, Ann Arbor, Michigan, and Columbia, Maryland; in the Netherlands city of Emmen and its suburb of Emmerhout; and in the Japanese town of Nose), or even a larger-scale U.S. government sponsored demonstration (in Haddonfield, New Jersey), serve a variety of purposes.[6–9] None are addressed solely to the service of the unserved, but rather reflect a variety of joint services such as the access of commuters to a local rapid transit station (Haddonfield), the overall transit service of a modest-sized town (Regina), or a premium service competing with a regular bus service (Ann Arbor). Surely the economics of the situation argue for trying to combine a service that meets the needs of the otherwise unserved and, as well, is attractive enough to gain discretionary riders. Whether this will be possible on service grounds alone is far from clear at this point.

Much needs to be learned from these and other demonstration installations of such services about costs, patronage,

4.2 Demand bus services. Top three figures show demand bus in Nose Town near Osaka, Japan; lower left figure shows dial-a-ride van in Ann Arbor, Michigan; and lower right figure shows Buxi demand bus in a suburban housing area of Emmen, the Netherlands.

and inter-modal transfers to other services in the urban areas. On the other hand, not much needs to be learned about the technology, which is relatively straightforward, at least in the first, relatively simple, demonstrations. Variously called Dial-A-Ride or Dial-A-Bus, these services combine the telephone communications of a person desiring a ride calling in to a dispatching office which is in radio contact with a fleet of small buses (10–20 passenger size usually), much as a taxi dispatcher is in contact with a taxi fleet today (figure 4.2).

Such buses operate on general routes through areas on more or less fixed headways, but they have the flexibility to deviate from their basic routes to pick up and drop off passengers on demand, hence the term demand-responsivle.

In areas where telephones in homes are near universal, making the service work is clearly a transportation problem, but where home telephones are still a relative luxury it becomes a communications problem as well.

Deciding whether the service is to be a so-called many-to-many service or a many-to-few service is another issue. An example of a many-to-many service is one which picks up passengers over a relatively dispersed area and delivers them around a similarly dispersed area. Complete area coverage of a small town or suburb might fit this model. Many-to-few service, on the other hand, offers pickup at one end of the trip over a relatively wide area, but delivery to only one or several particularly important central locations, such as a rail transit stop, railroad or bus station, or shopping area.

The more extensive the area served, the longer and more circuitous the routes, the larger the buses, and the more destinations served by a single route—the more likely is a move away from personalized services. All these are tendencies toward reducing the quality of service to those taken out of their way to service the needs of others. But they are also tendencies toward reducing the costs of operation and increasing the area of at least some service. The difficulty and confusion of these services today is in deciding just what it is that they are trying to provide and on what grounds they are to be evaluated.

ACTIVITY CENTER CIRCULATION

Activity center circulation refers to those services specially oriented to movement entirely within a small, but high-density portion of an urban area. Here surface congestion is extremely intense, pedestrian activity is prevalent and often difficult, parking may be in relatively short supply, and pollution effects of vehicular traffic are particularly measurable.

Such conditions can extend over quite large portions of the central cities of the dozen or so largest-sized urban areas in the world, but even there they are localized in peaks of higher intensity some places than others. And in some of the more spreadout urban areas without a single, dominant focus of downtown activity, a number of such activity centers may exist some distance apart. Besides central city commercial and employment centers, more and more specialized centers of activity—such as universities, airport terminals, and even recreational areas—are experiencing difficulties of intense localized congestion, such as pollution, parking, and lack of pedestrian amenity.

Special services employing buses are possible which can help in circulating persons within such centers, distributing passengers from autos in peripheral parking or from transit stations to more central locations, and reclaiming some of the existing street and parking space for higher quality land uses. But such bus services are only a small beginning in dealing with the special problems of such areas. Much must be done as well with auto and land-use policy before substantial changes can be observed.[10]

4.3 Minibuses in use for activity center circulation in downtown Los Angeles, California.

Some of the needed bus improvements are linked up with priority treatment of surface transit vehicles in central cities, as will be discussed in the section following this one. But there are quite localized operations that can be considered, and these have already been implemented in a variety of activity centers.

Special mini-buses have made their appearance in the downtown areas of a variety of cities at one time or another. These buses, operating on simple shuttle routes or short loops in high density areas, have largely been viewed as pedestrian aids for circulation within the center, although sometimes they link up with parking facilities as well to complement the auto trip to or from the center itself. Such services are operating in the downtown areas of Los Angeles, California (figure 4.3) and Minneapolis, Minnesota, with quite small buses, frequent service, and low fares.[11]

The same type of small buses operating along a special mall in Washington, D.C. were deemed both a success and a failure a few years ago. It would appear, at least in part, that the success precipitated the failure. The buses did attract patronage with a combination of their special shuttle service through a shopping area, their low fare and frequent visibility, and their novelty and amenity. But they were unable to meet the demands placed upon them when patronage increased above comfortable levels, they were costly to operate, and they declined in amenity and novelty as time wore on.

Now that they have been discontinued in operation, a more extensive service, using more and longer routes, and operating larger, yet not full-size, buses, is under serious consideration to replace them. This service is modeled on the London Red Arrow bus service, and was in fact recommended as one outcome of a promising international consulting mission of the London Transport Executive. The Red Arrow buses are a medium-sized bus with relatively few seats and abundant standing room for special routes in London from the commuter rail terminals of British Rail to the activity centers of commerce and employment in the downtown. Routes are short and direct, travel times are short, and fares are relatively low for this service which runs mainly during morning and evening rush hours.

The history of the Washington, D.C. mini-bus service illustrates the importance of flexibility and careful staging of new transit services, at least in the beginning stages of activity center development or redevelopment. By beginning with limited but high quality bus service, initial demand can be met and perhaps additional demand generated in the same or like areas. This additional demand can be met with larger, or more frequent, or more extensive, bus services until the point where perhaps exclusive right-of-way systems can be justified to take over this service. That is the strategy, in part, of the Los Angeles downtown circulation system. Longer-term plans call for several lines of exclusive guideway transit systems to serve a major redevelopment project in the downtown, neighboring car parks, and much of the rest of the several square mile area.[12] But before these systems are ready, familiarity with a specialized activity center circulation service is being obtained through the use of mini-buses as an interim measure.

Special malls created out of existing streets for "pedestrian precincts" are sometimes shared with low-speed, specialized buses or elephant trains to handle the longer trips still within the same activity center. Together this increases the amenity of the space thus reclaimed from street traffic and curb parking. It also facilitates walking as a mode of urban travel, sometimes mechanically assisted when time, weather or distance might be an inhibiting factor. The subject of pedestrian precincts itself has been taken up in other O.E.C.D. work, particularly by the Urban Affairs Division in the past few years.[13]

CENTRAL CITY OPERATIONS

Where surface vehicular congestion is much more geographically extensive than in a few, rather well-defined activity centers distributed about the urban area, additional measures to give priority treatment to buses apart from other street traffic are promising.

4.4 Central city bus mall in Minneapolis, Minnesota with pedestrian amenities, minibus circulation, and downtown portions of regular bus routes combined.

Special bus malls have been created in some cities. A particularly successful example can be found in Minneapolis, Minnesota. There, a major commercial and office street in the downtown was turned over to the exclusive use of buses and taxis for much of its length, with limited cross-traffic allowed to interfere with the progression of the buses down the mall at intersections (figure 4.4). These buses are completing or beginning the downtown portion of extended trips throughout the central city in a congestion-free setting which is also high in pedestrian and passenger amenity, with frequent bus shelters, benches, lights, information kiosks, public telephones and landscaping.

This particular mall combines several activity center circulation functions in a portion of the downtown: a pedestrian facility, a fleet of mini-buses operating over short downtown loops, together with portions of the center city operations of the regular bus fleet. Proposals exist for similar "pedestrian transitways" in Madison Avenue, New York City and Oxford Street, London. In London, a half-mile long stretch of the busiest portion of Oxford Street has been closed to all traffic except buses and taxis.

Where an entire street cannot be given over to bus operations, a smaller allocation of space, such as a single, exclusive lane, may be. Bus lanes on surface streets, as distinguished from the high-speed busway of bus rapid transit, may be short-priority stretches of only a city block or two to avoid a specific traffic bottleneck, or they may be an extensive system of auto exclusion from much of the congested area during peak hour operations. London offers an example of the former, quite limited use of this concept, while Paris contrasts with a more extensive allocation of such facilities in many of the central portions of the city.[14–16]

Exclusive bus lanes may be either in the direction of normal street traffic, where it is often difficult to enforce the exclusion of other traffic, or they may run against the normal flows and thereby inhibit such encroachment by private vehicles. The operation of such lanes is much more developed in Europe than it is in North America, although experiments have been introduced in Washington, D.C. in conjunction with the downtown portion of the Shirley Highway bus rapid transit demonstration discussed earlier.

Priorities can also be assigned to buses through traffic signal controls and other measures restraining flow. Much work is now going on in the United States with traffic signal timing to sense and then assign priority to traffic flows in proportion to number of persons carried rather than to number of vehicles. Much of the work in Washington, D.C., had been intended as the site of a demonstration of this technique. A computerized traffic control system was contemplated for installation at 112 signalized intersections in the downtown area. This program, however, was curtailed after an initial period because of the complexity of traffic flows and the management of multidirectional vehicle flow. Where such techniques have succeeded have been along arterial streets where buses have their own lanes and can pre-empt signals at intersections. Much work remains to be done on urban traffic control systems and much work continues in progress on traffic control system development.

Other work is going on in the United States in metering the flow of vehicles onto freeways at ramps to keep the flows uniform and less congested. Priorities for buses over autos could be assigned at such freeway entry points to increase the flow of people carried by the facility.

A simple version of such a scheme has recently been introduced on the Oakland-San Francisco Bay Bridge, a heavily traveled commuter corridor between two portions of this large, water-separated urban area in northern California (figure 4.5).

4.5 Car pool and bus priority lanes on the San Francisco-Oakland bridge, showing location of lane signs and sample car sticker.

This is a toll bridge facility where autos, buses, and trucks have to queue up to pay a toll at a barrier before proceeding in the westerly direction toward San Francisco. Priority passage for buses without stopping to pay tolls, together with a special bus lane leading to and extending far beyond the toll barrier, both were introduced in April 1970 to speed up transit service across the bridge. Late in 1971, an experiment was tried, giving the same priority treatment to automobiles carrying three or more persons—"carpools"—during the peak morning period from 6:00–9:00 A.M. Within one week the number of carpools per day jumped from 1,000 to 2,000 where it has held relatively steady since.[17] This experiment in giving "people-preference" at points of traffic flow control has proved sufficiently successful that it is no longer an experiment, but operating as regular policy for the foreseeable future.[18]

Other possibilities for sensing and controlling the location, passenger content, and operations of transit vehicles and private automobiles on critical street facilities are gradually being developed, and would seem to promise major improvements in traffic operations to the potential benefit of surface transit. Automatic Vehicle Monitoring (AVM) is a means of sensing, at frequent intervals, the location of vehicles carrying special electronic equipment which communicates a signal back to a central control facility by a variety of means differing in technical sophistication and cost. Experiments with a variety of systems are going on in the U.K., France, and the United States.[19–20]

One place where AVM is already in operational practice is Hamburg, Germany, where it is used to keep track of a significant portion of that city's fleet of buses. When a bus equipped with this system begins to fall behind its schedule it is immediately apparent in the central dispatching control center as a discrepancy between actual location and scheduled location on an automatic plotting device. Information can then be transmitted to the driver of such an off-schedule vehicle to speed up or delay as the case may warrant.

In its simplest form such a system can give some assistance in improving the marginal performance of surface vehicles in regular operations. But the larger benefits of the AVM concept come when data generated from it can be used to identify the recurring points of troublesome street operations, keep track of passenger loads in individual vehicles, and otherwise vastly increase the sophistication of knowledge of what is happening in the complex operations of a traffic system in which transit and private vehicles are mixed.

For example, it is the hope of the Hamburg Transit Authority (HHA) to gather sufficient data to identify and document through AVM those portions of the Hamburg street network where delays are hampering surface operations, and thereby to be able to propose to city authorities a comprehensive system of exclusive bus lanes and traffic priorities to improve transit operations.[21]

Such on-board electronic equipment, coupled with sensors and communications equipment located in the streets, is also part of an experiment being sponsored jointly by the General Motors Corporation and the city of Rochester, New York to increase the flow of buses on city streets. By platooning buses—running them close together in groups, rather than spreading them out in more or less equal separation down a street, or exclusive lane or mall—delays of loading and unloading can be significantly reduced. And the flow rate along whatever surface facility is being utilized can thereby be increased.

To gain these time-savings, however, it is necessary to load and unload a number of buses at one time in roughly the same area. To organize the queuing of passengers at such a multiple stop it is necessary to know which buses are coming when and exactly where they will be stopping. Electronic signboards, like the familiar time and temperature indicators, indicating which bus is arriving at which portion of the bus stop, can alert passengers to exactly where to board as each successive platoon of buses approaches. The information on which bus is where is transmitted from the bus to the sign through sensors located somewhat down the street. The increased capacity provided by a single street lane utilizing such a platooning concept can be quite valuable in congested central city areas where street space must be used at its most efficient.[22]

Together, the combination of bus lane priorities, traffic signal timing, turning and ramp access advantages for transit vehicles, AVM data sources and operating controls, platooning of bus arrivals, and other possibilities inherent in the management and operation of surface street space in central cities, all offer great potential for the improved efficiency and service quality of bus operations in the not-too-distant future.

5 TRAMWAYS, RAIL RAPID TRANSIT, AND SUBURBAN RAILROADS

All of the systems discussed in this section are basically rail systems, operating on more-or-less exclusive guideways and serving moderate to high density volumes of traffic. As we will see, the simple distinctions that used to separate them into three different types of rail service are rapidly disappearing.

In the not-so-distant past a tramway was characterized as a street railway that shared street space with other forms of traffic, its rails imbedded in the normal surface of major urban streets. In fact, it was often the interference of trams with other traffic that led to their removal from a number of cities, both because of a rush to do everything possible to facilitate motor travel and because of the declining performance of trams themselves when mixed with the flow of more and more congested surface traffic.

In contrast to the tram, rail rapid transit was an elevated or underground urban railway separated spatially from surface traffic on its own exclusive right-of-way. Functioning without the interference of cross traffic it could operate more rapidly and reliably than trams, but at a higher cost since it required more specialized facilities.

To be economically justifiable rail rapid transit thus needed to serve higher volumes of traffic and yet could not offer as geographically extensive a service as the tram. Its operations were consequently in the likeness of a high capacity railroad, with long trains of coupled cars running at moderate to long intervals on a relatively few linear urban routes limited to the geographic boundaries of a major city such as New York, London, or Paris.

The suburban railroad was simply a commuter railroad shuttling between suburbs and central city railway terminals shared with the normal, longer-haul national rail passenger services. It offered a specialized service of home-to-work and back travel from residential communities to points of distribution relatively close to employment centers, principally at rush hours in the morning and evening.

Often the suburban railroad tied in with the rail rapid transit system and/or trams at the central city terminals for further distribution of passengers to their specific destinations. But it basically shared the facilities of the regular railroads, and thus created only incremental costs where, or as long as, a strong national passenger rail service functioned.

Now these former distinctions are becoming hard to hold to. The tram, where it has flourished and been retained, is tending more and more to exclusive guideways both on the surface and in tunnel. Rail rapid transit is moving far beyond the limits of the central city into the former province of the suburban railway. And suburban railways, where not displaced by technologically more advanced rapid transit systems, are tunneling beyond the passenger rail terminals at the periphery of the city center, right to the very heart of its core, and thereby giving far more direct service. Often, too, the trams are getting longer and the trains shorter to increase the efficiency and capacity of the one, and the service frequency and flexibility of the other.[1]

In their most modern applications all of these systems may: use an exclusive right-of-way; be found on the surface, elevated or in tunnel; flexibly operate in single cars or coupled in trains; and be automated in some of their operations. The best manner to distinguish among these various evolutionary, exclusive guideway, urban area rail systems now, and on into the future, would seem to be in terms of their capacities and the geographic areas served.

The upper range of these capacities range from 10,000 per hour for a line of existing technology tramway up to 40,000 per hour for rail rapid transit.[2] But what is becoming critical for the more spread out and lower density cities of newer vintage, as well as for the more modest-sized cities of several hundred thousand population, is how low the capacities for an exclusive guideway system can be and still remain economically viable in meeting transportation needs.

Research and development programs are now actively underway in a number of countries, most notably France, the United Kingdom, and the United States, to obtain a mini-tram or modest capacity exclusive guideway system with basically the same service characteristics as today's higher capacity rail systems, but with substantially reduced capital and operating costs. As for the geographic areas served, these are ranging from relatively small areas such as airports and downtowns, through moderate-sized urban areas of a few hundred thousand, to extremely large urban regions of many million inhabitants. The technologies may differ somewhat in these variously sized areas, but at the heart of many of the ventures are incremental or evolutionary advances to existing rail services.

In examining what is happening right now in the improved design, operations, and environmental compatibility of tramways, rail rapid transit, and suburban railroads, it is useful to look at three aspects of these systems. These are: the nature and location of the exclusive right-of-way; the type of service provided; and the incremental staging of the technology.

THE EXCLUSIVE RIGHT-OF-WAY

The concept of the semi-métro in Europe ranks with that of bus rapid transit in the United States as a major breakthrough in the use of existing technology to accomplish urban transport goals in an efficient and effective manner. Only a few years ago the principal choice in public transport operations was between (1) low capacity and slow, but relatively inexpensive, surface modes mixed with normal traffic (in other words, trams and buses) and (2) rail rapid transit at high cost and capacity. Lately, however, these two innovations in operations have begun to close the great cost and capacity gap between them.

The semi-métro is essentially a system of tramways with an exclusive right-of-way for much of its trackage. Although a good deal of this exclusive way is at surface grade, the tramway goes into tunnel under the streets and behaves as if it were a classical subway in critical portions of the urban area where surface space is at a premium and requirements for amenity are high.

It is the tunnel portion of the right-of-way that is most publicly visible and characteristic of the new tramway investment in European cities. But improvements in rolling stock, in automation of some functions, and in exclusive surface rights-of-way are part of the picture as well.

The semi-métro represents a long-range commitment to the tram technology as being adequate in capacity and performance to meet a given city's needs. This is not necessarily true for a shorter-range, temporary installation of the same technology, termed a pre-métro. In the latter case, tunnels and stations have to be built to accommodate larger rail rapid transit trains, and when travel demand justifies expansion to higher capacities.

The semi-métro already exists in a number of cities and is being planned for others. Among those with installations today are Brussels, Vienna, Rotterdam, Frankfurt, and Cologne.

Planning is now underway for this type of installation in the middle-sized British urban area of Newcastle-upon-Tyne, and funding is available for its implementation. Although the idea is not particularly new, its growing usefulness as a cheaper step short of full métro investment seems particularly current in practice.

Upgrading and electrification of the existing rail network now used by British Rail Diesel Service will eventually constitute 26 miles of the track layout. Only 8 miles of the system will be entirely new construction. This 8-mile stretch, however, is critical because it includes north-south tunnels running under central Newcastle and south of the River Tyne. About one-quarter of this work had been accomplished by early 1977. Once the rail system is operating, the existing bus system will be reorganized to support the rail service.[3]

In fact, such an underground facility for a portion of the Boston streetcar system has existed for some time now; and in San Francisco the trams are finally going under its main commercial thoroughfare, Market Street, after some seventy years of recurring interest and planning studies.

Figure 5.1 shows what is now a street railway system (The San Francisco Municipal Railway, or "Muni") as it will appear when undergrounded in conjunction with BART, the new Bay Area Rapid Transit system, for a portion of its route. A pedestrian concourse is at the first underground level, Muni at the second, and BART at the third.

The strongest incentive for installation of semi-métros has come in those countries that maintained their tramway systems more or less intact during the past few decades while others were retiring theirs. Since the surface trackage investment in outlying areas has been preserved, together with rolling stock, operating know-how, and patronage, the incremental investment in these places for new downtown rights-of-way underground is somewhat less than would be the costs of starting an entirely new system from scratch elsewhere.

It is the incremental and opportunistic assimilation of bits of trackage, street and highway lanes, and modest technological advances that characterizes much improvement of urban transit systems today. The semi-métro concept in practice is evidence that there is great promise in this direction.

5.1 Artist's cut-away drawing of three-level underground transit facilities beneath Market Street, San Francisco, California. Here the regional rail system (BART) on the third level interlinks with the municipal tram system (Muni) on the second level and a pedestrial concourse at the first level underground.

The Skokie Swift in the Chicago metropolitan area is an American example of the same phenomenon, albeit less comprehensive or visible. There, an abandoned railroad right-of-way extending into some of Chicago's northern suburbs was adapted to commuter shuttle service by what were essentially tram cars; and ties to suburban parking and urban rail rapid transit at each end were established in one of the Urban Mass Transportation Administration's first federally aided demonstrations.[4]

Elsewhere in the United States, efforts to make use of abandoned rail lines in urban areas are beginning to follow some of the European practice, although belatedly and in fragmented fashion because tramways and their rights-of-way were not preserved.

The building of new underground, or otherwise exclusive, rights-of-way for rail services is not restricted to the tram in tunnel. In Germany some of the suburban railroads are being converted to more direct and useful service as regional railroads by tunneling new rights-of-way for them into and through the very hearts of cities.

Two major examples are the constructions now completed in Hamburg and Munich. These new S-Bahn facilities connect existing railroad terminals on two sides of each community through underground links, with several stops directly under the city centers. They should greatly reduce the problems of transfer and distribution for rail commuter passengers, and thereby save both time and energy for potentially large numbers of regional trips.

At the central, Marienplatz, station the interchange between S-Bahn and U-Bahn will be about as shown in figure 5.1 for BART and Muni in San Francisco, except that the positions of the regional rail system and urban rail systems will be reversed. The U-Bahn is on the lower level in Munich.

And in Chicago, a goodly portion of that community's new investment in rail rapid transit will soon be going into undergrounding the old elevated "Loop", which has had a somewhat notorious impact on the amenity of the downtown. An interesting operating innovation that will go along with undergrounding the rapid transit system will be a special distributor subway serving just the few mile square downtown area.[5]

Thus, from regional rail systems to downtown distributors, there is a pressure to get existing rail technology on its own exclusive right-of-way and underground as well.

LINE-HAUL OR AREA-WIDE SERVICE?

Whether a transit service is line-haul or area-wide in nature is an important characteristic to distinguish. The same technology can often serve for both, but differences in the geographical extensiveness of the facilities provided and the types of trips involved make for a substantial variation in the most desirable features of operation of the system.

Line-haul service has traditionally meant simple shuttle operations along a single corridor or variety of corridors, as with radial commuter rail lines converging on a center city, often even arriving at different stations, as seen in the service of British Rail in London, or S.N.C.F. in Paris. For trams, the same type of line-haul service is often present in radial, non-intersecting lines extending from the city center outward through less dense, residential areas. Much of the argument today in favor of the concepts of semi-métro and bus rapid transit is based on the ability of these systems to provide capacities somewhere in between those of trains and surface transit for the line-haul service function at in-between costs as well.[6]

Area-wide service, on the other hand, is a more homogeneous and dispersed type of operation in which considerable intersection of lines allows a grid-like coverage of an urban area or part thereof. The interior portions of the London Underground, Paris Métro, and New York Subway systems provide this kind of coverage for much of the densest parts of those cities through easy transfers from one line to another.

This kind of service gives access throughout an area instead of focusing on a point or a few points strung out as beads on a string. The ability to serve modest volumes of demand economically is the key to the extensiveness of service such a system can provide. It is in this function that the private auto on surface streets often performs most brilliantly and at which transit systems so often fail.

For line-haul service it is the maximum capacity that should be served that is the important sizing variable for the technology or the options utilizing that technology. But for area-wide service it is the *minimum* demand that can be served economically that provides the equally critical sizing variables, both for the technology and for the operations of a system.

Obviously sometimes a single transit system provides both services over different portions of its extent, again taking as an example the Paris Métro, which serves a line-haul function at its radial extremities and an area-wide function in its interior network of intersecting lines. But since many urban areas and their various activities are becoming distinctively less focused on a single center, the line-haul function of traditional transit operations is giving way gradually to a need for a stronger area-wide function, both for newer installations of existing technologies and for entirely new systems.

Thus part of the appeal of smaller capacity systems, at least when reduced costs go along with them, is two-fold. In moderate-sized cities which do have a suburb-to-center commute problem, but insufficient demand for any high capacity rail systems (such as Strasbourg or Grenoble in France, or Leeds or Bristol in Britain), lower capacity line-haul installations like semi-métros and bus rapid transit fill one type of pressing need. In other, larger but spread out, cities which might have the option of a limited rail rapid transit network but then only along one or a few corridors (such as Los Angeles or Minneapolis-St. Paul), the option of the lower capacity systems gives the choice of a more extensive network that can carry out area-wide functions as well.

Much of the technological research to be described in chapter 7 has, as its aim, the development of automated transit systems capable of extensive area-wide services. These systems, generally categorized as Automated Guideway Transit (AGT), have been extensively reviewed by the United States Congress, Office of Technology Assessment in 1975, and categories of systems identified on the basis of vehicle size and operating characteristics. None of the systems implemented thus far provide an area-wide network capability to serve any large-scale urban area. Those generally categorized as group rapid transit provide a smaller scale of highly automated line haul service.

The key example of this in the United States is the Morgantown demonstration project in Morgantown, West Virginia, where an automated system involving 5.4 miles of single-lane guideways and 45 vehicles is now in operation, linking two separated campuses of the University of West Virginia. A second phase, extending the system to 5 passenger stations, 8.2 miles of single-lane guideways and about 78 vehicles, is now under development.[7] While this system, in its installation in Morgantown, does not have an area-wide capability, further developments of this automated technology could evolve in that direction.

The concept of a mini-tram—or a smaller capacity and cheaper version of a mass transit system on its own right-of-way than even the current tram technology represents—is receiving much support among industry and transport ministries in Britain, France, Japan, and the United States. In both France and the U.K., where those trams, that once existed were removed rather than preserved some time ago, the new research, development, and demonstration programs seem basically aimed at solving a line-haul service problem in moderate-sized cities.

While the linkage to tram-like service is well-recognized, at least in Britain where the term seems to have originated, much more is claimed for the new systems elsewhere. Therefore, there exists considerable confusion generally about what is actually new and worthwhile about such medium-capacity systems as the Westinghouse Transit Expressway in the United States,[8] URBA in France,[9] or VONA in Japan,[10] to name just three which are illustrated in figure 5.2.

5.2 Three examples of medium capacity guideway systems. From top to bottom: Westinghouse Skybus (U.S.), URBA (France), VONA (Japan).

To the degree that these technologies are presented as medium-capacity line-haul systems, they should find their true test in the existing transit marketplace, principally on economic grounds in comparison with bus rapid transit and modern tramway systems of a more conventional nature. Showy vehicle designs, technically interesting propulsion and suspension systems, or elevated rights-of-way should not be allowed to befuddle the basic question of what service is provided at what cost. Most of these technologies are available today in prototype form, often running on experimental or even semi-public test tracks.

But the full-fledged urban applications and regular operations that would begin to yield reliable cost and performance data for potential customers are still largely missing. These should begin to emerge in the next few years, as the Transit Expressway may or may not begin operations in Pittsburgh, VONA gets a new town tryout in Japan, and other installations of such technologies are considered in France.

An additional confusion, however, is the characterization of some of these mini-tram technologies as small vehicle systems for activity centers or as personal rapid transit systems of near-individual mobility. Both of these services are different from the line-haul function that these systems were originally designed for, but both may be additionally possible as portions of a *family* of transit systems designed from similar building blocks of the basic technology. This overall family of systems is what has been analyzed in detail in the "Automated Guideway Transit" assessment referred to earlier.[11]

The jobs are different, however, and would seem to represent substantially different problems of design, cost and implementation. These will thus be treated in chapter 7 as activity center systems and personal rapid transit.

In addition to the considerable investment being made in rail infra-structure and new automated technologies, considerable effort is being invested as well in modernizing tram or light rail vehicle technology. Widespread efforts are being made to upgrade the image of a well-established transit mode through an attractive car design, seating placement, graphics and automation.

The fact that an effort has been made to distinguish between line-haul and area-wide transit services does not mean that conventional technologies cannot serve area-wide needs, if consciously applied with that end in mind. As has already been pointed out, that is precisely a portion of the service carried out by the intersecting lines of the Paris Métro and other central city bus, tram, and rail rapid transit lines through passenger transfers.

Efforts are now underway in a number of cities to make these transfers physically and economically easier and more extensive. Not only are extensions to the rail rapid transit networks of cities, such as London and Paris, extending out into new territory within the central city or its suburbs to offer new line-haul services, but they are providing a denser, more interconnected area-wide service throughout the central areas as well. The same is also true with the transfer possibilities of the new downtown S-Bahn services in Germany and the regional scale rail rapid transit systems of Paris (Reseau Express

Regional/RER), the San Francisco Bay Area (Bay Area Rapid Transit/BART), and Stockholm.

STAGING AND SPECIALIZATION OF SERVICES

Few cities now constructing, or contemplating the construction of, new rail systems have been entirely without some form of rail service in recent years. Many of the largest commitments of resources have been in urban areas where the dependence on tramways, rail rapid transit, or suburban railroads for urban transportation has been well established by the structure of the central city and the patronage of its existing services.

Thus the new investments can be considered incremental—albeit large—extensions or conversions of previously more limited services. Limited, that is, either in geographic extent or quality of service. It is far less risky, from a city's point of view, to know something from existing development and behavior patterns about the utilization and impact of new transport services.

For this reason, the staging of services in time and geography is an important aspect of decisions regarding new transit investment, particularly the fixed guideway kind, where vehicles cannot simply be rerouted to other areas of the city or sold—as with buses—if demand proves uneconomical on a given line or overall area.

The pre-métro, as mentioned earlier, is an important concept in staging transit services, allowing some time for a wait-and-see approach to public acceptance for urban rail service before more costly investments in rail rapid transit technology are made, if at all. Such is the strategy being adopted by some communities, such as Stuttgart, where the growth of demand is somewhat uncertain.

Although the pre-métro name refers specifically to tram technology operating in tunnels capable of accepting rail transit trains, the concept of staging services is a more general one. For example, in those American cities where no recent history of rail operations exists, bus rapid transit using existing freeways might usefully precede decisions about whether to invest in rail or other fixed guideway systems.

But the key to gaining reliable information from such an exercise is to make sure that the quality of service being rendered initially is sufficiently good to adequately reflect consumer preferences. Patronage on a poor performance surface street bus system is no indicator of the potential market for regional rail service, for example.

Staging in geography is not only a transit investment strategy, but an urban development one as well. How this can work successfully is demonstrated by the present practice of investment in the Paris region.

In the Paris region new investments in conventional métro line extensions, an entirely new regional rapid transit system—the RER—and suburban rail extensions to a variety of new towns even beyond the regional service network of the RER are staged to (1) bring existing quality rail services (in the case of the métro and the suburban railroad) to more extensive areas to keep pace with urban growth or to help encourage it, or (2) to provide a sufficiently better service of the same type (but with increased speed and fewer stops) that will help alter the geography of the region in terms of time and space in order to decentralize certain business and governmental functions out of immediate physical proximity to each other in overcrowded central Paris.[12]

None of the transport investments is thus radically uncertain in terms of departures from past experience, yet each will have major impacts on the growth and development of the region.

When the plan is completed, each of the different rail services in the Paris region will have its own specialized function to perform. Such also is the case with the trams, rail rapid transit, and railroads in the Munich region when its ambitious plans, too, are implemented.

But whereas in Paris the essentially new level of service—the RER—is intended to take over a geographic function not previously performed by any existing system, in Munich the situation is somewhat different. There, the newest mode introduced into the region is a rail rapid transit system that will replace a number of existing tram lines in serving the downtown. These trams, rather than being discontinued, will be reoriented to new lines feeding various rail transit stations, initiating a new specialized service to make the best use of the equipment available.

The concept of specializing the functions of the various systems available carries over to street systems as well. A single point is raised by Jack Eisen in his excellent paper prepared for the Economic Commission for Europe, Seminar of 1976, referring to the integrated planning and development of a variety of transit services in both Munich and San Francisco.[13]

6 INSTITUTIONAL INNOVATIONS IN PLANNING AND ADMINISTRATION

The changes possible in the operations of existing modes of urban public transport, as discussed in the previous two chapters, open vast realms of possibility in terms of improved service to the public. But for the impact of such operational improvements to be substantial enough to make a real difference in the overall transportation balance between private and public mobility, much must be done to increase the levels of investment allocated to these public modes and also to integrate the separate modes so that they can perform comprehensively as an urban or regional system.

The many cities visited in the course of data collection for this report differ widely in their existing investments in public transport and in their physical and administrative structures as well. These differences will of course influence the potential amounts of new investment in such modes as well as the degree and nature of institutional change needed to bring about an adequate administrative system of public and private transport.

If the financial and political commitment is not already present to provide the resources necessary for at least a moderately good public transport system in a given urban area, the incremental advances in transit operations talked about earlier, or the planning and administrative reforms to be talked about in this chapter, make no sense.

But the task set out for this report is not to discuss in detail the adequacy of investment in public transport today in various urban areas. It is rather to concentrate on the possibilities of improved performance by public transport where and when such investment is considered. A major premise of the report, and particularly of this chapter, is that substantial new investment in public transport infrastructure will be required by almost all major urban areas, before incremental improvements in operations, planning and administration can have any significant effect on the overall problem of mediating the impact of the automobile on the environment.

In most of those urban areas to be discussed here, substantial investments *have* been made, or are currently in the process of being made, in such public transport infrastructure. And planning or administrative innovations have been found to be key elements in these places in gaining maximum utilization or impact of the transport facilities involved.

These innovations are advanced in this chapter, in the form of brief case studies, as possible suggestions for other urban areas to consider in looking at the non-technical side of public transport improvement. Together with adequate levels of financial investment in public transport, evolutionary operational improvement to existing technologies, and comprehensive policies of automobile restraint in certain urban areas or portions thereof, such institutional innovations in planning and administration can contribute major change in the adequacy and balance of both present and future urban transportation modes.

Four separate, but related, issues of planning and administration are discussed in this chapter: administrative coordination of public transport; coordination of transport and land use planning; complementary planning of highways and transit; and economic innovations in finance and pricing transport services. The examples cited in each case are simply illustrative of present practice in a variety of cities sampled, not specific recommendations for other cities to necessarily consider following.

ADMINISTRATIVE COORDINATION OF PUBLIC TRANSPORT

While competition between suppliers of a given service is often encouraged in relatively strong sectors of an urban economy as an effective means of keeping efficiencies high and prices low, there are few today who would claim that urban public transport fits within the category of services for which internal competition is a cure rather than a curse.

The modifier "internal" is important here, however, as there are many concepts afoot for improving urban transport by opening competition more directly (1) between private road use and public transport use by imputing real marginal costs to urban highway users through forms of road pricing, and (2) between taxi operators to increase the number of vehicles available, provide for shared ridership, and reduce passenger costs.

But in terms of different modes of public transport, there are a variety of efforts now going on aimed at strengthening their economic and service characteristics through administrative and financial coordination of formerly competitive private firms or public agencies.

Perhaps the most famous is the reorganization of bus, train, and rail services in Hamburg, Germany, which took place a few years ago,[1] and which is being adopted in similar form for the Munich region as well. In both of these instances private suburban bus firms, city transit agencies and regional operations of the national railway system (Deutsches Bundesbahn) have been coordinated so that complementary services, rather than competitive or disjointed ones, are provided.

The tasks and revenues for the various modes and operating agencies are apportioned centrally by a governing body made up of representatives of these agencies, forming a transit alliance which thus coordinates schedules, fares, and routes, but leaves the specifics of operations in the hands of the individual participating agencies.

In the United States, regional transport agencies are coming into existence largely as planning bodies, as comprehensive and coordinated planning for transit modes is becoming prerequisite for obtaining federal grants for transit system capital expenditures. As more and more transit systems are put back on their feet, economically, so that a potential for possibly harmful competition starts to exist, these agencies may be called on to perform functions similar to those in Hamburg and Munich.

The introduction of service by the Bay Area Rapid Transit System in a portion of the San Francisco urban region was the first instance of this kind of coordination in recent years in the United States. The existing patterns of service and economic viability of several other high quality public transit agencies in the area (Alameda-Contra Costa Transit and the San Francisco Municipal Railway) will be much influenced by the way in which new service patterns, fares, and transfers between systems are handled. A major thrust of the regionwide Metropolitan Transportation Commission has been toward the integration of transit services, fares and funding. While some progress has been made, much of the potential for coordination remains as yet untapped by this recent new planning and coordinating body.

Since a varied array of transit services, from small-area demand bus operations to regional-scale rail transit linkages, will need to operate within the same regional territories for a major revitalization of urban public transport capabilities, this problem will appear more and more frequently in cities not yet as richly endowed with transit services as Hamburg.

If a city or region is open to the opportunities of coordination, not only in its existing operations but for future ones as well, a number of ad hoc possibilities may arise. Lausanne, Switzerland offers an example of this in the planning recently carried out for a new rail service which would benefit from the coordination of several existing services and the construction of a new facility as well.[2]

There, a relatively short rail transit link of almost a funicular sort already links the central portion of the city with its recreational frontage on Lake Geneva, and several intervening areas of business and residential importance. And a suburban railroad, which now rather poorly connects some areas of potential residential growth with portions of the commercial center, has its operating franchise about to come up for review.

By combining a goodly portion of the railroad's existing right-of-way with newly constructed right-of-way to bring rail service more directly through other important portions of the community and into the center where linkage with the existing shuttle system is possible, it may be possible for the Lausanne region to rather inexpensively obtain the first line of what could now be upgraded to a rail transit system.

But to capitalize on such opportunities, which also must exist in many other communities, coordination and imagination are required. For example, such an opportunity involving an abandoned railroad right-of-way running into downtown Milwaukee, Wisconsin, provoked the serious consideration of exclusive busways as a transit solution for that community. And a recent bill placed before the California State Legislature seeks to institutionalize this search procedure for new opportunities in abandoned rail rights-of-way by requiring the State Public Utilities Commission to consider the usage for public transport of rail lines considered for abandonment in urban areas. In Portland, Oregon, the presence of numerous abandoned tram rights-of-way has led to a surge of interest in light rail as an alternative regional transit technology for the metropolitan area.

COORDINATION OF PUBLIC TRANSPORTATION AND LAND USE PLANNING

The concept of coordinated planning for public transportation and the urban land uses that it serves is certainly not innovative in and of itself. Cities were planned or grew up around public transport services when these held a near monopoly on the means of transport for the bulk of the urban population.

Private fortunes were made on real estate promotions

built up around suburban rail and tram lines, for example, when these opened up new areas for residential development at the beginning of the present century, particularly in the United States. And currently in Tokyo and other cities in Japan, privately owned commuter railroads discharge many of their passengers into the subterranean arcades of large department store complexes which they in turn own. And thus they capitalize on the ready market offered by such a concentrated daily mass of customers.

But the notions that public transport creates public land values and development opportunities that are meaningful in an era of wide-spread private automobile ownership, that land uses can be planned in detail to maximize the benefits from public transport access, and that the timing, as well as the location, of urban activities might be subjected to planned intervention by public agencies—all of these are relatively untried elements of urban transport policy in the great majority of cities.

The Value Capture concept of taxing increased land values around transit stations has recently been injected in the United States public transportation policy, and the Urban Mass Transportation Administration is encouraging communities to leverage the value of their transit investments through concentrated planning of land use in impact areas adjacent to transit systems.

Stockholm has a long and successful history of building its rail transit facilities and its suburbs together, a line and a town at a time. Theirs is perhaps the purest example of what can be done to build-in through mutually-supportive policies of a single government agency or closely coordinated agencies both the transit ridership and the urban development that public transit permits (or encourages).

For cultures and urban economies less centrally planned than Stockholm's, other possibilities present themselves as well. In San Francisco, for example, additional floor space for high-rise office construction is offered as a financial inducement to firms who will locate adjacent to a BART station and link their premises directly to the system through underground concourses.

This bonus is specified as part of a zoning ordinance adopted by the city government to encourage the use that will be made of BART and to reduce the dependence of downtown businesses on automobile access.[3] Through such an intervening law, which acts to organize private incentives toward collective gain, another major set of alternatives is open to economies where urban development is largely conducted by the private sector.

The French government, with its new towns development plan for the Paris region, is taking major strides toward a mixed system of incentives that will influence the location of future industrial and commercial activity in not just the central areas of Paris but in much of the region as well.

As an incentive to the locational decisions of private firms, major public transport investments in rail service are planned with their potential impact on land-use accessibility relative to business activity directly in mind.[4] Some government agencies are themselves being located in respect to these accessibility factors and as seeds of growth for the new communities.

However, it should be noted that major road construction, which is also part of the same regional plan, plus the phasing and amount of highway construction, will certainly influence the benefits that the transit element has on the overall economic and environmental consequences of the implemented plan.

The concept of staggered work hours is just one example of the types of social and organizational issues that are now being discussed in an effort to adjust to the inadequacies of transport investment and spatial organization in cities. Under this arrangement the absolute peaks of commuter travel could be reduced in volume and spread out in time if employers in a given urban region, and even in a given office building, would shift the working hours of their employees from an 8 A.M. to 5 P.M. (or any other general set pattern) to a variety of starting and quitting times, spanning anywhere from 6:30 to 9:30 A.M. to 3:30 to 6:30 P.M., so that arrivals and departures by both transit and private automobiles would no longer be as bunched as they presently are in most cities.

Limited efforts in this regard have already been made; the U.S. Department of Transportation, for example, has taken the lead in Washington, D.C. in staggering the starting times of its various agencies. Similar traffic spreading effects are also possible if a four-day work week, with some staggering of days, should become fairly common practice within the next few years for other than transport reasons.

COMPLEMENTARY PLANNING OF HIGHWAYS AND TRANSIT

The matters of planning and finance touched on in this and the following section offer perhaps the most important areas of change and improvement in urban transport that we have mentioned thus far. And this fact will doubtless hold true for the future as well, even if new modes of urban transport are developed and introduced.

But central to these matters are major political questions of how resources are raised and distributed in urban, regional, and national economies under a variety of political cultures too numerous and too complex for analysis in this sort of quasi-technical report. Therefore the examples and concepts mentioned here are intended only as an indication of the importance of these policy areas for urban transport.

The complementarity of planning between urban highways and public transport is perhaps the weakest link of all between agencies which should be able to mutually reinforce their activities towards improved public transportation. In many countries the political gap between the supporting interests for highway building and for transit operations has caused the administrative agencies to operate in isolation from each other, and it is the bringing together of these political interests under a more general concern, such as "the urban environment" or the "regional economy," that is needed before anything other than fragmented, ad hoc forms of cooperation can emerge.

Chicago has made complementary planning of highways and transit work through the very special case of rail rapid transit sharing rights-of-way with the city's newest expressways. The first instance of this came in the mid-1950's when Chicago's Congress Expressway (now Eisenhower, I-90) needed a right of way about 550 feet wide extending the full width of the city in an east-west direction from the central business district. An elevated transit line with a right-of-way up to 75 feet in width already existed in that general location and became the prime land acquisition parcel for the future expressway.

Rather than relocate the transit line elsewhere after construction of the expressway, it was relocated in the median, and most expressways in Chicago have been planned with a similar philosophy in mind, following the lead of this initial ad hoc response which brought highway and transit planning together over a jointly needed right-of-way.[5] Figure 6.1 shows a transit station in the center of one of these expressways.

In London in 1972, transit planning as well as highway planning has come under the general responsibility of the

6.1 Aerial view of rail rapid transit right-of-way and station in freeway median, Chicago, Illinois.

Greater London Council. Prior to that, the two operated more or less separately, with the London Transport Executive reporting to its own board and dealing with the national government directly on transport matters, and highways being planned separately under the Council's leadership.

The controversy surrounding transit/highway emphasis in a volatile public inquiry over the Greater London Development Plan is most often attributed to the fact that the traffic surveys and initial transport planning carried out by the Council were completed in the early 1960's when roads were the only responsibility of the Council and its predecessor agency. Now, with both road and transit responsibilities joined, a major organizational step has been taken toward resolving any transport conflicts through a more general purpose political entity. Major ring roads surrounding inter-London were abandoned during a change in government in 1973. New strategies aimed more at transit development and some traffic restraints for the central area have been implemented instead.

The United States may also soon be attempting to create more multimodal agencies at state and regional levels to overcome some of the competition between roads and transit which has become damaging to both. The U.S. Department of Transportation is itself a fairly new governmental creation, and departments with similar merged responsibilities are rapidly coming to life in many states.

But it is at the level of the urban region that the need for coordination and cooperative planning is the greatest. The regional transportation agency, the Metropolitan Transportation Commission, which has operated since 1970 in the nine-county San Francisco Bay region of California, attempts to function as a planning body for the several disparate transportation agencies planning separately for their modes in the region. More such agencies have recently been created for other metropolitan centers in California and other states.

And the Paris region, as cited earlier, is planning its new highway construction as well as its transit construction, to complement the overall development goals of the region with its new towns, growth centers, and the like. All of these efforts to break down the separations between highway planning and public transport planning for urban areas bear close watching, but it is far too early yet to judge long-range results.

ECONOMIC INNOVATIONS IN FINANCING AND PRICING TRANSPORT SERVICES

In the choice between public and private modes for a given trip, one psychological barrier that has frequently been noted is that of perceived cost. When one has an auto readily available, with its gas tank full and inexpensive or free parking at most destinations, the cost of any given urban trip may be perceived as marginal or nearly zero. In contrast, when one uses public transit a fare is always charged, a practice which reinforces the notion that transit has a personal cost each and every time that it is used.

A variety of things are now being tried, or nearing an experimental stage, that seek either to change the real cost experienced by an individual for auto and transit travel, or to alter the perceived cost in a way more favorable to transit usage.

An example of the former is road pricing, or congestion pricing, a favorite remedy of economists to the fact that personal automobile travel in congested times and places costs more in terms of both externalities and demand for capacity than does travel at off-peak hours or in the countryside.

An example of the second sort is the ticketing policy of the Bay Area Rapid Transit system in which large value tickets (say $20) are to be sold which can be used over and over again with the trip costs deducted by automatic devices until the value of the ticket is exhausted. It is hoped that those people who choose to buy the multiple ride tickets will come to consider transit usage a monthly cost rather than a per trip cost, and that this will be of some importance in encouraging regular commuter ridership.

While road pricing has long been talked about by the economists for its beauty in theory, it has seldom been implemented, and then usually for intercity toll roads rather than for congested urban facilities. The technical problems associated with road pricing according to geographical location and time of day have not been the major barrier to implementation; rather politics has been and continues to be the largest stumbling block.

Because many people have long adjusted to the notion that roads and streets are "free," at least in the per trip sense, it will be difficult to make this type of shift. But the temper of times and places change, and the rising general concern for the urban environment, together with a growing public faulting of the private automobile for many environmental ills, may well be moving general opinion to the point where road pricing becomes politically feasible.

Such road pricing as now exists in urban areas is largely restricted to bridge and tunnel pricing. And that until very recently, has not been tied to overall transport policy, but rather has been used as a means of paying for particular facilities. Recent raises in bridge tolls in New York City and free ridership for car pools of three or more across the San Francisco Bay Bridge are the first evidence of a change in this policy, at least in the United States.

Together with the types of freeway metering operations and bus priority systems talked about in Chapter 4, road pricing adds to both the complexity and the potential that would seem inherent in viewing highways as systems to be operated, and not just built and maintained. Among the most advanced experiments in the operation of highway facilities is one now going on in Los Angeles, California under the sponsorship of the California State Department of Public Works.

There, a 42-mile triangular loop of three freeways is being constantly monitored electronically, and visually (by heli-

6.2 Freeway operations control station in Los Angeles, California, showing aerial television monitor, automatic traffic condition indicators and control consoles.

copter based television), and operationally (by highway patrol cars and emergency vehicles). Traffic is denied entry to the system at on-ramps until flow conditions permit easy entry and smooth overall traffic performance.

This means delays at entry but better overall performance for individual vehicles once entry is gained.[6] Figure 6.2 shows the control center for the system in operation under rush hour conditions. While this experiment is not contemplated at the moment as one in which road pricing will be involved, the growing citizen resistance to further freeway building in Los Angeles and other cities may bring this technique of facility rationing into consideration in the not too distant future.

Two recent innovations in the pricing of public transport services are consolidated fares (in Stockholm) and free fares (in Rome).

In October 1971 a single monthly ticket went on sale in Stockholm which allowed unlimited use of, and transferring between all of the public modes in the Stockholm region. A single card (as shown in figure 6.3) available for roughly $10 brought the suburban railroads out to a certain prescribed distance from the city center, into a common fare system with urban transit for the first time. Thus one could board a commuter railroad in those suburbs not now served by rail transit, travel downtown, and transfer freely to either rail transit, bus or both in completing a morning work trip, all for the fixed monthly charge. Much more recent activity has been carried out in the area of fares as described in the Eisen paper referred to earlier.[7]

6.3 Suburban rail services included in Stockholm area monthly ticket plan, with sample 50-kroner ticket.

And in Rome much publicity was afforded to an experiment in free fares in areas where ridership was increased. However, auto traffic was apparently not measurably reduced by the no-fare policy. Since detailed reports of neither of these experiments have yet been seen, it would be difficult to draw very meaningful conclusions from them.

In Chicago there is some talk of making the new distributor subway in the downtown free for certain local trips within its service area. Coupled with the financial innovation of having the one-third local share (municipal) of the system cost paid for out of a special property tax levied on the immediate area served, the no-fare possibility goes a goodly step toward reshaping what had been the conventional view of transit paying for itself into the emerging view of consideration of it as a general public service.

This general subject was the theme of a portion of the Fourth International Symposium on Theory and Practice on Transport Economics, held by the Conference of European Ministers of Transport in the Hague.[8] The general tone of the meeting seemed to recognize a shift from previous years toward a policy of service rather than profit, a point raised as well by the London Transport Executive in its testimony before the Greater London Development Plan public inquiry.[9]

But if public transport does not necessarily have to be profitable, or financially self-sufficient, then large questions of how it is to be paid for, and by whom, are raised. There are a variety of practices afoot, ranging from general municipal property taxes, and employers' fees, to sharing formulas involving the gas tax support of existing highway systems. For the purposes of this report, and at this time, they are too rapidly changing and too dependent on the local political cultures of individual nations, and even political subdivisions within nations, to address in any particularly helpful or general way.

But the importance of this financing issue warrants underlining again. It is a critical basic determinant of how existing transit services are to survive and grow, and how fast and how far future transit innovations are to be developed and implemented.

7

AUTOMATED GUIDEWAY TRANSIT: Small Vehicle Systems in Localized and Areawide Service

Widespread interest in new forms of automated transit systems was generated in the late 1960's. One of the initiators of this interest was a request by the United States Congress for a study of the potential of new systems of urban transportation to alleviate many of the problems of existing urban mobility. While the results of the study accomplished under that congressional request, *Tomorrow's Transportation,*[1] did not indicate a single technology which would easily remedy problems of urban transportation, it did indicate there were wide-ranging varieties of technological activities under discussion for actual research and development, principally in the United States and Great Britain.

Since there was no one solution commonly agreed upon, the Congress and the incoming Federal administration chose not to proceed with a coordinated research and development program in this area. Rather, the report was translated into several foreign languages and circulated widely in the international technological community. The results of these were a sudden interest in Japan, France, Germany, Sweden, and Canada. Many initiatives were undertaken in the early 1970's in all of these countries with various degrees of technical success and some economic international hardship. Thus, at this writing in 1977, there is a track record of mixed success in the field of automated guideway transit, which has some potential for the future, but in far more modest ways than was conceived of in the early 1970's.

The most recent analysis of the entire field has again been carried out by the United States Congress, through its new Office of Technology Assessment, created specifically to analyze for the Congress major technological issues of importance to federal policy making. The report was prepared in June 1975 at the request of the Senate Committee on Appropriations to help guide their transportation subcommittee in determining future funding for technological research development and demonstration in this area for the Urban Mass Transportation

Administration. Rather than summarizing that report here, I should indicate that it is "must reading" for anyone interested in the recent status of this entire field. Certain definitions developed in that report now are widely used in international discussions of the automated guideway transit field, and I would like to repeat them here:

Shuttle-Loop Transit (SLT). (Example: Tampa International Airport.) This is the simplest type of AGT system. Vehicles move along fixed paths with few or no switches. The vehicles of a simple shuttle system move back and forth on a single guideway, the horizontal equivalent of an automatic elevator. They may or may not make intermediate stops. Vehicles in a loop system move around a closed path, stopping at any number of stations. In both shuttle and loop systems, the vehicles may vary considerably in size and may travel singly or coupled together in trains.

Group Rapid Transit (GRT). (Example: AIRTRANS, Dallas/Fort Worth Airport.) These systems serve groups of people with similar origins and destinations. The principal differences between GRT and the simpler SLT are that GRT tends to have shorter headways and a more extensive use of switching. GRT stations may be located on sidings off the main guideway, permitting through traffic to bypass. GRT guideways may merge or divide into branch lines to provide service on a variety of routes. Vehicles with a capacity of 10 to 50 passengers may be operated singly or in trains. Headways range from 3 to 60 seconds.

Personal Rapid Transit (PRT). (Example: Cabinentaxi in Germany is a prototype; there are no systems in passenger service.) The term PRT, as used in this study, is restricted to systems with small vehicles carrying either one person or groups of up to six usually travelling together by choice. Plans for PRT systems typically include off-line stations connected by a guideway network. Under computer control, vehicles switch at guideway intersections so as to follow the shortest uncongested path from origin to destination without intermediate stops. Most proposed PRT systems call for vehicles to be operated at headways of three seconds or less.

The previous version of this report, which was developed for OECD in 1972, had separate chapters dealing with activity center circulation which applied to horizontal elevators, high-speed conveyors, and people movers. It also contained a separate chapter on personal rapid transit defined as small vehicle

7.1 Technical details of prototype vehicle and conveyor chain in VEC conveyor system.

systems in area-wide service. The discussion of activity center circulation is closely related to that of shuttle loop transit and group rapid transit and somewhat overlaps the previous discussion of chapter 5 relating the tramways and rail rapid transit, particularly where reference was made to the Morgantown System, Westinghouse Transit Expressway, URBA and VONA. The most important aspect of the shuttle loop transit and group rapid transit categories is that they fulfill a very specialized function of circulation within the specialized activity centers, such as central business districts, major shopping centers, airports, universities, and the like. A number of these have been installed and others are under active development.

Technological proposals for activity center mobility are being advanced today in basically two forms: (1) as discrete—small or large—vehicles moving on separate guideways, much as other scales of separate right-of-way transit systems do, but automated at modest speeds over relatively short distances of perhaps one-half to two miles or so; and (2) as continuous, conveyor belt-type platforms operating at similar speeds, modest in comparison to normal vehicular systems, but quite high-speed in terms of familiar passenger belt and escalator systems in use today.

Neither idea is particularly unique. After all Form 1 may be only a scaled-down version of the more extensive mini-tram or PRT systems discussed elsewhere, and Form 2 a scaled-up version of the moving pavements in use at the turn of the century at exhibitions in Paris, Chicago and elsewhere.[2] But a combination of the speeds at which they would have to operate, the capacities they would have to comfortably accommodate, and the types of land use they would have to complement are distinct from anything that has preceded them in actual use, apart from an amusement park or exhibition setting.[3]

Research and development on a few systems of each type is currently underway in several countries, under a variety of sponsors, and at relatively low levels of funding. Some application studies have been made, and working demonstrations at half- and full-scale already exist, or will shortly for half-a-dozen variations of the basic concepts.

In addition, the number of technological variations being discussed by individual inventors lacking development funds is legion. These latter proposals will be ignored for the purposes of this report, in order to concentrate on those systems reasonably expected to be operable in the next few months or, at most, years.

Activity center systems promise to yield the first functional innovations in public transport available through new technologies because they are cheap in comparison to other innovations, not only in terms of research and development, but in terms of demonstration cost and installation cost as well. Also, they can be tested in a limited configuration, causing a minimum disruption to surrounding land use and other transport services, and can easily be built into specialized land uses particularly suited to showing them off well; airport terminal facilities being the prime example.

Moreover, they are virtually unopposed by auto/highway interests, since it is becoming generally accepted that limits on the private automobile in dense centers are in the interests of many, including the motoring public.[4] And finally, the discrete vehicle systems may be developed and tested in this more modest form before advancing in cost and sophistication toward either medium-capacity line-haul rail, or area-wide Personal Rapid Transit, functions.

Four such automated systems were demonstrated at a major marketing exposition on transportation technology (TRANSPO '72), arranged by the U.S. Department of Transportation at Dulles Airport from May 27 to June 4, 1972.

Although these systems were billed there as Personal Rapid Transit (PRT) systems, they were operating under relatively modest performance requirements for network complexity, headway distance between vehicles and control functions. It is considerably more appropriate to consider all of the installations at the Dulles exposition as early versions of discrete activity center systems, some of which may be later developed into PRT systems with much more rigorous performance specifications.

The four systems which were chosen for demonstration at TRANSPO, out of 13 proposals submitted, were constructed by the Bendix Corporation (Dashaveyor), the Ford Motor

7.2 "People-Mover" test track and general exposition setting at TRANSPO '72, Washington, D.C.

Company (ACT), the Rohr Corporation (Monocab), and Transportation Technology Incorporated. Each operated in a relatively small area of demonstration space during the exposition and was available for public ridership on either a simple loop or shuttle line.

The systems continued to undergo testing following the exposition, and at least one of them may be chosen by the federal government for more extensive demonstration in a true city operating environment.

Consumer preference information, as well as technical operating data, was collected while the systems were in public usage, and will be analyzed extensively. The TRANSPO setting of the four demonstrations is shown in figure 7.2, with one of the test tracks shown in the foreground.

The Bendix System is one originally developed by the Dashaveyor Company, an independent enterprise only recently purchased by Bendix's Aerospace-Electronics Group. Its origins were in work begun for fully automated materials handling systems, one of which now operates at the White Pine Copper Company Mine at White Pine, Michigan.

The Ford ACT[5] System was quite similar in concept to that of Dashaveyor, with a 24-passenger vehicle operating under electric power on rubber tires in a concrete channel. The 64¼ ton cars accommodate 12 persons standing and will travel at a cruising speed of 30 miles per hour.

The Rohr entry at TRANSPO was an overhead, suspended cab system produced by Monocab, Inc., a recent corporate addition to Rohr's other transportation activities, which already included the Flexible Bus Co., the U.S. franchise for the French high-speed rail Aerotrain system, and the production order for the Bay Area Rapid Transit system rail transit cars.

The Monocab system for Dulles consists of 2,000 feet of guideway, two 6-passenger vehicles, and one, two-gate passenger transit station. There had already existed a demonstration track of similar length on corporate property near Dallas, Texas. The switches are on-board activated by the guideway. As with all other TRANSPO demonstration systems, this one is designed to operate in demand responsive and scheduled modes.[6]

Air cushion suspension and linear induction motor propulsion made the Transportation Technology, Inc., entry technologically distinct from the other Dulles entries, yet the transportation functions, controls and modes of operations were basically similar. A variety of sizes of vehicles are possible with this technology, as are a variety of speeds. But for TRANSPO, relatively small, low-speed vehicles were used, guided in a concrete track which contains what is essentially one winding of a stretched-out electric motor. Rather than steel or rubber wheels providing the support for the vehicle in running on the guideway, a thin film of air blown through pads on the vehicle itself provides this function.[7]

This list of four firms did not exhaust the number of those claiming capability to produce, at that time, a system capable of meeting the specifications set for the Dulles exposition. But it was representative and that is what the U.S. Department of Transportation was trying to achieve in allocating $6 million equally among four contractors for this demonstration. Each of the successful firms has invested money of its own as well to meet the tight deadlines that this project entailed.

Now, after considerable hiatus, the U.S. Department of Transportation is again encouraging the further development of such systems in urban deployments. A new program, administered by the Urban Mass Transportation Administration, termed the "Downtown People Movers" (DPM) program is underway, and, as early as 1970, several cities have been awarded engineering funds for the development of applications for such technologies in limited deployments.[8]

There is obviously a crucial gap between demonstration of a new transport concept or system on a company test track, or even in the semi-artificial public environment of an exposition, and the "real" physical, social and political environment of a city. That gap has remained unbridged in recent years for a variety of systems developed to the prototype stage, or even somewhat beyond. Examples include the Habegger mini-rail and TELECANAPE systems introduced at the 1964 Lausanne Exhibition,[9] the SAFEGE monorail at Chateauneuf-sur-Loire, WED-way "People Mover" at Disneyland, California, and perhaps several others.

The credibility of such technological innovations as everyday urban transport systems is apparently not established in an artificial or laboratory setting when major public investments must be decided upon. Therefore, how and where the new systems are actually first applied in an urban setting will be one of the most crucial aspects of that further development.

The Dulles exhibition was simply an introduction to the public of some technological possibilities. As such it was vastly important as an initial step toward a technological revolution in ground transportation. But from the standpoint of application it remains simply as a laboratory.

The plan to install one, and possibly more, of the technologies demonstrated at TRANSPO '72 in a real city had been formulated in general terms prior to the exhibition, although the city and the exact nature of the application has not yet been announced, but Congressional reluctance to fund such an urban demonstration in recent reviews of the UMTA budget has threatened to curtail this line of public experimentation. A failure to follow through with some sort of such "real world" demonstration would be quite limiting to the short-term progress of advancing such systems toward actual use.

In Canada, a quite systematic and deliberate effort to demonstrate a new form of urban transportation in application along the Toronto waterfront was once in the planning and technological screening phases. Although this demonstration was not aimed at being so much a practical demonstration of an activity center system in an urban context, as it was a partial demonstration of large-scale systems for either line-haul or PRT uses, it was intended to perform as such initially.

Sponsorship of the demonstration project was by the Department of Transportation and Communications of the Province of Ontario. Nine firms from the United States, Canada and Europe were invited to submit candidate system descriptions by March 17, 1972, for screening in Phase I. These were Alden,[10] Ford, TTI, Uniflo,[11] Krauss-Maffei, Bertin Aerotrain, URBA, Tracked Hovercraft Limited, and Hawker-Siddeley Canada Limited.

Two to four systems were due to be selected for site-specific proposals in Phase II after the Fall of 1972. One system would then be chosen for pilot-scale construction and testing around the first of February 1973. Krauss-Maffei of Munich, Germany, was selected and was well on their way toward developing a system when technological difficulties and changing priorities of the German Federal Government reduced the level of governmental subsidy for overall system development. On such a complicated international note of technological and financial reversals, the project was scrapped mutually by the Canadians and the Germans.

It is also useful to describe the intra-airport installation of the Vought Aeronautics Corporation "Airtrans" system in the 18,000-acre Dallas/Fort Worth, Texas, aviation facility as a leading example of the types of equipment that are beginning to be introduced in such locations. This is a vehicle and guideway network that will move passengers, employees, baggage, mail, and trash within the airport complex using rubber-tired, steerable vehicles that will be propelled automatically on concrete U-shaped guideways by electric motors.[12]

The initial routing of the 40-passenger vehicles will provide service between specially designed stations at four terminals, two remote parking areas, a hotel and Federal Aviation Administration office and a maintenance area—a total distance of 13 miles. Depending on system demand, the system will operate with 68 vehicles.

Average trip time—travel time and waiting time—between any two stations in the system is less than eight minutes at speeds of 17 miles per hour. For future development, the Airtrans System offers the advantages of being dual-mode—automatic in the guideway or, with modifications, driveable off the guideway.

It is premature to call the application of any of these technologies in urban context within the next few years more

7.3 LTV "Airtrans" vehicle showing possible dual mode capabilities.

than either medium-capacity line haul systems (discussed in Chapter 5 as evolutionary extensions of the tram on its own right-of-way) or activity center systems. The more advanced concepts of area-wide, personalized transit which may emerge from considerable refinement of these technologies will take much larger and complex demonstrations to show their possibilities and problems. Thus, the critical thing to look for in applications of new systems within the next few years is not what they are called, but what service they actually perform.

The need to demonstrate civil technological programs to the public in forms that they can readily understand; to overcome widespread doubts in the transit industry that anything technologically innovative in urban transport is possible or even desirable; to begin to spend real money in this field to keep industrial interest alive; to improve the image of a tired and dull public service in its political competition for scarce resources—all these are valid and important reasons for a bit of boosterism to spur interest, support and future expectations. And for these reasons (as well as others less understandable) efforts are being made to simplify and to short-cut expensive research and development programs and to bring PRT systems—or their less exacting cousins—to limited performance demonstration soon.

But the importance of recognizing a thing for what it is—not for which it is called or wished to be—has been a central theme of this inquiry until now. It must continue to be so if the report is to be truly helpful to the discussion of serious alternative policies for urban transportation. Therefore, much of what is currently being said publicly about PRT in a variety of countries is in need of critical analysis, to clear the air about a subject of immense importance to the future of urban transportation.

Without such critical review early on, the potential for disappointment and political backlash when costs arise, complexity increases, and failures occur becomes a very real and great danger perhaps unrealized in many technical quarters today. A few such disappointments in the not-so-distant past have already occurred. It is with this perspective in mind that we offer the following optimistic, yet critical interpretation of recent developments in the field of personal rapid transit.

Reportage in the popular press and even in the technical literature to the effect that personal rapid transit was already here in the early 1970's needed to be treated with considerable skepticism. Pressures on individual firms and governmental agencies to make such claims were quite understandable in the short run, yet very dangerous in the long run, and it is only now, after a gap of some five years that the Urban Mass Transportation Administration is once again promoting the installation of advanced technology systems through its modest Downtown People Movers program. Even then, many of the advanced technologies being preliminarily tested on test tracks on the Colorado desert and elsewhere have either been scrapped or cut back substantially due to lack of congressional enthusiasm for such programs. At least one other major United State firm, the Rohr Corporation, deeply committed to advanced technology and urban transportation, has backed almost entirely out of the field.

Despite the present climate of skepticism over the more advanced concepts of personal rapid transit, some efforts do continue in the field. A recap of the personal rapid transit developments over the last decade is perhaps useful at this time.

TRANSIT SYSTEM CONCEPT

The core concept of an area-wide transit system with personalized service marks a radical departure from most thinking about the public sector of transport in the urban transportation field. There, for years, the dominant ideologies and technologies have been oriented to line-haul movement of masses of people, hence the term "mass transit" as a synonym for public transportation in cities.

In previous sections of this report dealing with incremental advances in bus and rail operations the area-wide versus line-haul distinction was made, and ways of reorienting some existing transit technologies to area-wide services were discussed. But these technologies still use large "mass transit" vehicles and require passenger transfer and inconvenience to allow an intersecting network of essentially line-haul technologies to function as an area-wide service. New technology designed expressly for personal transport without transfers over a truly area-wide network would thus enable a major breakthrough in transit service, a service impossible with today's technology and incompatible with today's equation of public transport with mass transport.

A basic description of PRT, as contained in the pioneering *Tomorrow's Transportation* report, bears repeating here:

> It would consist of small vehicles, each carrying about the same number of persons as an automobile. These vehicles would travel over an exclusive right-of-way or guideway network, either over standard routes, or else automatically routed individually from origin to destination at network stations.
>
> Personal rapid transit would provide travelers the important advantages of minimum waiting time at the origin station, and private, secure accommodations. At the heart of the concept is the premise that personal transit would serve a metropolis, except perhaps for its lowest density outskirts, with a network or grid of lines, each perhaps a mile or two apart.
>
> Empty passenger vehicles or "capsules" would be available at each station on the network. The riders would enter one, select and register their destination, and then be transported there automatically, with no stopping. The average speed would be essentially equal to the vehicle speed. The station spacing on a guideway network for the system would have no influence on speed of travel. Passenger demand and station costs would dictate proper station spacing.
>
> Empty vehicles would be recirculated automatically to maintain an inventory at each station, and passengers could be routed past stations without stopping until they reached their destinations. Ideally such a system would give travelers the same privacy as a private automobile, although during peak periods in cities with particularly heavy corridor movements a traveler might have to share a vehicle with two or three other passengers.

The guideway network covering the metropolitan area is the essential ingredient of the personal rapid transit system. Without a network of guideways the system could hardly avoid conventional heavy dependence on work trips and a radial orientation to existing central business districts. Thus it could not provide adequate transportation alternatives in large metropolitan areas with a wide dispersion of trip origins and destinations. No matter how sophisticated the technology, transit which operates without some sort of network service pattern almost certainly will remain a marginal service in the movement of urban populations.[13]

An argument is often made, particularly in American cities, that a new rail transit system would carry only a small fraction of urban trips, perhaps 5 percent at the highest, and therefore could not offer a viable solution to such difficulties as congestion, air pollution or even the general transportation needs of people without access to automobiles. Buses, the argument then goes, are the only real alternative to total auto dependence in cities of less than one million population as well as the major choice outside of high-density corridors even in the larger cities. For they at least have the flexibility to go anywhere over the widely distributed existing road network.

That argument may well be largely true—at least until there is the potential of an exclusive guideway technology that can be more extensively distributed than high-capacity train-like technologies. And it is this crucial point that is ignored most frequently in transit/highway debates which make claims of discussing the future, rather than the present.

Thus the central issue and advantage of personal rapid transit as an innovative transit service with major, not minor, potential influences on urban transport is represented in the last sentence of the quotation above: "No matter how sophisticated the technology, transit which operates without some sort of network service pattern almost certainly will remain a marginal service in the movement of urban populations."

Continuing the description of PRT found in the *Tomorrow's Transportation* report:

Network systems of personal rapid transit would perform economically with travel demand ranging from 1,000 to 10,000 persons an hour in a travel corridor—the medium to lower density conditions in which mass transit systems today usually perform inadequately. Yet these corridor travel demand levels prevail in most metropolitan areas. The network system, moreover, could have average speeds of 50 to 70 miles an hour, a substantial improvement over average urban freeway speeds.

The roadbed or guideway for personal rapid transit might consist of rails or surfaces for air bearings; the vehicles could use steel or rubber wheels or air pads. Propulsion could be in the vehicle or in the guideway itself. Each guideway would be about 5 feet wide and could be a single-lane over substantial portions of its length. The narrower and lighter structures should require less land. They also could be more attractive than many urban freeways. All these options are open to the design engineer; no particular solution has yet been shown to be outstanding.

A personal rapid transit system having these performance characteristics is an important element in a viable urban transportation system for a number of reasons:

• An *exclusive right-of-way* is essential if public transportation is to be automated and if it is to escape the congestion of general street traffic. Forced to compete with automobiles on crowded streets, other forms of mass transportation are inherently at a speed disadvantage.
• *Automation* can make transit service more competitive with the automobile, since it is the only safe and efficient way to operate a system using numerous small vehicles.
• *Small, individualized vehicles* avoid the chief delays of present rapid transit: Stops at intermediate stations for other passengers and waiting or dependence on a schedule at the origin station.
• Additionally, *use of small lightweight vehicles, with the quiet suspension and propulsion mechanisms* which can be developed, and the less massive elevated and station structures such systems would permit, would minimize the impact of the system on the environment.[14]

While elements of the advanced technology that can lead in this direction are already under development in research laboratories around the world, it is less the technology and more the overall systems concept that is at issue now. It is a harmless enough expediency for an automated tram such as that operating at Morgantown, West Virginia to be called a PRT, while development continues toward more sophisticated networks, vehicles, guideways, and operations. But if it becomes widely believed that such an installation truly represents personal rapid transit, and further research and development is thus curtailed at a relatively low level of service capability, then the mislabelling represents a major disservice to all of those who hold out some hope for major technological and service improvement in urban transport.

Instead, it should be possible to keept the concept of PRT intact while incrementally bringing the technology to higher and higher levels of sophistication. The Dulles TRANSPO demonstrations could have some influence in that regard. But the crucial question there is which kinds of characteristics of the small scale transit systems will be emphasized in the next demonstrations, if any, and possibly be slated for real cities, following technical tests and public trials at the exposition?

At some more distant, evolutionary point in system development, the sophistication of the needed control technology—perhaps the efficiency of the power-consumption requirements, the maintainability of the constituent system components, and other factors—will be central technological issues that must be resolved before really major commitments can be considered for PRT systems. But for the present and very near term, existing technologies are being utilized for as much of system hardware as possible by many of those companies promoting capabilities in this area.

That is, rubber tired, automotive-type suspensions, conventional electric motors for on-board propulsion, small scale computers for controls, and other such components are being lashed together in relatively crude, but imaginatively adequate first-generation personal rapid transit systems. These will probably be adequate for small scale, cheap public demonstrations soon that can be used—or misused—to powerfully convey an image of potential and progress in urban transit research and development.

Keeping a perspective on the difference between where PRT is today and where it must develop in the future to become more than an exposition or test track novelty will be crucially important in the next few years for industry, governments, and the public. It will be, that is, if initial visibility is to be turned into adequate political and financial support to carry out the necessary research, development, and demonstration work to make real urban PRT installations possible in the long run. This point should be kept in mind while considering the material contained in the following two sections on technological research and development, and on application studies.

Technological Research and Development

The research and development activities that must precede the actual delivery of an extensive, complex, safe, reliable, and unobtrusive system of personal rapid transit in an important urban environment have recently begun to get underway in several advanced industrial countries.

The technological aspects of the research and development, at least as far as getting a single prototype into simplified operation, probably appear quite simple and straightforward to industries comfortable with meeting the high technology

7.4 Scale-model PRT guideway layout at Aerospace Corporation headquarters, El Segundo, California.

challenges of aerospace and electronics systems. And indeed, individual firms and government laboratories with strengths in these areas are quite noticeable in the present flurry of technological activity surrounding PRT concepts in every country now taking an active interest in the field.

But to existing urban transit operators and to the lower technology, yet mass market oriented, automotive industry, as well as other skeptics of whatever stripe, the issues of mass production, reliability, maintenance and operating costs, and consumer acceptance are among the major concerns. And these issues cannot be grasped well at the early stage of prototype demonstration.

In addition to the technological aspects of research and development, there are a host of political, social, and environmental issues that will appear only at the stage of actually attempting to demonstrate a system in an urban setting. Then other values also will be at stake and failure could be costly to the community as well as to the transport system proponents themselves. Dealing with such issues requires a different type of research and a different view of development and demonstration than that used to deal with technological, or even economic and production, concerns.

Most of the R and D now underway is of the technological prototype sort, or its analytic precursor, systems analysis. But a limited effort is also being devoted to application studies that seek some indication of the general functioning and acceptability of such PRT systems in real urban situations. In addition, simpler types of systems that have some of the PRT features and can be readied now or in the very near future are serving as partial tests of the technological components, operating requirements, production costs, urban design requirements, and consumer acceptance of the more complex and extensive systems that should gradually follow.

With the potential for incremental technological innovations in urban transit spread as widely as it is over trains, and trams, and activity center systems, it is quite sensible and likely that research and development identified with these systems types will constitute much of the research input to personal rapid transit programs as they evolve. However, in this section, and the one following on application studies, we will concentrate on research and development clearly identified as PRT in nature.[15]

In the United States, where it is generally acknowledged that serious interest in personal rapid transit began, a variety of relatively small industrial firms began doing their own systems design and prototype development, following the systems analysis performed for the *Tomorrow's Transportation* report nine years ago and refined elsewhere since. Most visible among these are Alden, Transportation Technology, Dashaveyor, Monocab, Carveyor, and Uniflo.

But operating at relatively low levels of financial investment, and in a climate of skepticism at both urban and federal levels of government, these firms generally failed to establish sufficient credibility to move their designs beyond their own test tracks into serious consideration for urban applications. And without the prospects of either production contracts for their relatively rudimentary initial systems efforts or substantial infusions of public or private investment to increase the complexity of the designs, the whole field seemed stymied. There was no way that a small, private firm could, by itself, break through into a field that must, by its very nature as a public service, be supported by public funds.

Only with the agreement of the Urban Mass Transportation Administration to sponsor demonstrations of four such systems at a special public exposition, did these individual firms obtain the opportunity to establish the needed public credibility to move privately initiated, small vehicle transit systems forward in the United States. Even at that, the relatively cheap investment of only $6 million was divided equally among the four systems that were selected from some 13 applicants. Much of the cost that then went into design of the systems for demonstration at TRANSPO was private investment, leveraged by the federal involvement and "seed" investment.

A significant change in the industrial composition of the private firms now active in the United States PRT field has followed from this decision. Almost immediately following the sorting-out process involved in the TRANSPO competition, Dashaveyor was purchased by Bendix Corporation, the Monocab division of Varo was purchased by Rohr Corporation and the Ford Motor Company became a serious entry in the field. The other final participant, Transportation Technology, had been already affiliated with the Otis Elevator Company for some time. Thus, with this qualitative shift in the financial makeup of the private sector, and the federal credibility afforded to the overall enterprise, it can be expected that the pace of development of new transit technologies will increase in the United States.

Another variation on the processes involved in developing a new, small vehicle transit technology—one which recruited and employed high technology laboratories of government and the aerospace industry—is also being carried out in the United States as part of the Morgantown project previously mentioned in Chapter 6. In this instance the systems analysis was first carried out, not by a producer company or a government systems laboratory, but by a university and local government, who together had a real transit problem in need of solution.

Once given the national importance now afforded the Morgantown system, the Jet Propulsion Laboratory, one of the quasi-governmental systems facilities with a good record in managing space programs, was brought in by UMTA to do the systems design and to direct the overall implementation of the project. Awards for actual development of the vehicles and controls were subsequently awarded to the Boeing Company, a major aircraft manufacturer, and the Bendix Corporation, another aerospace and electronics enterprise.[16]

None of the three proprietary systems that had been studied by the joint university/city team in their own initial systems analysis and systems design studies was selected for development as had been originally envisioned, although the Alden

Self-Transit Systems Corporation which had been the preferred choice of the university/community team was included as a subcontractor of Boeing on vehicle development. After the systems design phase had been completed the Jet Propulsion Laboratory withdrew from further managerial responsibilities. Project management, in addition to the continuing vehicle development work was awarded to Boeing.

In the Morgantown experience, systems analysis, systems design, and prototype development have been sharply separated in both timing and responsibility. But the effort has brought even more high technology firms into the potential PRT industry and given the Urban Mass Transit Administration some good experience in the management and delivery of advanced technology.

In all of these efforts, plus a number of developments in Canada, France, and Japan, the early promise appearing in 1972 has fallen on rather hard times. The Morgantown system has gone through fits and starts of both technical and potential problems and now does seem destined for completion of the second phase, although in somewhat limited form compared to its original concepts. Many of the original firms participating in the TRANSPO demonstrations have failed to find other clients and have barely held on to their interests in the field.

It is still too early to tell the ultimate value of these projects, but the TRANSPO and Morgantown demonstrations are recognized as measures which at least started the whole field of new technology moving without necessarily delivering a workable PRT technology (as it has been defined in this report).

In Germany the PRT situation is less ambiguous, organizationally more advanced, and technically progressing at a somewhat steadier pace than in the United States. There a consortium of firms has been funded by the central government to carry out an entire research, development, and prototype demonstration program leading to an initial version of a PRT technology, and at a relatively low cost and within tight time budgets.

The two industrial firms which are participating in this program are ones of considerable technical diversity and financial strength. Demag is a producer of heavy machinery and steel products located at Duisburg in the Ruhr, while Messerschmitt-Bölkow-Blohm is an aerospace and aviation-based firm headquartered near Munich.

These firms had separately been surveying previous work in the field of transit technology as a potential market opportunity in 1970–71, and had been carrying out preliminary systems analysis of PRT-type concepts, when they discovered their overlapping activities. They subsequently decided to join forces in exploring the potentials of small, automated vehicle technology for urban transit applications.

Demag's interest can be broadly identified with the energy and steelwork aspects of such systems, while those of MBB run to the controls and vehicles. Together, they are identified with a substantial systems study performed about a year ago with the city of Freiburg, Germany as the example application used for data and detailed analysis.[17] Since that initial study they have continued their systems analysis and design work with travel data and urban environmental constraints from a variety of real settings, including portions of Munich and the town of Hagen in the Ruhr.

The project—called CABIN TAXI or CAT for short—is sponsored by the German Ministry of Education and Science in conjunction with in-house funds of MBB and Demag. A letter of intent to this effect was transmitted to the firms early in 1972 indicating that some DM 15 million would be available for an initial two years, with 80 percent government funding and 20 percent from the individual firms themselves. A development schedule had the following targets:

- Testing of all essential components of this personal rapid transit system on test stands is planned for 1972 and has already commenced to a considerable extent.
- A test track for the prototypes will commence operation in 1973.
- Completion of a first larger experimental network to study user acceptance is planned for 1974.
- The first public network is expected to start operation in 1976.[18]

Figure 7.5 is an artist's rendering of the small passenger cabins loading and discharging passengers in a station environment. Briefly, a few of the technical characteristics of the CABIN TAXI system include: a cabin seating two adults and one child, plus storage space for luggage and with passive safety devices; propulsion by two asynchronous linear induction motors; a design speed of 36km/hr.; and an elevated track system that allows vehicles to run in opposite directions above and below, thereby saving on width of right-of-way that otherwise would be devoted to double tracks. Switching is controlled by the vehicle, not the track.

Automatic control is effected through a three-level hierarchical system of data flow which splits the control requirements between (1) the vehicles, (2) discrete points of the network

7.5 Artist's conception of MBB-DEMAG CABIN TAXI system, concept in station and street environments.

such as stations or switches, and (3) overall supervision of the network. This particular pattern of control is intended to reduce data flows over what would be the case with centralized control, as well as to insure redundancy and reliability in the network control even in case of a network management computer breakdown.[19]

A decision on whether or not to go ahead with installation of a public network, as opposed to merely an experimental one, will have to be made after the results of the first two years of the project are evaluated. The 1976 data for opening operations of an installation in a real urban setting was thus contingent on successful performance during previous stages, and on additional funding becoming available beyond that just committed.

The test tracks have successfully performed a limited application of the technology being installed in a hospital complex in Germany, but larger installations have, thus far, not come to fruition.

In France, the single firm of Engins Matra—an aerospace, automotive, and communications conglomerate of sizeable proportions, headquartered in the Paris suburbs at Velizy—has been selected by the national government to examine the prospects of PRT-type technology from systems analysis on to prototype demonstration. The origin of French interest in such technology stems both from domestic innovations, such as the novel coupling/decoupling concepts in an earlier proposal called AT 2000, and from the general level of activity in the field elsewhere in the world.

A twenty-eight month contract that commenced in March of 1971 was negotiated between the firm and the French government. As a research and development grant, it carried Frs. 7.3 million of government commitment, and called for another Frs. 2 million of investment by Matra itself.

Systems studies of a range of technologies, with somewhat different performance characteristics but all having small vehicle systems which are generically called ARAMIS, are being carried out. The emphasis is on quick delivery of a system using off-the-shelf technologies for its components. Data for analysis and designs has been obtained from a variety of French urban settings, such as medium-sized communities like Nice and Strasbourg, and the suburban communities ringing Paris to its south. Close headway tests of the French vehicle have been run, but no installations are currently proposed. A more limited product of this firm is operating as an automated guideway transit system in a new community of Lille-Est, but is more along the lines of shuttle technologies discussed earlier.

British efforts in developing PRT technology were well along on a system called Cabtrack by Spring of 1971, but were slowed by the Department of the Environment which oversees transport in the U.K. The decision came just short of a contract award to the industrial firm Hawker-Siddeley to carry the systems studies closer to hardware production.[20] The effect of this position is to take Cabtrack back to the drawing boards for comparative study with minitrams and more conventional technologies, with an admonition to watch carefully the costs and environmental impacts of each alternative. No major technological moves in this direction have developed in Britain in recent years.

And in Japan, serious efforts to develop a PRT system under the name CVS (Computer-controlled vehicle system) are well advanced under the sponsorship of the Japan Society for the Promotion of Machine Industry (JSPMI), and the technical

7.6 Computer-controlled vehicle system (CVS) in 1/20th scale as displayed at the 1971 Tokyo Motor Show.

guidance and instructions of some members of the University of Tokyo and the Ministry of International Trade and Industry (MITI).

A New Machines and Systems Development Center was set up in April, 1972 within JSPMI. Through this center the project team, consisting of university scholars and technical representatives of a variety of participating industrial firms, is now designing and testing the first full-scale vehicles, tracks, and control systems for an automatically controlled, small-vehicle system.

Previous to this effort, feasibility analyses, systems studies and construction and operation of a 1-to-20 scale model of a CVS network were carried out in 1970–71 to ascertain whether the project warranted significant support. The model with some 60 vehicles operating over a simulated portion of the central Ginza area of Tokyo, was prominently displayed at the fall Tokyo Motor Show in 1971. Figure 7.6 illustrates the model as it appeared to the public on this occasion.

With financial support assured for the next stage of research and development in 1972–74, plans were made to construct a test course for the CVS at the automotive test course of MITI at Higashimurayama, Tokyo. An initial 200 m. of this track was scheduled to be opened in September, 1972, and test runs of the first CVS cars made during the period of October to December 1972. A full test course of 4,700 m. was completed in 1973, and full computer operation of a 100 vehicle test system carried out from February to October 1974.[21] Successful tests have been run, but lack of funding for installation in a real community has led to dismantling of the very extensive test track facility.

The system itself consists of two-to-four-passenger vehicles, operating on a separate, overhead network of guideways, automatically controlled by a hierarchical computer control system. Rubber-tired vehicles—driven by electric motors with energy obtained from a trolley wire within the guideway and guided by a guide wheel inserted in a groove along the guideway—will operate at speeds of from 40–60 km./hr. at short headways, yielding a capacity of some 6,000 vehicles per hour for a single lane of traffic. Frequent stations and a rather dense network of guideways of two types (a high-speed superway network with 1 km. spacing and a slower-speed path network with 100 m. spacing) have been contemplated for urban applications in the systems studies thus far performed.[22]

No plans or schedules have yet been announced for an urban demonstration. However, numerous studies are underway as to possible application sites, principally involving new community development outside major urban centers. Considerable institutional ambiguity between the Japanese agencies responsible for technological development for urban planning and for road construction have caused some of the uncertainty in this program.

Application Studies

The vast majority of personal rapid transit efforts at the present time is of a technological and system analysis nature. Precious little work has been done in the area of studying the possible social and physical effects of PRT systems as they might be applied in urban settings.

In some ways it is certainly too early to attempt to predict or assess the consequences of a major commitment to an innovative urban transport mode such as this one, but it is exactly because of the unexpected social and physical consequences related to the implementation of another urban transport technology—the private automobile—that interest was initially aroused in PRT as an alternative. High quality application studies of the potential effects of personal rapid transit systems in widespread usage are thus a desirable and important part of any nation's research and development activities in this field.

Urban studies have been made by virtually all industrial firms and technological laboratories with programs in new transit system development, but the type of studies carried out almost always are of the nature of a travel demand forecast, for later use in designing the location and capacity of hypothetical system vehicles, guideways, and stations. While such studies are valuable inputs to the design process, they fall far short of the types of social, architectural, and planning studies that would be beneficial to cities, citizens, and national governments in appraising the merits and demerits of these technological research activities.

Three suggested actions placed before the United States government by a special committee of the National Academy of Engineering are extremely pertinent to this point:

1. Federal urban transportation programs should focus increasingly on providing better quality of urban life, not just better transportation.
2. The increasing focus on the quality of urban life clearly calls for a better understanding of the interactions and relationships between urban transportation systems and the functions of metropolitan areas. This, in turn, requires an enhanced program of analysis and real-world experimentation.
3. The proper design of urban transportation experiments and the implementation of more effective investment programs also call for an increase in supporting social science thinking and analysis.[23]

The thrust of these suggestions would appear to be (1) that a good deal more attention needs to be paid in the future to the roles of transport service in urban life, not just to its engineering and economic features; (2) that conscious thought about how, when, and where technologies are applied in demonstration and initial application deserves additional emphasis; and (3) that broader areas of professional competence ought to be drawn upon earlier in the systems analysis and design process and on throughout the demonstration and implementation phases of transport system development. While these points were meant to apply to an entire spectrum of urban transport-related programs, they are especially important in the case of a service concept as significant as personal rapid transit could become.

Three application studies are worth mentioning since they have gone somewhat outside the limited engineering perspective of industrial laboratories. As such, they are indicative of

7.7 Excerpted figure from a systems study of PRT applied to Gothenburg, Sweden, showing service area within 20 minutes of city center.

additional sources of evaluation for new industrial products like PRT, and of application studies of slightly broader scope.

In Gothenburg, Sweden a study of possible application of personal rapid transit to the future needs of that city of approximately 500,000 has been underway for several years now. It began as a separate study of one alternative to the proposed rail rapid transit program for the city, but has become an integral part of a comprehensive town planning study of several transport alternatives and their potential influences on city development.

Here, the city government is itself carrying out the study. And since there are no potential PRT system suppliers active among Swedish industries today, the inquiry has been able to look worldwide for comparisons of potential technologies. Final results are not due for several years yet, but interim progress has been reported.[24] Figure 7.7 shows one sample PRT network analyzed for Gothenburg and the geographic coverage of transport service within 20 minutes of the city center. A decision not to go ahead with further pursuit of a personal rapid transit system was made in 1973, although the application study was the first significant one done for an entire community.

A somewhat similar study—this one carried out by a university research group—was undertaken in Minneapolis-St. Paul, Minnesota with the partial support of the Minnesota Legislature. Although not part of a comprehensive study of alternative modes of transport for the Twin Cities, it has played a part in transit decision-making there along with studies of rail rapid transit and bus improvements.[25]

And in the U.K., an architectural study performed for application of Cabtrack, has been the most detailed environmental assessment of PRT yet attempted.[26] The results of the study showed such a system to bring mixed blessings in a complex urban environment. In fact it is a valuable contribution to the assessment of transit system impacts in certain architectural surroundings and to the learning environment in which such interdisciplinary collaborations will have to be conducted in the future.

Figure 7.8 shows the type of photo-montage used in the study as a visual assessment of the potential impact of guideway siting in various locations. Although the findings were that such an installation would perhaps be harmful in a local environment such as shown in the picture, and therefore ought not to be considered further, these findings got distorted in the context of discussions that were held after the pictures were released to the public. There are, therefore, certain dangers involved in conducting such studies—but the dangers of going ahead without them may be far greater.

Supplementary Services and Dual Mode Systems

The questions of how extensive a PRT network can be, the frequency and location of stations, whether guideways must be on new rights-of-way or might reclaim portions of existing street and freeway networks—these and a variety of other concerns will accompany major future commitments to PRT-type technologies in real cities.

7.8 Photo-montage architectural study of cabtrack PRT system visual impact on a residential street, London.

Whatever practicable limits are placed on the extent of the basic PRT network, it is certain that there will be some need for supplemental transport services to get to and from stations to the individual origins and destinations of passengers. Whether this is by elevator access to stations actually within important buildings (a planning and architectural design issue as much as a technical issue), or by pedestrian activity along streets or malls, or by personal auto or public transit conveyance, will depend on where in the urban area the stations are located. Of interest in this section is the particular question of access in lower density areas where distances would be too great for pedestrians to walk.

If private autos had to be used for station access it is doubtful that large numbers of persons would find advantage in leaving their own car at a station for a trip across the city, only to find that they again needed the missing car at the other end to complete the trip comfortably. For commuter-type trips to high density areas where parking is difficult and a walk to the final destination is reasonable and comfortable, as with a rail transit system or suburban rail system serving a center city, auto access at only one end of the transit system perhaps does make sense. But for the much more scattered trips that a PRT system hopefully would be serving, there will be a need for local mobility at either end.

For those able and willing to drive, trips which required transfers at both ends probably would not be made by transit; autos would continue to meet these ends. But for those unable to drive, supplementary systems such as the minibuses or Dial-a-Ride services already discussed in Chapter 5 might be valuable.

And, quite possibly, a rental service of small urban public autos might make sense for feeder service if the desire for personalized mobility carried throughout the trip. Such a public rental service has been talked about for some time now, and a few experiments are going on with variations of this possibility.

The small-scale system of "autos banalise" in Montpelier, France is one such scheme,[27] and the extensive system of rental cars operating in most countries which is focused on business travel from airports and major hotels is a longstanding version of another. Studies of such rental vehicles as one means of feeder service to BART in some of its outlying areas have been done by consultants and other such studies may well exist.[28]

Beyond the notion of supplementary services to allow a PRT system to be useful in serving low density areas as well as medium and high density ones is the more advanced technological concept of dual mode systems. Such systems would allow the operation of small transit vehicles both on a captive guideway under automated, systems control, and off the guideway on local streets, under individual driver control.

Such a concept would further blur the distinctions between autos and transit to the point where similar flexibility was offered without transfer in both public and private systems. And at even later stages, perhaps privately owned vehicles could operate on guideway portions of the system as well as on the streets.

Attempts to grapple with such a concept have come from at least three directions. As much as a decade ago, Detroit auto manufacturers were interested in automating highways and personal automobiles to gain increased safety, capacity, and speed out of the auto/highway system. Since then a number of things have been done to bring automation and controls to highway operations, such as those discussed under "Central City Operations" in Chapter III on bus improvements. But efforts at automating automobiles are not being as actively pursued as they once were; at present the emphasis is on automating the street and freeway operations for manually controlled vehicles.

A second direction is in the dual mode bus studies and technological development being pursued somewhat halfheartedly in the U.K., United States, and perhaps elsewhere. Here, the prospect is one of automating the operations of the bus on a guideway, then putting it under manual control on surface streets. This is distinct from the non-technological dual mode bus operations which have a driver and a special lane on existing freeways, as in the Shirley Highway Demonstration discussed at length previously.

The recently-cancelled Throughways bus project in the U.K. was the most advanced conceptually in this automation regard, although it was difficult to see much advantage to an automated bus except as a technological precursor to automated, dual mode personal vehicles, without a salaried driver. A major review of dual-mode possibilities and past activities has been completed for the U.S. government,[29] and funds were requested of Congress to begin a dual-mode research program at a level of $16 million in fiscal '73.

A dual-mode bus program in Japan currently remains a high technological priority there. The overall field of automated guideway systems has had rather searching review during the mid-70's. While many technological tests were successful, no country successfully has overcome the impediments to even modest scale installation of more than simple shuttle systems in urban applications. The most extensive test installations have been those of the Japanese in their 1975 Sea Expo in Okinawa and the Morgantown project in the United States. One will have to wait and see how successful the Downtown People Movers project will be in the United States before seeing where the field will go here. In Japan, the state of the economy and the organizational impediments will have to be substantially improved before a new round of progress can be made in this new technological field.

8 GOVERNMENT AND THE PACE AND QUALITY OF URBAN TRANSPORT EVOLUTION

A good deal of this report has been devoted to emerging attempts in O.E.C.D. member countries to produce innovations in transit service through imaginative technological and administrative means. Through the international dissemination of comment about these local and national initiatives, it is hoped that a heightened awareness of the possibilities for public transport innovation will be produced among transportation and urban planning professionals and government officials.

If the comments have been at times skeptical of the optimistic expectations of the promoters of specific innovations, or somewhat critical of the performance of government agencies in selected instances, they have been made with the cautious hope that knowledge of difficulties as well as of success can be of substantial help in the further development of urban transportation.

But there is much more than might be attempted by such a document as this, and by the deliberations of such a body as the O.E.C.D. Despite the many and varied attempts at innovation observed in the course of this report, the prospects for realistically implementing much of the promise of such innovations, particularly those of substantial technological novelty, are at present quite marginal. And the possibilities for undue delay and embarrassing failures in applications are significant, given the present array of organizational and political factors bearing on the whole structure of institutions involved in urban transit development.

These dim prospects do not lie in technological problems concerning new systems, nor in the transportation characteristics of existing ones. They lie rather, in the gaps of communication and responsibility between those institutions with specific transportation missions and those with differing tasks, such as enhancing the quality of the urban environment, or promoting the application of science and technology to human affairs.

As described in previous sections of this report, the potential for bringing new urban public transport services into the overall context of enhancing the urban environment is a challenge that exists for the taking. It must be confronted successfully on a number of levels if the emerging environmental and transportation needs of major urban areas are to be met in a reasonable time and at a reasonable cost.

The discussion of this final chapter is addressed to some of the roles that government, largely national government, can play in accelerating the pace and insuring the quality or relevance of innovation in urban public transport. There are possibilities as well, however, for both local governments and international cooperative bodies to contribute to this effort. The principal focus at each level of potential government action is in overcoming the gaps in present institutional arrangements that limit the understanding, the resources, or the responsibilities necessary to deal with urban transportation policy in the context of overall environmental quality and innovative transport service possibilities.

Among the needed changes are:

1. *A wider appreciation of the policy context in which urban transportation problems and opportunities should be viewed.* In other words, an understanding that transit acts as environmental constraints and development tools influencing the quality of life of entire urban regions, and not just as matters incidental to the movement of people and goods, is an important objective at all levels of government. The recognition of the automobile's contribution to urban air and noise pollution; the decline of economic and service viability of public transport for those urban dwellers without direct access to private motor vehicle usage; the land use impacts on parkland and housing stock, on scale and amenity of historic city centers, and on dispersion of urban activities—the inclusion of all such external factors in costing transportation and land use alternatives, in arriving at planning decisions, and in organizing the functions of government agencies with urban transport responsibilities will be significant in initiating a consciousness of this broader context among transport planners.

2. *A focus on the complementarity of existing transportation systems in planning and operations to replace competition or independence in the policies of urban transportation agencies with responsibilities for alternative modes of travel.* This includes not only the type of coordination achieved between existing public modes as in such cities as Hamburg,

Stockholm and Munich, but also between automobile and transit modes as is beginning in London and San Francisco, and in the planning for entirely new modes still to be developed and introduced in the urban environment.

This type of complementarity is most important at the level of governance of urban regions, which may differ significantly in terms both of constraints and desires, but it could be reflected as well in funding and administrative flexibility for support of the various transport modes at the level of central government.

3. *An enhancement of the relationships between government and industry in supporting and evaluating the possibilities of more innovative, technological modes of urban public transport.* At the level of central government this would mean closer exchange of information on desired characteristics of new systems for application and on technological possibilities and costs of alternative approaches. The type of cooperation evidenced in Germany by the creation of the SNV (Studiengesellschaft für Naheverkehr) as a national agency to study urban transportation problems, jointly under the sponsorship of industry and urban transport agencies, is one prototype for such continuing consultation between producers and consumers of potentially innovative technologies.

4. *The working together of local transport planners and industrial developers of specific technologies long in advance of the availability of a specific hardware*, both to inform the local area as to the nature and credibility of future possibilities and to inform the industrial producer of realistic urban needs and constraints.

An effort by the French IRT (Institute de Recherche des Transports) to promote this type of local cooperation in application studies in such cities as Lille, Rouen, Grenoble, and Strasbourg is significant in this regard. The burden for such application studies should not have to fall entirely on the industry or the local government, but could have the support of central government, or even international agencies, since the exploratory findings should be of use to a wide variety of participants in the innovation process, and not just to the specific locale which might well choose not to utilize the particular technology even if it became available, nor to the industry which might be developing a technology for general production by other firms as well.

5. *The encouragement and coordination of comprehensive studies of the impact of urban transportation on major urban*

regions. This is a key possibility perhaps available to O.E.C.D. at this point in time for assisting in the consideration or implementation of the kinds of possibilities cited in the four paragraphs above. Such studies might attempt to analyze the varied impacts of major existing or planned investments in urban public transportation in the context of personal mobility, urban physical and social development, environmental quality, transportation policy making arrangements, and prospects for the nature and amount of additional transport investment in the next several decades.

This possibility for research arises now because substantial new investments in public transportation are being undertaken or seriously debated in a variety of cities in O.E.C.D. member countries which will doubtless provoke some measurable changes in not only urban mobility, but urban structure as well. Serious study of the impacts of these investments could lend considerable impetus to the goal of widespread understanding of the varied roles of public transport in urban environmental policy and of the demands for and possibilities of innovation in the service capabilities of urban public transport operations, planning and technology.

APPENDIX IIA
Service Characteristics of Existing Modes of Urban Transport

SERVICE CHARACTERISTICS OF EXISTING PUBLIC MODES

At least five characteristics of urban transport operations differentiate the services provided by the several generic modes now available. Certainly other descriptors could be added for greater detail in categorizing the technology of the individual varieties of a given mode, in specifying speeds, waiting times, and personal comfort or in differentiating the spatial location of the traveled right-of-way, but these are of secondary importance in categorizing the basic transport service provided. For the primary differentiation of these modes one may helpfully specify: (1) the scale of urban geography served, (2) the size of the vehicle, (3) the degree of exclusiveness of the traveled right-of-way, (4) the flexibility of the routes and schedules, including transfers and (5) the manpower requirements for operating personnel. While successive waves of economic, technological and social changes have altered the quality and acceptance of the various services rendered by these public modes over their long histories, the types of service actually provided have altered little in all this time.

The basic unit service required of a transport system is to get a passenger from an origin to a destination, making up a single trip. An urban population may be very heterogeneous, as may be the nature of their origins, destinations and times of travel. The potential densities of travel between various general locations are influenced by the distribution of land use activities (commerce, industry, services, residences, recreation, etc.) and the income and other individual characteristics of the inhabitants, but the presence or anticipated presence of at least a minimally adequate transport service influences the locational choices as well. The densities of travel service thus either required or generated, as well as the personal characteristics of the travelers variously located, as aggregated from individual persons, origins, destinations and trips, become the demand characteristics for the transport system of an urban area.

Where travel densities are high and incomes low, one kind of service becomes operationally viable; where travel densities are low but incomes high, another, sometimes quite different, service becomes viable, perhaps both in economic and operating terms. At such service extremes the public transport options are well defined and set out limits on the range of services that need to be provided in a heterogeneous physical and social urban environment. Relatively cheap, high- or moderate-density, line-haul transport service along a fixed route and with a fixed schedule has been the dominant style of mass public transit for as long as most of us can remember. On the other extreme there has been the elite public transit option of the taxi, an individualized transit device with a driver for hire to go almost anywhere, flexibly at one's bidding, for a price few were willing or able to pay.

Within these limits much is possible, although until very recent years almost nothing was done to fill the very broad gap in services with lower density, higher quality mass transit or lower cost, nearly individualized transit. The move to fill this gap is the most exciting aspect of the incremental advances in public transport operations which are overtaking the long-established public and private modes. Lower residential densities, rising incomes, growing environmental awareness, and a number of other factors are causing the distinctions between mass and individualized transit expectations to fuzz considerably. Which distinctions, then, are those that are "fuzzing," and which are remaining fixed? Let us look to the descriptions of existing modes in operational terms to find out.

A preparatory warning is in order concerning terminology for existing modes of urban transport. Beyond the usually expected language differences in terms from country to country, there are confusions within a single language based on differing origins of technical and popular terms, moveover, the absence of some adequate terms at all in a given country for a service, system, or technology absent there but perhaps widely in use elsewhere. Thus a combined glossary/dictionary of transit terms in English, American, French and German is provided as Table 1 to help sort out the extended variety of terms for a basically simple, and limited, set of existing modes.

Bus

Buses are generally considered to be moderately large-sized single vehicles operating on shared rights-of-way with other street traffic and subject to the rules and regulations of such traffic with the exception of special stops for entry and exit of passengers. They operate on fixed routes and schedules which are subject to change only intermittently when new

Table 1 Glossary/Dictionary of Transit Terms

English	American	French	German
Bus Motorbus Trolleybus	Bus	Autobus	Omnibus Autobus
Tram	Street Railway Trolley Cable Car Streetcar	Tramway	Strassenbahn
Tram in Tunnel		Semi-métro Pre-métro	U-Strassenbahn Unterpflaster-bahn
Underground Tube	Rapid Transit Subway	Métropolitain Métro Métro Classique Métro Urbain	U-Bahn (Untergrund-bahn)
Regional Railroad	Suburban Railroad Regional Rapid	Chemin de Fer Métro Suburbain	S-Bahn (Stadtschnell-bahn)
Taxi Cab	Taxi Cab	Taxi	Taxe

routes are deemed viable or others discarded. They are operated manually by a driver, sometimes assisted by an on-board conductor to collect fares and by curbside dispatchers to regulate the flow of vehicles and passengers and to dispense information or change. There has historically been a distinction made between buses with individual, on-board power sources (motor bus) and those with electrical power pickup from an overhead wire (trolley bus), but that is a technological distinction which has little influence on service, except to reduce somewhat the flexibility in routing to meet changing travel demands.

Street Railway, Tram, Trolley, Cable Car

A considerable amount of confusion exists in the world at large, but particularly in the United States, over terminology for a rather clearly definable mode of urban public transport. This is the street railway, or system of vehicles, sometimes individually operated and sometimes coupled in trains, which run on their own set of rails in city streets. Street railway is the usual American term, while tram is the British term for the same concept. Since such street railways in the United States were often powered by electricity gathered from an overhead wire, following the earlier days of horse traction, they gained the popular terminology of trolleys, even though that term technically refers only to the small, grooved metal wheel which rolls in contact with the electrical conductor overhead. In

Britain trolley more frequently refers to the trolley bus, so that the potential for confusion in using the term in an international report such as this one is serious, as is the further possibility of confusing special terms which distinguish street railways in terms of the location of their rights-of-way in other countries. Thus, street railway and tram will be used interchangeably, and "trolley" will not be used in its narrowest, technical meaning. The famed cable cars of San Francisco, the funiculars of Lyon and other cities, and the revived horse trams of Disneyland-type amusement parks are all to be considered technological variations on the street railway.

The degree of exclusiveness of the right-of-way for tram or street railway operation is a matter of great variability. Historically the right-of-way was shared with other street traffic and the rails were merely imbedded in a more generally usable street surface, at least within cities. Exclusive rights-of-way were provided for interurban tram operation which, where and while it existed, provided the equivalent of regional rail service. While the basic definition assumes operation in the street, and thus a shared right-of-way, more and more tram operations depend on an exclusive right-of-way for at least a portion of their operations.

Where this degree of exclusiveness goes as far as abandoning even street-level operation in favor of a tunnel over a critical portion of its route a new set of terms has been invented of late. While technically a tram in a tunnel in English, the French terms *semi-métro* or *pre-métro* have been used in almost all countries including Britain. Surprisingly no special term has been used for this tram variation in the United States, despite such operations in Boston for a considerable period of time, and new construction to this effect in San Francisco for some of the surviving downtown tram lines (sharing the same tunnel with the new regional rail rapid transit system, BART). The distinction between a *semi-métro* and a *pre-métro* is supposed to be whether or not the tram technology is intended to be a long-term transit solution or only an interim solution until the demand for full rail rapid transit (*métro*) technology has developed to a point where the trams can be displaced by this higher capacity rail equipment. Often, however, the words are used interchangeably. We will try to keep the distinction as clear as possible.

Routes and schedules for trams are fixed, and drivers and conductors are used in roughly similar proportions to buses except where the articulated operations of some tramlines allow fewer operators than for a similar density of bus service. Individual tram cars, when operated separately, are usually similar in size to buses.

Rapid Transit

Rapid transit, usually referred to as rail rapid transit, is basically railroad technology applied to an urban setting. Vehicles are rail cars coupled in trains running over exclusive rights-of-way, quite often restricted to the political limits of a single major city, such as Paris or New York. The traveled right-of-way may be elevated, at grade, or underground; but it is the extensive underground segments of the systems in dense urban centers that give the popular terms subway, underground, tube, *Untergrundbahn*, *U-Bahn* to these urban passenger railways. The French term, *métropolitain*, refers to the urban aspects of the rail service, as distinct from regional or national railways. Since the investments in exclusive rights-of-way are costly, especially for those underground sections serving urban cores, the altering of routes is nearly impossible, although there is some flexibility in scheduling to meet peaks and valleys of demand over time. Since large volumes of passengers are involved in such high-capacity systems, most manpower requirements for ticketing, information, and the like are taken care of at stations, rather than on board the vehicles. Usually only a single operator is required for an entire train, even though each car unit may be separately powered, thereby reducing the manpower costs of operation over small-unit systems such as buses and trams.

Regional Railroad

Regional railroads have traditionally been passenger trains for commuters operating on conventional railroad rights-of-way, time sharing some or all of their routes with freight service and long distance passenger trains as well. They have encompassed much more extensive geographically urbanized areas than have rapid transit systems, functioning to link suburban, residential communities with the commercial, service, and industrial activities of the regional center city. Not only are the routes fixed by the railroad right-of-way, but the schedules are fixed to offer service almost exclusively to match peak morning and evening employment commuting hours. On-board manpower requirements are roughly comparable to those for minimal passenger-train operations elsewhere. Newer versions of regional rail services have taken on close similarities to urban rail rapid transit, as a diversity of urban activities has spread to the former exclusively residential suburbs and beyond, making trips between suburb and city more frequent, more varied, and longer. The commuter railroad is thus beginning to yield both a new kind of operation and a new array of functions in regional rail service. The partially completed *Reseau Express Regional* (*RER*) in Paris, the improved operations of the *Stadtschnellbahn* (*S-Bahn*) in a variety of German urbanized regions, and the Bay Area Rapid Transit system (BART) in San Francisco are all examples of this changing concept of regional rail service in addition to the continuing presence of commuter rail systems as well.

Taxis, Jitneys, and Car Pools

In its purest form the taxi is an individual means of public transport, with a fleet of drivers and cars available for personal hire, using the shared public street and highway system, for trips flexible both in route and time throughout an urban area. Whether hailed in the street, boarded at a special taxi rank, or arranged for by telephone, the vehicle and driver are available to the unique travel requirements of the passenger requesting the service. But in addition to operating in this purely individualistic mode, the taxi in some communities is allowed to combine the different trips of a small number of separate individuals going in similar general directions, yielding a kind of collective transport with personalized origin-to-destination service characteristics. When such a service operates a little like a bus, over more or less fixed corridors regularly, but with individualized origins and destinations, it is often called a jitney. And when such a collective taxi service is informally arranged among the passengers themselves using one or more of their own automobiles, with one or all of their members in turn as the driver, it becomes a car pool.

Other Modes: Air and Water Transport in Urban Areas

In addition to the ground transport systems previously defined, there are both air and waterborne modes that serve various functions similar to the ground modes in special cases of geography or affluence. Where extensive bodies of water create either barriers to surface modes or special opportunities for rights-of-way, ferries have provided special urban or regional transit links, often of high amenity and outright pleasure for the passengers. Such ferry operations have long existed as a part of the Hamburg Transit Authority's operations, and they have been recently restored as regular commuter services on San Francisco Bay by the Golden Gate Bridge and Transportation District. Further, in some urban areas with widely dispersed centers of activity and largely lacking in any comprehensive system of high-speed regional ground system, air-shuttle services using either helicopters or relatively small planes have emerged to serve some special, highly elite, business traffic. But the general occurrence of either air or water modes is limited in most large urban areas; and the functions performed are only variations on the predominantly ground modes already identified, whether operating in a rapid transit, or even a taxi, mode.

SERVICE CHARACTERISTICS OF THE PRIVATE AUTOMOBILE

Some descriptive words, too, are in order concerning the service characteristics of the private automobile in terms of similar descriptors to those applied above for public modes. These are an extension of the remarks in Chapter 2 concerning specialization and complementarity of transport functions.

It should not be forgotten that for the automobile to operate with all of the personal freedom and flexibility ascribed to it, more is involved in this system of private urban transportation than merely a personalized *vehicle*, privately owned. There must also be a *driver* or operator, whether or not it is the specific passenger desiring transport for the given trip. There must be an extensive and relatively uncongested network of traveled public *rights-of-way*. There must be a comprehensive technical, social and legal system of *operating controls* governing the operation of the entire traffic system. And there must be *storage facilities* provided throughout the urban area for these private vehicles when not in use.

Although the automobile is often idealized as the perfect, and ultimate, form of transport because of its seeming ability to take anyone to anywhere he might choose, at a time compeltely within his control, and at a cost competitive with other transport options, it frequently fails to meet these criteria of performance in certain types of urban applications. Even with certain questions of externalities set aside, the private automobile fails to perform certain urban transport functions well, while performing others brilliantly. It is necessary to distinguish these instances with some clarity in the context of transport operations. If public transit is to be complementary with the private auto and if the reciprocal relationship is to hold as well, we need to identify those service gaps in the operations of the private automobile and its supporting system of shared rights-of-way, traffic controls, and storage facilities which other modes might better fill, or might fill as well, in order to plan more comprehensive urban transport policies.

When there is no driver or private vehicle available, for whatever reason, then the rest of the system offers no service to a potential passenger unless it is shared with transit vehicles as well. While the auto performs its direct origin-to-destination service brilliantly well in the relatively uncongested surroundings of low-density suburban land uses, it does so only for those with access to a driver and a car.

In much higher-density areas in central city and extended urban areas as well, congestion in the form of just too many autos for the street or roadway efficiently to carry may slow movement to speeds far below the optimal envisaged in a comprehensively adequate system of roadways. Likewise, storage facilities in such high-activity locations may impose limits as well, either on the number of vehicles that can be accommodated, on the amount of activities that can be fostered, or on the amenity of the area itself.

And additionally, where the conjunction of numerous trips in a given geographic corridor, even with their diverse origins and destinations, gives rise to an excess of autos over roadway capacity, for important links of the overall urban transport network at specific times of day, failures and inadequacies of the automobile system prevail. If these congestion difficulties cannot be avoided by lower density land uses or additional roadway construction at reasonable economic, social or environmental cost, then the private automobile system in that specific application must be judged as inadequate, and not indeed capable of fulfilling the ultimate promise of universal mobility idealized above. Thus it, too, may be relegated to fixed routes and schedules, may only directly serve certain land uses, albeit the vast majority of them without complementary transit usage, and, in lower density areas, may require wider availability if an overall urban transport policy is to work properly.

APPENDIX IIB
Organizations Contacted in the Course of Study

Canada

Department of Transportation and Communications, Province of Ontario, Downsview, Ontario.

France

Engins MATRA, Velizy.

Institut d'Aménagement et d'Urbanisme de la Région Parisienne (Paris Regional Planning Institute), Paris.

Institut de Recherche des Transports (Institute of Transportation Research), Arcueil.

Société d' Economie Mixte du Métro de l'Agglomération Lyonnaise (Lyon Rapid Transit Planning Agency), Lyon.

Germany

Hamburger Hochbahn Aktiengesellschaft (Hamburg Transit Authority), Hamburg.

Kraus-Maffei AG, Munich.

Krupp Forschungsinstitut (Krupp Research Institute), Essen.

Messerschmitt-Bölkow-Blohm GhmH, Ottobrunn bei München.

Stadtwerke-Verkehrsbetriebe München (Munich Transit Authority), Munich.

Japan

Ministries of Construction, Industry and International Trade, and Transport.

Hankyu Bus Company, Osaka.

Isuzu Motors Limited, Research and Experiment Department, Fujisawa.

Japan Center for Area Development Research, Tokyo.

Japan Society for the Promotion of Machine Industry, Tokyo.

Japan Transport Economics Research Center, Tokyo.

Japanese National Railways, Technical Development Department, Tokyo.

Universities of Kyoto, and Yokyo, and Tokyo Technical Institute.

National Police Agency, Traffic Police Bureau, Tokyo.

Tokyo Metropolitan Government, Bureau of Capital City Development, Tokyo.

Netherlands

Centrum voor Vervoersplannen (Center for Transportation Surveys), Utrecht, Netherlands.

Sweden

Chalmers Tekniska Högskola (Chalmers Technical University), Gothenburg.

Göteborgs Sparvägar (Gothenburg Transit Authority), Gothenburg.

Stockholms Läns Lansting-Trafikhontoret (Stockholm Traffic Department), Stockholm.

Switzerland

Battelle-Centre de Recherche de Genève, Geneva.

Institut de Technique des Transports, Ecole Polytechnique Fédérale de Lausanne (Institute of Transport Technology), Lausanne.

United Kingdom

Department of the Environment, Ministry of Transport, London.

Greater London Council, London.

Imperial College of Science and Technology, London.

London School of Economics and Political Science, London.

London Transport Executive, London.

National Research and Development Corporation, London.

Tyneside Passenger Transport Executive, Newcastle-upon-Tyne.

United States

The Aerospace Corporation, El Segundo, California.

Alden Self-Transit System Corporation, Milford, Massachusetts.

Bay Area Rapid Transit District, Oakland, California.

The Bendix Corporation, Aerospace Division, Ann Arbor, Michigan.

Boston Transportation Planning Review, Boston, Massachusetts.

State of California, Division of Bay Toll Crossings and Division of Highways (District 4, San Francisco; District 7, Los Angeles).

Chicago Transit Authority, Chicago, Illinois.

Department of Transportation: Urban Mass Transportation Administration, Federal Highway Administration, Office of the Secretary, Washington, D.C.

Ford Motor Company, Transportation Research and Planning Office, Dearborn, Michigan.

General Motors Corporation, Transportation Research Department, Warren, Michigan.

Jet Propulsion Laboratories, Pasadena, California.

Massachusetts Institute of Technology, Cambridge, Massachusetts.

Metropolitan Transit Commission, Saint Paul, Minnesota.

Office of Science and Technology, Executive Office of the President, Washington, D.C.

Regional Transportation District, Denver, Colorado.

Southern California Rapid Transit District, Los Angeles, California.

Transportation Technology, Inc., Denver, Colorado.

University of Minnesota, Minneapolis, Minnesota.

West Virginia University, Morgantown, West Virginia.

International

European Conference of Ministers of Transport, Paris.

O.E.C.D. Road Research Program, Science Directorate, Paris.

SELECTED BIBLIOGRAPHY

Koza Amano, "Toshikotsu No Kadai To Chokiteki Na Kaiketsuhosaku Ni Tsuite" (The Problems of City Traffic and their Long Term Solutions), *JSSC* Vol. 8 No. 73, 1972.

J. E. Anderson, J. L. Dias, W. L. Garrard and A. L. Kornhauser, *Personal Rapid Transit* (Institute of Technology, University of Minnesota, April 1972).

Lyle C. Fitch, "Improving Urban Transportation," in *The Future of American Transportation*; edited by Ernest W. Williams, Jr. (Englewood Cliffs, N.J.: Prentice-Hall, Inc., 1971).

S. Olof Gunnarsson and Yngve Westerlund, *Nya Transport-System*, Byggforskningens informationsblad B 18: 1971 (Statens institut för byggnadsforskning, Stockholm, 1971).

G. Hupkes en Anderen, *Transport in de Toekomst* (Amsterdam: Pantoskoop, 1970).

Institut d'Aménagement et d'Urbanisme de la Région Parisienne, *Bulletin d'Information de la Région Parisienne*, No. 2, "Les Transports" (Paris, March 1971), and No. 4, "Les Transports Nouveaux" (October 1971).

Institut d'Aménagement et d'Urbanisme de la Région Parisienne, *Les Technologies Nouvelles de Transport*, Cahiers, Volume 26, l'ère partie (Paris).

Institut de Recherche des Transports, *Transports Nouveaux dans les Villes Moyennes*, Rapport de Recherche #2 (Arcueil, France, October 1970).

Japan Committee for Urban Transportation, *Tokyo Oyobi Sono Shuhen Ni Okeru Kosoku Tetsudo O Chushin Tosuru Kotsumo No Seibe Zokyo Ni Kansuru Kihon Keikaruni Tseuite* (Concerning the Basic Plan to Improve the Transportation Network Mainly of High Speed Mass Transit in Tokyo and its Suburbs), Report No. 15, March 1, 1972.

Japan Ministry of Transport, Transportation Policy Committee, *Daitoshi Kotsu Ni Okeru Basu Takusi Ni Kansuru Toshin*

This list includes only references of a general nature. Technical reports dealing with different specific transport systems or application studies are cited in the body of the text as footnotes.

(A Report on Buses and Taxis in Large City Traffic), August 20, 1971.

Japan Ministry of Transport, Transportation Technique Committee, *Shin Kotsu Sisutemu No-Kaihatsu* (Development of New Transportation Systems), May 11, 1971.

J. R. Meyer, J. F. Kain and M. Wohl, *The Urban Transportation Problem* (Cambridge, Mass.: Harvard University Press, 1965).

National Academy of Engineering (U.S.), *Urban Transportation Research and Development* (Washington, D.C.: 1972).

O.E.C.D., Consultative Group on Transportation Research, *Improvements and Innovations in Urban Bus Systems* (Paris: 1969).

O.E.C.D., Road Research Programme, *Optimisation of Bus Operation in Urban Areas* (Paris: Research Group T3, October 1971).

Wilfred Owen, *The Metropolitan Transportation Problem* (Garden City, N.Y.: Doubleday & Co., 1966).

Regional Transportation District, *Transportation Systems and Technology—Issues and Trends* (Denver, Colorado: December 1971).

Brian Richards, *New Movement in Cities* (London: Studio Vista; New York: Reinhold Publishing Corp., 1966).

Wilbur R. Thompson, *A Preface to Urban Economics*, Ch. 9, "Traffic Congestion: Price Rationing and Capital Planning," and Ch. 10, "Interactions among Problems: The Problems of 'Solutions'" (Baltimore, Md.: Johns Hopkins Press, 1965).

U.S. Department of Housing and Urban Development, *Tomorrow's Transportation—New Systems for the Urban Future* (Washington, D.C., 1968).

U.S. Department of Transportation, *An Analysis of Urban Highway Public Transportation Facility Needs* (Washington, D.C.: April 1972).

U.S. Department of Transportation, *Metropolitan Transportation Planning Seminars* (Washington, D.C.: 1972).

Urbane Information, 4, 70 "Urbaner Verkehr, Zukunft des Verkehrs, Neue Verkehrssysteme, Umweltschutz, Verkehrssoziologie," (Urbanistic Verlag, Munich, August 1970).

Working Group on Bus Demonstration Projects (U.K.), *Report to the Minister of Transport* (London: April 1970).

NOTES TO PART II

Notes to acknowledgments

1. O.E.C.D. Conference Proceedings, *Better Towns with Less Traffic*, Paris, France, April 14–16, 1975.

2. United States Congress, Office of Technology Assessment, *Automated Guideway Transit*, June 1975.

Notes to chapter 3

1. The first metropolitan railway system was constructed in London in 1863, using steam engines and running in open trenches. Hackney-coaches appeared in London from 1633 onward as the beginnings of the taxi trade. Carosses à cinq sols—the forerunners of buses—were inaugurated in Paris briefly in 1662 as carriages for public hire on fixed routes, with such service reappearing there in the 1820's under the name of omnibus. And the first recorded street railways or tramways occurred in the United States in about 1830 in Baltimore and New York, with horse traction. For an interesting and detailed history of such modes see for example Charles S. Dunbar, *Buses, Trolleys and Trams* (The Hamlyn Publishing Group Ltd., Feltham, Middlesex, England, 1967).

2. Certainly other descriptors could be added for greater detail in categorizing the technology of the individual varieties of a given mode, in specifying speeds, waiting times, and personal comfort or in differentiating the spatial location of the traveled right-of-way, but these are of secondary importance in categorizing the basic transport services provided.

Notes to chapter 4

1. Wilbur Smith and Associates, *The Potential for Bus Rapid Transit*, (Automobile Manufacturers Association, Detroit, February 1970), pp. 40–44.

2. U.S. Department of Commerce, National Bureau of Standards, *The Shirley Highway Express-Bus-On-Freeway Demonstration Project*, Washington, D.C., August, 1971.

3. Donald A. Morin, "Status Report on Highway-Related Public Transportation Projects in the United States," *Transportation*, Vol. 1, No. 1, May 1972.

4. U.S. Department of Transportation, *An Analysis of Urban Highway Public Transportation Facility Needs* (Washington, D.C., April 1972), pp. 30–32.

5. This is not, however, always the case. Implementation of a modified demand bus system in Japan has been initiated in an effort to reduce costs, partially by going from large two-man buses to smaller, one-man operations in low-density areas.

6. Daniel Roos, "Operational Experiences with Demand Responsive

Transportation Systems." Paper delivered to the Highway Research Board Meeting, Washington, D.C., January, 1972.

7. *Collected papers of the Third Annual Conference on Demand-Responsive Transportation Systems*, Ann Arbor, Michigan, June 12–14, 1972.

8. Geurt Hupkes, *Buxi: Demand-Responsive Bus Experience in the Netherlands* (Highway Research Board, Washington, D.C., January, 1972).

9. T. Tanizawa, I. Tsuji, J. Kaneshige and Y. Sakano, *Demand Bus Implementation in Nose Area*, Hankyu Bus Company, Isuzu Motors, Ltd., and Fujitsu, Ltd. (Japan, November 1972) and I. Kaneshige, Y. Funaya, Y. Sakano, and T. Ogawa, *Demand Bus System for New Residential Area in Japan* (Society of Automotive Engineers #720218 Detroit, January 1972).

10. Victor Gruen, *The Heart of Our Cities* (Simon and Schuster, New York, 1964).

11. Southern California Rapid Transit District, "To Help Improve Downtown Mobility," (Los Angeles, California, September, 1971).

12. "Central City Los Angeles 1972/1990: Preliminary General Development Plan" (Los Angeles, April, 1972).

13. C. Kenneth Orski, *Vehicle-Free Zones in City Centers*, OECD Symposium on Road Traffic Restraints, Cologne, October 1971.

14. E. R. Ellen, "Ways of Helping Buses in Urban Areas," *Revue de l'UITP*, Vol. XVIII-4/1969.

15. M. Neizonnet, "Practical Experience with Use of Bus Lines in Paris" in *Improvements and Innovations in Urban Bus Systems*, (OECD Consultative Groups on Transportation Research, Paris, 1969).

16. P. Engelbrecht, "Freie Fahrt dem Stadbus!", *Der Bayerische Bürgermeister* (#5, 1969), pp. 121–125.

17. State of California, Department of Public Works, *Feasibility Study on an Exclusive Lane for Buses and Carpools on the San Francisco Oakland Bay Bridge*, April, 1971).

18. Although buses do not stop to pay tolls, they are counted, and the transit authority billed later. It is the time-saving of not having to stop and pay the toll that is the advantage, not a financial one. However, in car pool operations the additional benefit of avoiding a 50¢ toll adds a financial incentive. Once the decision was made to keep the experiment as a permanent operation a small charge was made for passes for car pool cars, but only to meet some legal requirements for the bridge's bond financing. Even this nominal charge was enough to reduce the number of car pools per day to 1,600.

19. The MITRE Corporation, *Automatic Vehicle Monitoring Technology Review and Benefits and Costs of an Automatic Vehicle Monitoring System.* U.S. Mass Transportation Administration (Washington, D.C., August and September, 1971).

20. Institute for Public Administration, *Urban Vehicle Monitoring Technology, Economics and Public Policy* (U.S. Department of Housing and Urban Development, Washington, D.C. 1972).

21. Dr. Ing. Hans Tappert, *Real-Time Bus Control Improving the Reliability of Bus Transportation*, Hamberger Hochbahn A. G. (February, 1969).

22. Jerold W. Scheel and James E. Foote, *Bus Operations in Single Lane*

Platoons and Their Ventilation Needs for Operation in Tunnels (September, 1968), and *Comparison of Experimental Results with Estimated Single Lane Bus Flow Through a Series of Stations Along a Private Busway*, (Warren, Michigan: Research Laboratories, General Motors Corporation, May 1969).

Notes to chapter 5

1. Dipl.-Ing. Arnold Mies, *Hamburg's Idea for Automation of Rapid Transit Lines—Basic Reflections on Train Movements* (Hamburg: Gesellschaft für Verkehrsberatung und Verfahrenstechniken MBH).

2. P. Scherrer, *Nouveaux Métros en Allemagne Fédérale* (Lyon, France: Association "Lyon-Métro," November 1970), pp. 3-6.

3. *Public Transport on Tyneside—A Plan for the People*, Tyne and Wear Passenger Transport Executive, Newcastle upon Tyne, September 1973.

4. Chicago Transit Authority, *Skokie Swift—"The Commuter's Friend"* (Chicago, Illinois, May 1968).

5. City of Chicago, *Transit Planning Study—Chicago Central Area* (Chicago, Illinois, April 1968).

6. J. R. Meyer, J. F. Kain and M. Wohl, *The Urban Transportation Problem* (Harvard University Press, Cambridge, Mass., 1965), pp. 169-249.

7. *"Personal Rapid Transit,"* University of West Virginia, Morgantown, West Virginia (University brochure).

8. MPC Corporation, Transit Expressway Report (Port Authority of Allegheny County, Pennsylvania, 1967).

9. Maurice Barthalon, *L'Urba—Nouveau Moyen de Transport de Villes de Demain* (Vauves, France: Cie. d'Energetique Lineaire, 1969).

10. Mitsui & Co. and Nippon Sharyo Seizo Kaisha, VONA *Vehicles of New Age* (Japan, 1972).

11. United States Congress, Office of Technology, op. cit.

12. A technologically more advanced, high-speed rail connection between the new town of Cergy-Pontoise and the business center of La Defense was actively under cause by the French Government several years ago, but was scrapped when the political intervention of a small community between the two ends of the line insisted on a stop there as well. As infrequent stops are required for a high-speed service, its ability to perform the region shaping function is greatly diminished and the addition of that stop, coupled with the death of the Aerotrain inventor caused this system concept to be abandoned. This is just one more example of the institutional blockages to technological innovations in the transportation field.

13. United Nations Economic Commission for Europe, Seminar on the Role of Transportation in Urban Planning, Development and Environment, background paper, *The Changing Role of Urban Transportation*, June 1976.

Notes to chapter 6

1. Charles N. Barnhard, "Commuter's Dream City (Hamburg)," *The Lamp*, Winter 1971 (Standard Oil Company, New Jersey).

2. Commission Intercommunale d'Urbanisme de la Région Lausannoise, *Avant-Projet d'une Nouvelle Liaison Ferroviaire Gare CFF-Centre Ville-Romanel* (ITEP/EPF-Lausanne-CETP, Juin 1971).

3. City of San Francisco, *San Francisco Downtown Zoning Study, Final Report*, December 1966.

4. Andre Taure, "L'Innovation dans les Transports," *Bulletin d'Information de la Région Parisienne*, October 1971, pp. 10–11, and Pierre Merlin, "Une Même Politique: Urbanisme et Transports," *Bulletin d'Information de la Région Parisienne*, March 1971, pp. 11-16.

5. George Krambles, *Expressway Rapid Transit*, Chicago Transit Authority, July 1971.

6. Department of Public Works, State of California, *Los Angeles Area Freeway Surveillance and Control Project* (January 1972).

7. Eisen, op. cit.

8. Conférence Européenne des Ministres des Transports (C.E.M.T.), *Choice Between Private and Public Transport* (Topic I) and *Economic Criteria for the Maintenance, Modification or Creation of Public Transport Service Which May Not Necessarily Be Profitable* (Topic II), Fourth International Symposium on Theory and Practice in Transport Economics (The Hague: 5–7 October 1971).

9. Greater London Development Plan Public Inquiry, *Comments by the London Transport Executive*, GLDP Inquiry Proof E12/2, November 1970.

Notes to chapter 7

1. United States Department of Housing and Urban Development, *Tomorrow's Transportation* (Washington, D.C.: U.S. Government Printing Office, 1968).

2. Brian Richards, *New Movement in Cities* (London: Studio Vista; New York: Reinhold Publishing Corp., 1966).

3. A number of systems that could actually do a reasonably good job of activity center circulation have been in limited use in Disneyland and Disney World amusement parks in the United States and at recent exhibitions at Lausanne, Montreal, and Osaka. The unreality of their apparent working environment to most people seems however to be a stigma these systems cannot shake.

4. For example, the Automobile Club of Southern California has supported and even helped initiate the Los Angeles downtown minibus service that will precede an exclusive guideway circulation system there, and has supported the concept of activity center transit systems elsewhere in the Los Angeles area (see *Toward a Better Tomorrow*, Highway Engineering Department, Southern California Automobile Club, 1971), and the Ford Motor Company has begun development of such a system to enhance the value of some of their own land holdings which would benefit from high-density, activity center construction in Dearborn and Detroit, Michigan.

5. ACT is an acronym, began in 1971 by referring to Activity Center Transportation System and by mid-1972 had curiously become Automatically Controlled Transportation System.

6. Monocab, Inc., *Stratoline* (Garland, Texas, 1970), Monocab, Inc., 2700 Oakland Avenue, Garland, Texas 75041.

7. *Transportation Technology, Inc.* (company and product descriptive brochure). Transportation Technology Incorporated, 4700 Holly Street, Denver, Colorado 80216.

8. *People Mover Profile, Technology Sharing, A Program of the U.S.*

Department of Transportation, Urban Mass Transportation Administration.

9. See Richards, Brian, *New Movement in Cities* (London: Stuio Vista, 1966), pp. 65-71 for a discussion of the Habegger systems and of the "Carveyor" system of Goodyear Tire and Rubber Co., existent since 1953 in a test configuration, but never installed for actual use.

10. Alden Self-Transit Systems Corporation, *STARRCAR System Description* (Milford, Mass., November 1971).

11. Uniflo Systems Company, *UNIFLO/PRT 70* (Minneapolis, 1970).

12. Vought Aeronautics, *Airtrans.* Ground Transportation, Dept. 2-18000, Vought Aeronautics Company, P.O. Box 5907, Dallas, Texas 75222.

13. United States Department of Housing and Urban Development, *Ibid*., pp. 60-62.

14. *Tomorrow's Transportation*, pp. 62-63.

15. It would be hard to identify all of the research going on in the development of component technologies that would be of specific importance to PRT systems in future years. It is clear, however, that changes are coming in propulsion, controls, modular construction (for guideways and for vehicle maintenance), suspension and other fields which may well have considerable importance in the evolutionary technological sophistication of these systems, as they need to increase in operational complexity and perhaps decrease in relative cost. And much of the component research of closest relevance to PRT is going on in firms and university or government laboratories with a keen interest in how this field is developing, but often without the resources or overall capability to attempt systems design and prototype construction. Work on linear induction motors, for example, is particularly widespread, as is vehicle automation.

16. West Virginia University, *PRT* (an informational pamphlet issued by the Office of University Relations/Publications, Morgantown, West Virginia, 1971). Also see Chapter 6.

17. Demag Fordertechnik and Messerschmitt-Bölkow-Blohm, *Systemanalyse Naheverkehrsmittel Cabinentaxi—CAT*, March 1971.

18. Richard Hesse, *German Experiences in the Planning and Development of an Automatic Cabin Taxi System* (Intersociety Conference on Transportation, Washington, D.C., May 31, 1972).

19. Hesse, *op. cit.* pp. 7-12.

20. Royal Aircraft Establishment, *Cabtrack Studies—Assessment of Autotaxi Urban Transport Systems*, U.K. Ministry of Technology, December 1968 (Vol. I), and January 1969 (Vol. II).

21. Japan Society for the Promotion of Machine Industry, *Computer-Controlled Vehicle System*, (Tokyo, 1972). JSPMI, 3-5-8 Shiba Koen, Minatoku, Tokyo, Japan.

22. Japan Society for the Promotion of Machine Industry, *A New Urban Traffic System CVS* (Tokyo, May 1972).

23. National Academy of Engineering, *Urban Transportation Research and Development*, (Washington, D.C., April 1972) pp. 11-13.

24. Sixten Camp and Rolf Oom, "Spårtaxi for Gothenburg," in *PRT op. cit.*, pp. 123-140.

25. Design Center, University of Minnesota, *Demonstration of Personal*

Rapid Transit in the Twin Cities (Institute of Technology, University of Minnesota, Minneapolis, June 1971).

26. Walace Russell, "Architectural and Environmental Studies of an Automated Public Transport System in an Urban Context," in *PRT, op. cit.*, pp. 59–122.

27. "Coin-Operated Taxis," in News section of *Transportation*, Volume 1, No. 1 (May 1972) pp. 100–101.

28. Schmidt, James W. and Robert K. Arnold and Stephen Levy, "Specification and Evaluation of Alternative Feeder and Local Transit Systems in a Suburban Area," 51st Annual Meeting of the Highway Research Board, January 1972, Washington, D.C.

29. Jet Propulsion Laboratory, *Technical and Cost Considerations for Urban Applications of Dual-Mode Transportation*, United States Department of Transportation, Washington, D.C. (May 1972).

III METHODS OF TRAFFIC LIMITATION IN URBAN AREAS

J. Michael Thomson

SUMMARY

In every city of the world the volume of traffic is limited, intentionally or unintentionally, by measures adopted by governments, both central and local. If these measures were relaxed, there would be more traffic; if they were strengthened, there would be less. In other words, the volume of traffic in a city is not something like the rainfall that has to be accepted; it is amenable to some control or influence by government. Furthermore, depending upon the methods used, control may be exercised selectively over particular classes of vehicle, in particular areas or streets, at particular times, and for particular purposes.

This paper reviews the methods that are being used or contemplated for the limitation of traffic. While not entering into arguments for or against traffic limitation, it necessarily takes note of the various reasons advanced in favour of limiting traffic and those put forward in opposition. Few people, of course, are wholly for, or wholly against, traffic limitation; most people want there to be some traffic but not too much. What constitutes "too much" is something which every city must decide for itself in the light of its own economic and geographical circumstances, its heritage, aspirations and values. This paper does not attempt to give any guidance on how far, if at all, cities should try to limit their traffic or traffic growth; but, recognizing the kind of reasons why some cities may wish to do so, it describes the methods available to them.

Some 35 methods are identified. They fall into three categories: *traffic restriction*, whereby people are prevented by physical or legal means from fulfilling certain travel desires; *traffic restraint*, whereby people are discouraged, but not prevented, from fulfilling certain travel desires; and *traffic avoidance*, whereby action is taken to avoid the desires arising in the first place.

Traffic restriction includes methods involving road closure, prohibitions, traffic management, vehicle regulation and restriction of driver licences. Pedestrian streets and environmental areas rely on methods of traffic restriction, which are now in use in the centres, at least, of most European towns and cities.

Traffic restraint methods alter the balance of attraction between using private motor vehicles and doing the best alternative thing. They operate by making private transport slower, more difficult or more expensive *relative* to the alternatives. In practice, congestion is a powerful restraint, which may be planned to occur in the least harmful places. Ways of reducing traffic speed, and other traffic management measures, can be used to discourage traffic from using certain streets and to discourage the growth of journey lengths. Allocation of more road space to buses, cycles and pedestrians can encourage the use of these modes while simultaneously restraining private motor traffic.

Traffic restraint also covers a number of ways of making private motor vehicles more expensive to own or to run. Toll-gates, daily licences and vehicle metering are all forms of road pricing that are discussed in some detail. In practice, the most common methods of deliberate traffic restraint are concerned with parking and include parking restrictions, controls and charges. A more popular, though less effective, method of making cars *relatively* expensive to use is that of subsidising public transport fares.

Traffic avoidance is a longer term way of limiting traffic: it relies upon planning control of land use and transport infrastructure. The possibilities are discussed of arranging a town so as to minimize the need for travel, especially travel by private motor vehicles. Improvements to the structure of public transport (by means of subsidized investment) and restrictions on roadbuilding in order to avoid traffic generation, are major aspects of traffic avoidance.

Finally, a number of examples are given of studies that indicate the effectiveness of various methods of traffic limitation.

The conclusion of the paper is that an effective armoury is available, from which cities may select the most appropriate instruments to limit the volume of traffic, if they so wish, with a considerable degree of selectivity.

9 INTRODUCTION

THE CONTEXT

The motor vehicle has brought many advantages to many people and will continue to do so on a growing scale. To attempt to stop it would be wrong and foolish. These unqualified assertions are an essential starting point for a study of methods of limiting the use of motor vehicles. For this paper is not an attack on the automobile but a review of just one aspect of the much wider problem of maximizing the net benefits obtainable from motor vehicles to society as a whole.

Unfortunately the benefits produced by motor vehicles are accompanied by costly and undesirable side effects, the magnitude of which is not a fixed proportion of the associated benefits. Hence, while the use of motor vehicles in general represents a great social gain, their use in particular circumstances may represent a social loss.

The undesirable effects of motor traffic are well known and are specially marked in urban areas. They may be regarded in four categories:

1. the cost and social impact of providing roads and parking space;
2. the nuisance caused by traffic in terms of noise, air pollution, accidents, danger, ugliness, etc;
3. the adverse impact on other forms of urban transport, i.e. public transport by road or rail, cycling and walking;
4. the adverse impact on the spatial form or structure of the urban area.

These undesirable consequences of urban motorization have often been described elsewhere.[1] They are all related to the volume of traffic. Therefore there are two possible approaches to the problem: to reduce the side effects directly or to reduce the volumes of traffic that cause them.

The former approach is the more appealing. Let us eliminate noisy, polluting engines, let us control land use and find ways of maintaining good public transport side by side with good private transport, let us devise ways of getting better value for money out of roads and parking space. Let us thus deal with all these problems. In principle they can all be solved, or nearly solved, by technological or administrative advances, or by unlimited expenditure.

All this is accepted. A major inventive effort supported by a generous budget must be applied to the whole problem of urban motorization. Only when this appears to be inadequate, within a reasonable period, is one justified in turning to the second approach and looking for ways of reducing, or retarding the growth of, traffic volumes.

This is the context in which this paper is written.

JUSTIFICATION

It is not the purpose of this paper to discuss the case for traffic limitation, but simply to review the methods which are being or could be used. Nevertheless there is some connection between the methods to be used and the arguments advanced for using them at all. There are three kinds of argument: moral, economic and pragmatic.

The *moral* justification, put forward by some environmentalists, claims that modern traffic represents an abuse of the traditional freedom of the highway. In most countries, if not all, the freedom to use the public roads was recognized as a basic right long before the invention of the motor vehicle. If motorists take the view that this right should apply without qualification to the use of motor vehicles, they will naturally regard every regulation and restriction as an interference with their freedom. Few motorists take this extreme view nowadays. Most would agree that there are limits to what should be permitted in respect of noisy, dirty, dangerous vehicles. Even in the days of the horse there was no "freedom" to use animals that were a menace to other road users. At the other extreme it is possible to take the view that there is no natural right, and never should be, to take motor vehicles into public places, and permission to do so should be regarded as a privilege, though few people would wish to withhold permission altogether. Depending on the basic viewpoint, therefore, the use of motor vehicles on public streets can be seen as either a right or a privilege, and those who take the latter view see no need to justify restriction; on the contrary, they see a need to justify the privilege.

The *economic* justification has been fully set out by many authors, including Beckmann, Walters, Roth, Stewart, Vickrey, Foster and many others.[2] Invariably the argument has been expressed in the context of road pricing, the conclusion being that higher charges for the use of congested roads could produce gains in economic efficiency; but the same argument can be used more generally to show that traffic limitation by other means, not only pricing, could also produce economic gains under certain conditions.

The argument may be put as follows. Consider an urban road network open to vehicles free of charge. Assume initially that all vehicles and drivers are homogeneous, and vehicle owners incur no charges, apart from their own operating costs, as a result of using the roads. They may pay taxes, such as vehicle license fees, purchase tax or income tax, which may be used to pay for the national road system, but these are assumed to be unrelated to any decision to make a particular journey. Financially, the motorist is unaffected by the costs that his journey imposes upon either the highway authorities, other road users or society at large, i.e. by the increment in *system costs* caused by him apart from his own operating costs.

System costs include all costs to the community arising from the operation and maintenance of both vehicles and roads during a given period. They include

1. vehicle operating costs, including the value of occupants' time;
2. road operating costs, e.g. police control, lighting, etc;
3. road wear and tear, i.e. items covered by road maintenance;
4. accident costs;
5. direct environmental costs, e.g. noise, dirt, fumes, danger, vibration and ugliness, suffered by both road users, including pedestrians and cyclists, and non-road users such as residents, workers, etc.
6. other external costs. The roads are not an independent system but part of the whole urban transport system; one should include changes in the cost and quality of rail transport caused indirectly by the road system.

As the volume of traffic on the network increases, all costs rise; some rise less than proportionately, others more so. Since the average speed of traffic declines as the volume of traffic rises, it follows that vehicle operating costs and the value of occupants' time rise more than proportionately. These relationships are well documented,[3] but not much is known about the precise relationship between traffic volumes and the other categories of system cost. As traffic increases, road operating and maintenance costs rise less than proportionately; so do noise and vibration; but dirt, fumes and ugliness probably rise more

9.1 Road system cost structure.

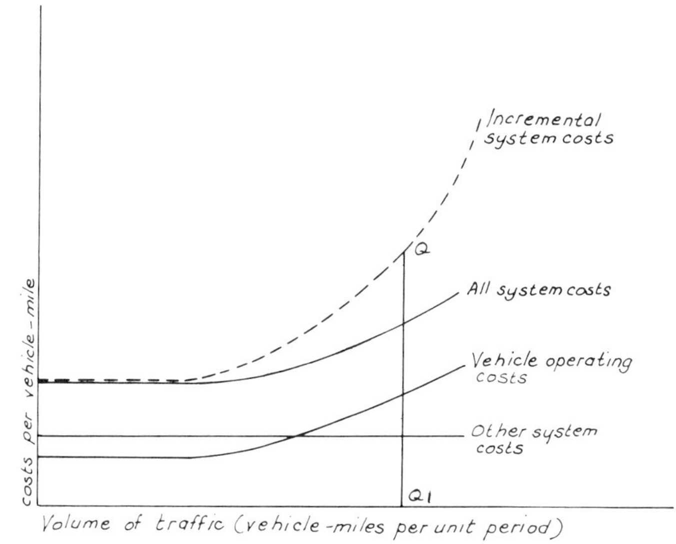

than proportionately. If one assumes that all these costs taken together, rise in equal proportion with the volume of traffic, the total cost structure of the system is as indicated in figure 9.1. Since the cost curve is rising, the incremental costs—i.e. the addition that each extra vehicle causes to total system costs—must rise more steeply; and the larger the volume of traffic, the higher they become. Under our initial assumption, each vehicle owner pays only his own operating and time costs, which fall short of the incremental system costs by a margin that rapidly widens as the volume of traffic increases. If, for instance, the volume of traffic is at QQ_1 where severe congestion is implied, the average vehicle operating costs are quite high but the incremental system costs are much higher.

It follows that decisions to use motor vehicles do not carry responsibility for all the costs they cause, and *some* vehicle journeys are therefore made which would not be made if the costs were directly chargeable to them. The economic efficiency of the system would be improved by the exclusion of vehicle journeys whose value is less than incremental system cost. In principle this could be achieved by charging an appropriate price or by other selective means of traffic limitation. According to an estimate in Central London, if 40 percent of cars could be thus restrained from entering the area the net benefits to the community would be worth about $10 million a year.[4]

This, briefly, is the economic justification. It is oversimplified, but complications such as fuel tax, which is a small charge related to vehicle use, and the heterogeneity of vehicles and drivers, which means that some vehicles cause higher system costs than do others, do not alter the conclusion: namely, that on congested urban networks, because of the structure of costs and the way in which roads are financed, some vehicle journeys are likely to be made which are not valued as highly—by those who make them—as the costs they impose on the system as a whole.

Finally there is the *pragmatic* justification for limiting traffic, sometimes put forward by planners and politicians, and used in development plans for cities like London, Cambridge, Bremen, Göteborg and Vienna. The view is taken that a laissez faire approach to the use of motor vehicles has been followed for a long time and has produced poor results, both in cities which have strived to satisfy the motor vehicle's appetite for road space and parking space, and in cities which have not. It has led *in practice* to highly undesirable developments, in the form of congestion, ribbon development, bad environment, collapse of public transport, decline of city centres, etc. This dissatisfaction with the sort of cities produced by free use of motor vehicles is accompanied by a commonsense belief that measures to limit the volume of traffic, in some places at some times, could do more good than harm.

OBJECTIVES

Traffic limitation means any kind of policy or programme adopted with the intention of reducing the volume of traffic below what it would otherwise be. There are, as we shall see, many ways of doing this. The methods most appropriate to a particular town will depend upon its particular problems, which need precise definition. Motorization produces the same problems in all cities but their relative incidence and the priorities accorded to them vary from one city to another. The dominant problem may be smog in Los Angeles, congestion in Lagos, noise in Rome, aesthetics in Amsterdam, or parking in Paris; and in London it has been claimed (by Sir Colin Buchanan) that the biggest traffic problem was accidents.

Similarly the benefits from motorization vary from one city to another. Whereas the motor car enables Americans and Australians to spread themselves in low density suburbs over their seemingly limitless acres, the same cannot be said of Hong Kong, Japan or most European nations. The effective mobility afforded by the car may be more important in Detroit, where most journeys are not well catered for by public transport, than in Hamburg where they are. Door to door travel, with a

Table 9.1 Possible Advantages and Disadvantages of Traffic Limitation

Possible advantages	Possible disadvantages
Short-term	**Short-term**
less congestion	inconvenience of using less desirable modes
less noise	adoption of less desirable destinations
less air pollution	adoption of less desirable activities
fewer accidents	more congestion on public transport
higher revenue for public transport	**Long-term**
lower demand for parking space	tendency to higher housing densities
easier conditions for pedestrians	higher investment in public transport
saving in vehicle operating costs	fewer economies of scale
Long-term	
less roadbuilding	
less road maintenance	
less provision of parking space	
better public transport	
less environmental deterioration	
less crime**	
less social divisiveness	
less depletion of irreplaceable natural resources	

* a transfer of passengers from cars to buses may not create greater congestion on the buses if the consequent reduction in traffic congestion permits higher vehicle-mileage from the bus fleet.

** the suggestion that motorization breeds crime is based on the break-up of urban communities and the social diviseness (into have and have nots) which are accentuated by the use of cars.

minimum of walking, may be more appreciated in the steamy Washington summer or the bitter winter of Montreal than in the mild weather of Europe's pedestrianised towns.

Before embarking on a policy of traffic limitation, therefore, a city should identify, in the light of its own circumstances and value judgments, the likely advantages and disadvantages. These will be short-term and long-term, and though the short-term effects will be foremost in people's minds and the main source of political interest, nevertheless the long-term effects may be the more important. A broad summary of short- and long-term effects is made in table 9.1; short-term effects are defined as those which occur prior to and independently of any consequential change in land use or investment. Table 9.1 is neither comprehensive nor universally applicable. The long-term effects depend on human reactions over time to the short-term effects and cannot therefore be predicted with certainty. It does not follow, for instance, that if traffic congestion is not relieved by traffic limitation it will lead inevitably to roadbuilding. Nevertheless the table may serve to emphasize the point that the effects of traffic limitation are not necessarily confined to the immediate obvious things; in the longer term, by weakening the impact of motorization, they may influence the physical and social structure of the urban area.

A clear definition of basic objectives will enable a city to specify more exactly the *functions* of a traffic limitation policy. There are five possible functions:

1. to limit the volume of traffic in particular streets, either permanently or at certain times. The objectives might be to aid pedestrian movement, to improve the amenity aspects, to reduce noise and fumes, etc.
2. to limit the volume of traffic in particular areas of the town, e.g. environmental areas or the town centre. The objectives might be to create quiet residential precincts or to stimulate the competitiveness of the town centre. It might be desirable to limit traffic in high density inner areas without necessarily doing so in outer suburban areas.
3. to limit the volume of traffic in the urban area as a whole. The objectives could be to reduce smog, to conserve natural resources, to ease the general pressure for more roads, to maintain a balanced use of public and private transport, etc.
4. to limit the use of certain types of vehicle, e.g. large cars, noisy motorcycles, lorries with low power-weight ratios. The objectives are self-evident.
5. to limit the use of vehicles for certain purposes. The commonest discrimination of this sort is against car commuters, the objectives being to relieve peak hour congestion and release parking space. Other examples are the limitation of through-traffic in favour of terminating traffic, and the limitation of long-term parkers in favour of short-term parkers; in each case the objective may be to encourage shoppers.

Logically a traffic limitation policy should be derived from a clear definition of objectives, followed by an analysis of where, when and what kinds of limitation are required to meet the objectives. This systematic approach will facilitate the selection of the most suitable among the many available methods, which we now proceed to describe.

THE THREE BRANCHES OF TRAFFIC LIMITATION

The idea of trying to limit the volume of traffic is likely to arouse fears of loss of freedom, of not being allowed to use one's car as one pleases, of being forced to stand in crowded buses and walk tedious distances in the rain. It is therefore important to realise that there are three general ways of limiting traffic, and they have different consequences for the individual's freedom of choice.

First there is *traffic restriction*, which includes all methods of preventing vehicles by physical obstruction or act of law from going where they would otherwise go. At first such restrictions are accepted without question. Nevertheless these are the methods that strike directly against the freedom of the motorist. How far they could or should be taken is a lively issue.

Secondly there is *traffic restraint*, which includes all methods designed, not to deny people their desires, but to influence their choice. A parking charge, for instance, does not prevent people from using their cars; it merely dissuades. A free bus service does not compel people to travel by bus; it merely encourages. Very heavy restraint would be regarded as prohibitive and therefore no different from straightforward restriction. But provided it is not excessive, restraint may serve to limit traffic without eliminating a realistic freedom of choice.

Thirdly there is *traffic avoidance*, which aims neither to prevent nor to dissuade people from fulfilling travel desires, but to avoid the birth of the desires in the first place. This is the ideal, where people have no desire to make journeys that are a burden on society. It is an ideal that requires the inhabitants of a town and their activity centres to be well located in relation to each other and to the transport system, which itself must be well located, designed and integrated.

In the long run one may look increasingly to traffic avoidance as the solution. In the short run one must rely more on traffic restraint and restriction. Restraint and restriction will themselves lead to traffic avoidance in the long run if people adjust their life patterns so as to avoid the restrictions and minimize the restraints. People can be expected to optimize their situations within the constraints bearing upon them. Until recently the lack of personal transport was a constraint upon the lives of most people, but they organised their lives in acceptance of this fact and apparently without feeling it to be a restriction. The availability of cheap automobiles opened up the vision of a complete removal of this constraint; and indeed the vision became a reality for the first families to acquire a car. But as the numbers have increased, city after city has concluded that unrestricted private transport *for everyone* is only possible, if at all, at an unacceptable cost, and that some constraint must therefore remain, at least until a superior technology permits it to be removed completely. For the minority who have enjoyed complete freedom to drive unimpeded and park freely anywhere at any time, this is a sorry decline; but in most cities those conditions have long since passed. The majority of citizens have never enjoyed them and never will. It should be easier for them to adapt to a city in which it is sometimes difficult, expensive or even impossible to make certain journeys by car. But even those who have become dependent on the car should eventually be able to reorganise their lives (by relocating their homes and other activities or by car sharing or "learning" to use new public transport services, and so on) so as to reduce their conscious desire to use a car where it is restricted or restrained.

The ultimate objective, therefore, is to create a city in which people no longer have a strong desire, nor feel it essential, to make such use of motor vehicles that they constitute a social problem.

The following three chapters discuss methods of traffic restriction, restraint and avoidance. The effectiveness of these methods is dealt with separately in chapter 13.

10 TRAFFIC RESTRICTION

The term traffic restriction is used for all methods of prohibiting or otherwise preventing vehicles, or particular classes of vehicles, from entering certain places at certain times. The general advantages of traffic restriction are simplicity and—superficially at least—equality; its disadvantage lies in its arbitrariness.

PHYSICAL RESTRICTIONS

No Roads, No Traffic

The simplest, cheapest and most effective form of traffic limitation is undoubtedly a physical obstruction. Extreme examples are in Venice and Zermatt, both of which towns have chosen to remain unconnected to their national road network and are virtually auto free. Of course, these towns are economically dependent on tourism, which is perhaps the only industry to derive a positive advantage from the prohibition of motor vehicles. Therefore the fact that both towns are prosperous is no indication that other towns engaged in other industries could prosper in such conditions. It is difficult to see how an industrial town could survive without road transport. The lesson to be drawn from Venice (and to a lesser extent Zermatt, though it is little more than a village) is that affluent people apparently can accept life without a car in return for the advantages of a auto-free city. Venice is exceptional, of course, on account of its canals and architectural heritage, but the engineering problems of building roads there are no greater than have been overcome in many other cities.

It is sometimes said that the surest way to avoid traffic is not to build roads. But most cities already contain dense road networks which, however inadequately, permit traffic to infiltrate almost everywhere. A few examples remain, however,o of large areas within cities that are auto-free simply because no roads have been built there. The Imperial grounds in Tokyo, Hyde Park in London, the Race Course in Lagos and Christ Church Meadow in Oxford are all large areas near the centres of congested cities where the "demand" for traffic routes is very high but has been rejected in favour of other things. The

10.1 Grid street pattern, Santa Ana, California.

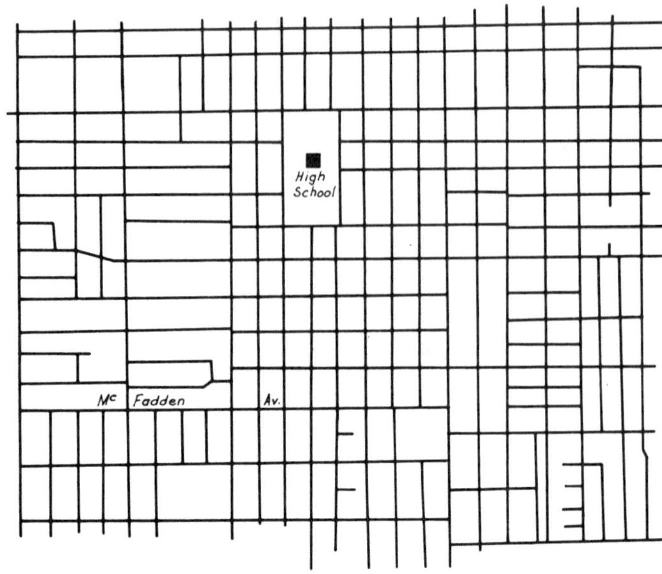

same cannot be said of Central Park in New York or the Bois de Boulogne in Paris.

The existence of large auto-free areas within a city forces vehicles to take long circuitous routes to move from one side to the other. This not only adds to journey times and operating costs but also creates more vehicle-miles, i.e. more traffic. Although this breaks a fundamental principle of traffic limitation—to minimize journey distances—there are so few major examples left of such areas that the traffic disadvantages are likely to be regarded as small in relation to the value of keeping the areas as they are.

Of more practical importance is the possibility of limiting traffic to the capacity of existing roads, by not increasing that capacity and not building new roads. We shall come to this question later.

Road Closure

Total road closure is easily implemented by the erection of bollards. A auto-free street is produced by the placing of bollards at each road access, as in some pedestrianised streets, but more often bollards are used to block access at one end to eliminate through-traffic. This simple technique can be used to transform the way in which an urban street network functions. As illustrated in figures 10.1–10.2, there are two diametrically opposed patterns of urban street networks (with intermediate variations). One is the grid pattern, in its extreme form designed to equalise the flow of traffic on parallel streets by giving motorists numerous alternative routes of equal length. This

10.2 Cul-de-sac street pattern, Sandhurst, England.

pattern is intended to maximize average traffic speed for a given lane-mileage. Whether it does so is a complex issue; but without doubt the grid pattern is intended to serve the interests of traffic movement. The other type of street network is the cul-de-sac pattern. The precise objective of this pattern is not clear; it could be to maximize the proportion of street mileage without through-traffic or to minimize the median traffic flow outside residential properties; but it is clear that this type of network is intended to serve an environmental objective.

The grid network will tend to fill with traffic as the size of the network expands and car ownership increases, but in the cul-de-sac network traffic growth is confined to the main roads. They may become congested and require widening, of course, but the purpose is to concentrate the traffic on to a few main roads rather than to diffuse it evenly over all roads. Bollards can achieve the same objective on a grid network or any other network, simply by blocking through-movements except on designated main roads.

A classic application of this technique was introduced in 1961 by Dr. Dorfler in Bremen as a means of reducing the volume of traffic in the city centre. As illustrated in figure 10.3, the principle is to divide the city centre into four "cells," the dividing lines being two roads reserved for buses and trams. Traffic is free to enter each cell but not to move directly from one cell to another; to do this it must go round by the ring road encircling the city centre. The technique has also been applied with five cells in Göteborg, and a two-cell version is planned in Liverpool. A good inner ring road is an essential feature.

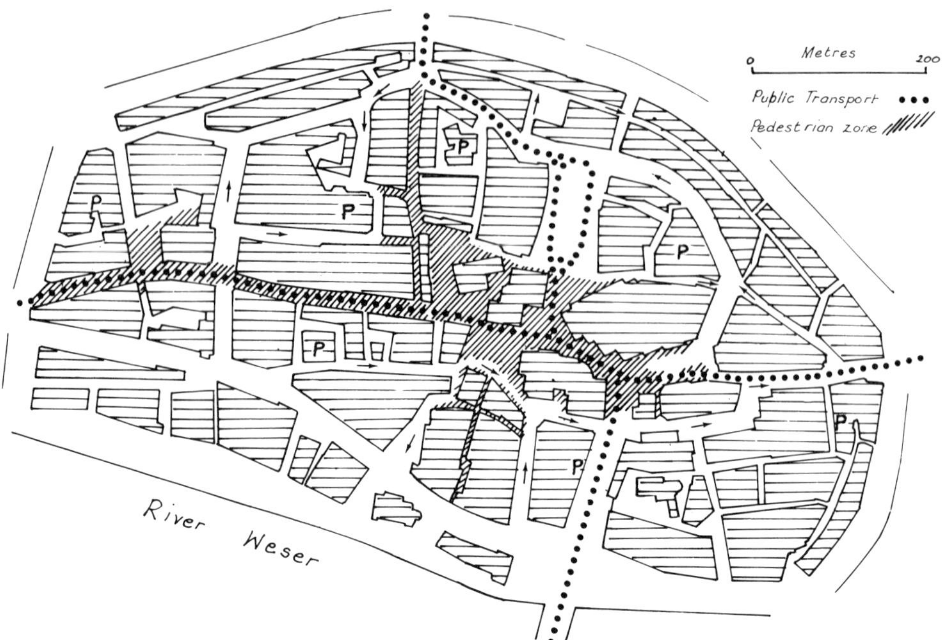

10.3 The Dorfler principle of traffic restriction, Bremen.

An unusual example of road closure has arisen at Windsor and Eton in England (figure 10.4). The traditional main road through these two communities crosses the river Thames by Windsor Bridge, which has a 17-foot carriageway. Historically and architecturally, it is a road of outstanding importance, which for years has been plagued by traffic. The opening of a major relief road in 1966, only 1100 yards away, failed to provide much lasting relief, and in 1969 over 13,000 vehicles per day were crossing the bridge. In April 1970, due to cracks in its cast iron ribs, Windsor Bridge was suddenly closed by Emergency Order to all traffic except bicycles, hand operated electric trucks and invalid carriages not exceeding 5 cwt. in weight. Traffic volumes fell by nearly 10,000 v.p.d. in Eton High Street and 4,500 v.p.d. in Windsor High Street, much of this traffic diverting to the relief road. The immediate benefits enjoyed by pedestrians and cyclists, and the improvement in the environment on both sides of the river, led to a demand to keep the bridge closed permanently, and in July 1971 the County Council decided not to rebuild it for motor traffic but to retain it for pedestrians and cyclists. Following a public inquiry into this proposal,[1] which was opposed by several organizations mainly on account of the difficulties caused to bus pas-

10.4 Principal streets, showing bridge closure, Windsor and Eton.

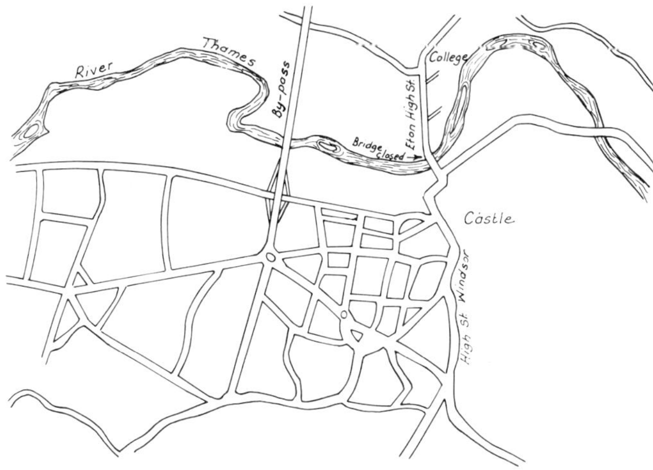

sengers, the inspector found that he could not approve of permanent closure until adequate arrangements were made to accommodate future growth of traffic by other routes.

Road blocks can serve to limit the use of back streets by through-traffic and simultaneously to increase the safety and capacity of the main roads by reducing the number of intersections. An example of this is the main Farnborough Road (in Hampshire, England) where a number of access roads have been permanently sealed off, initially by bollards but eventually by the planting of grass and trees across the carriageway. This is illustrated in figure 10.5.

Partial Road Closure

In many streets the total removal of traffic is required only at certain times, usually during the hours of peak pedestrian movement. At other times vehicles may be admitted, especially for the purpose of servicing premises with no rear access. It may be necessary or desirable to allow certain vehicles into a street which is prohibited to general traffic: e.g. ambulances, fire engines, telegraph and other maintenance vehicles, invalid vehicles, possibly residents' vehicles, and so on. The time-honoured device of a gate and commissionaire is widely employed, of course, to control the use of private roads and is found occasionally on public roads too, through parks and public estates. In a new residential development at Port Grimaud in France the access of cars is restricted by police-controlled gates; this high-class area was designed to be car-free partly in order to permit higher building densities on an attractive, but small, site. No example is known of this method being used to limit entry to a town centre or similar area. The commonly used methods are to leave some access roads open

10.5 Farnborough Road, showing road closures.

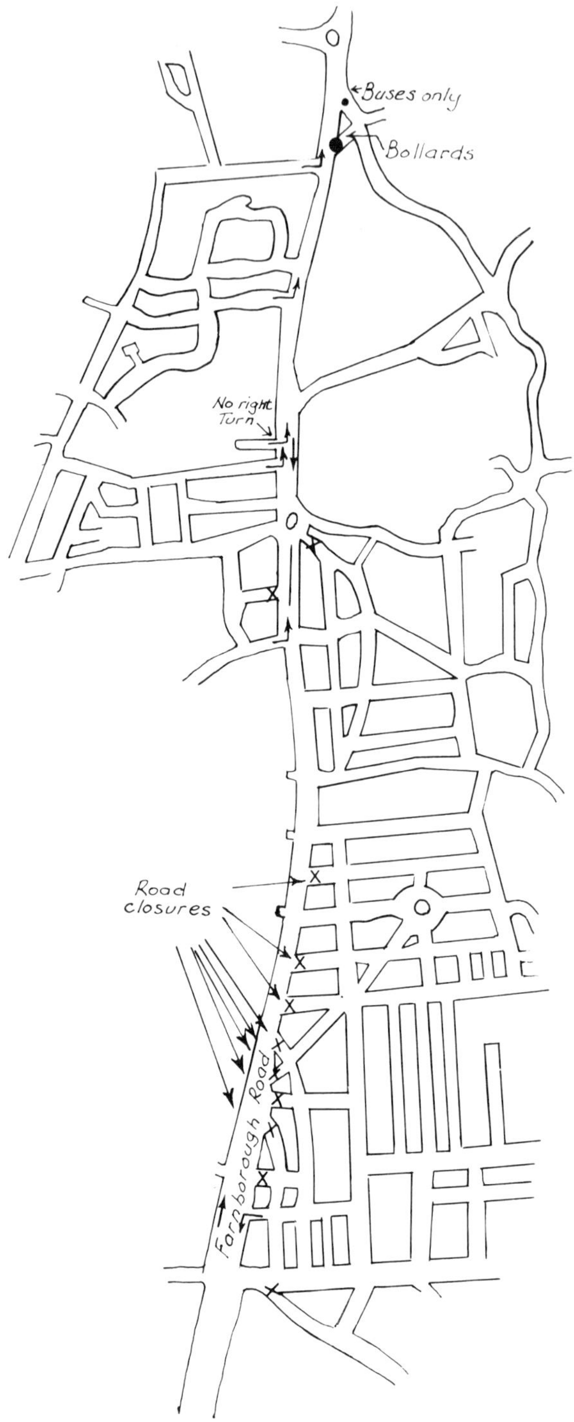

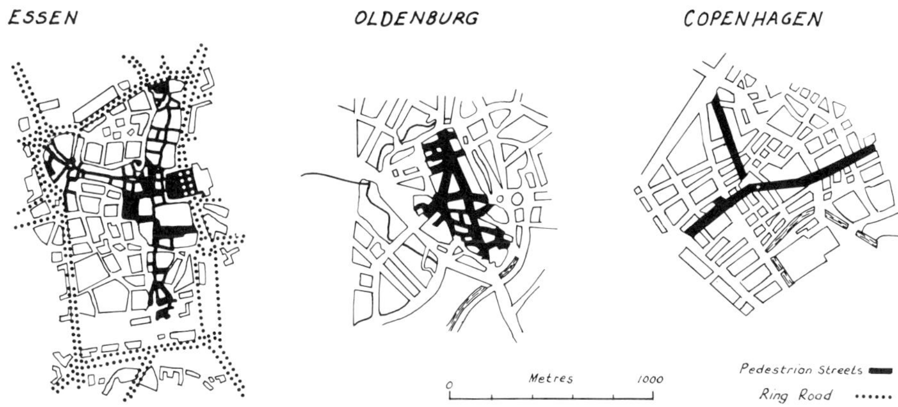

10.6 Pedestrian street patterns.

for emergency vehicles or for use by, say, commercial vehicles during specified hours, or to erect simple, removable barriers; these may be seen, for instance, in Yugoslav towns especially during the hours of the "corso" when it is customary for people to stroll in large numbers through the fashionable streets. Zadar is a good example.

LEGAL RESTRICTIONS

There are few urban streets where the total exclusion of traffic by fixed barriers is both desirable and practicable. Movable barriers too are not satisfactory if an appreciable flow of vehicles is allowed to enter. Regulations are more efficient, provided they are obeyed, but if policemen or traffic wardens have to be permanently engaged to enforce them they may be no more efficient than gates or controlled entry points.

Prohibition

Total prohibition (i.e. of all vehicles at all times) is rare and is invariably confined to small sections of street, because of the need for access by delivery, service and emergency vehicles. *General* prohibition, of all vehicles except essential vehicles requiring access, is now a common feature in the so-called "pedestrian" or "traffic-free" areas that have been created in more than a hundred European towns in the last ten years.[2] But the need for vehicular access appears to impose a fairly severe limit on the size of these areas, because the larger they are, the larger is the volume of exempt traffic having to pass through the periphery to reach the inner parts.

Figure 10.6 shows the size and shape of three important traffic-free areas. Essen is probably the largest such area in existence, but the most complete, in the sense of embracing the whole town centre, is in the smaller town of Oldenburg. The problem of access tends to force traffic-free areas into a narrow elongated shape, as was true in Copenhagen's Strget prior to 1975. Kühnemann and Witherspoon found that, "As a rule, traffic-free zones consist of a single pedestrian street sometimes extending into adjoining side streets."[3]

The difficulty in creating larger traffic-free areas appears to have impeded their introduction in the largest cities. In Paris, for instance, a dozen separate areas would be needed to give relatively the same effect as one area in Bonn or the Hague. Attempts to design a system of traffic-free areas in large cities, however, soon encounter an entirely different kind of problem: whereas smaller towns can create their areas by transferring most of the traffic to ring roads or nearby car parks, usually involving some construction, in large cities this is not possible. Some channelisation of traffic is possible, of course, by widening suitable main roads like Euston Road, London; and inner ring roads like the Route Peripherique in Paris can be used to take through-traffic out of the city centre. But most central area traffic in large cities is local traffic, and channelisation on a sufficient scale to permit the creation of a whole system of traffic-free areas could only be achieved by major new construction of roads and car parks within the city centre, or by "squeezing" the traffic, i.e. eliminating part of it.

Methods of Prohibition

The desire to reduce city centre traffic often leads to demands for the banning of all non-essential vehicles from the centre. There are minor examples like the island of Sark and the village of Polperro, but it has never been tried or even seriously studied in a major town. Traffic prohibition has only been employed on a small scale, in individual streets rather than whole areas.

The simplest and most easily enforceable prohibition is the straightforward "No entry" sign, but this of course does not achieve the selective discrimination required. It is therefore normally used only as a one-way restriction, and the likelihood of encountering oncoming traffic is possibly a reason why it is effective. *Discrimination by exemption* is often achieved by "No entry except buses" signs. Numerous variations of this technique are both possible and desirable. In Bremen, for instance, exemption is given to buses and trams. In Paris, taxis and motorcycles are permitted to share reserved lanes with buses. *Direct discrimination* against specific classes of vehicle, especially commercial vehicles, is also common. "No lorries over 3 tons unladen weight" applies in the City of London and similar restrictions are found in some streets in San Francisco. The Blue Zone system operating since 1955 in central Paris prohibits larger delivery vehicles in the afternoon but allows them to do their business in the morning. In Tokyo bicycles are prohibited in some streets, presumably in an attempt to concentrate them on other streets (whereas in New York the same objective is pursued by labelling certain streets as "bike routes").

Other specific prohibitions are effectively enforced for purposes other than traffic limitation. In Istanbul, dolmus buses are not allowed in the city centre. Although the main purpose in both cities is probably to protect the municipal bus companies from competition there is also a belief that the exclusion of small buses helps to relieve central area congestion.

Area Permits

Suggestions for banning large categories of traffic, e.g. all private motor vehicles, from large urban areas invariably founder on the difficulties and hardships that may arise for a few people who are particularly dependent on such vehicles. Doctors, salesmen, invalids, cabinet ministers,—these are some obvious cases where travel needs would not be adequately met by taxis and public transport alone. In practice, such is the complexity of life and movement in a city, a great many other cases would arise. Furthermore there is no need to impose such sweeping restrictions. Area permit schemes could therefore be used to restrict entry into a defined area to vehicles with a permit which must be applied for and granted by the appropriate authority. Permits could be granted annually to people with a permanent need for one, and for shorter periods to those with only a temporary need.

This form of control would not be difficult to enforce insofar as it was necessary to prevent vehicles entering the area without displaying a permit. It would have other considerable advantages: it would enable the authorities to control fairly exactly the volume of traffic in the area and to allocate the limited road space on the basis of need (or their interpretation of it).

There appear to be two major reasons why area permits have not been introduced on a permanent basis in any town. First, the power to decide who should be allowed a permit is one which no responsible civil servant would welcome. Secondly, abuse by permit holders using their permit for improper purposes could cause major enforcement problems and bring the whole scheme into disrepute.

Technically, however, area permits present no difficulties. Similar methods are used for restricting traffic on special occasions, such as coronations and military celebrations, and for restricting the entry and parking of vehicles at big sporting and festive events. But to use these methods successfully for the permanent restriction of traffic in cities would require an exceptional degree of impartial, sensitive and competent administration.

Vehicle Regulation

Insofar as the purpose of limiting traffic is to limit the noise, pollution and danger caused by traffic, this can be partly achieved by the prohibition—in urban areas at least—of vehicles that fail to meet rigorous standards. Although most countries now have such regulations, their effectiveness in dealing with *noise* and *air pollution* is not impressive. The standards demanded by governments are not high enough to do more than curb the worst excesses. The anti-air pollution regulations which were contained in the U.S. 1970 Clean Air Act were a major step forward. But in most European towns it is noise, not air pollution, that causes most nuisance, yet no country has dealt firmly even with the noise of motorcycles. In Sweden the Scania "quiet bus" was developed at a modest extra cost, but nowhere have bus companies or bus manufacturers been obliged by law to adopt the higher standards that are now technologically and economically feasible.

Not only are the required standards low in relation to what would be needed to achieve a big reduction in noise and air pollution, but the enforcement of these standards is ineffectual. Almost everywhere one can see numerous vehicles that appear to be exceeding the permitted levels of noise or exhaust emission. In the United Kingdom the number of prosecutions for such offences is so small that no one claims that the law is being enforced. A similar situation undoubtedly exists in other countries.

There are considerable technical difficulties, of course, in enforcing noise and air pollution standards, and this is an important reason why governments have not demanded higher standards. Far more progress has been made with *safety* regulations, in spite of many difficulties, and this is reflected in sharply declining accident rates in urban areas. In many cities, despite rapid growth in the volume of traffic, the number of accidents is being held down at a fairly constant level, or even reduced. The same cannot be said of noise and air pollution.

The growing criticism of motor traffic in towns, which has swept North America and Western Europe during the last fifteen years, is largely a reaction against the physical offensiveness of traffic. It would be sensible therefore to do whatever possible to restrict the physical offensiveness before restricting the traffic. Regulations against noisy, polluting and dangerous vehicles could go a long way towards this.

Another aspect of the physical offensiveness of traffic is *driver behaviour*, especially in relation to pedestrians. This is not just a matter of education. It is also a matter of regulation and road design. Broadly speaking, drivers tend to drive in the manner permitted by the physical characteristics of the road and the regulations appertaining to it. The hostility of pedestrians is aroused by inability to cross roads and by feelings of insecurity, which are consequences of road design and driver behaviour. Traffic engineers have gradually modified the street systems of many cities in the quest for fast, continuous traffic flow, thereby making pedestrian movement increasingly difficult and frightening. The attempt to assist motor traffic has thus promoted the conflict which has resulted in the rejection of traffic from many places, and the restitution of the pedestrian.

In situations where the pedestrian is pinned to the footpath for fear of his life, or waits interminably in the rain watching for an electrical gadget to give him permission to cross, he may begin to regard the motorist as a tyrant; but this role is often forced upon the motorist by the traffic engineering characteristics of the road. Traffic engineers have persistently sought to segregate pedestrians in order to speed up traffic. But segregation is often impossible or highly unattractive. More attention is then required to creating traffic movement which is *compatible* with pedestrian movement. The art of mixing traffic and pedestrians is particularly well developed in England, where the liberal use of traffic islands, zebra crossings, traffic signals and careful kerb design helps to create steadier traffic movement, less weaving and overtaking, slow cornering and shorter crossing movements for pedestrians.

Hence the regulation of vehicles and vehicle behaviour in urban areas can greatly lessen the conflicts which have led to demands for traffic limitation.

Route Restriction

It has been a common practice for a long time to restrict buses to certain routes and in particular not to allow them to travel down quiet residential streets. This is easily enforced on bus companies. Similar restrictions are becoming increasingly demanded for heavy goods vehicles. The maximum dimensions of goods vehicles have been permitted to grow, in order to reduce road haulage costs. The use of very large lorries has been greatly stimulated by the construction of national motorway networks but this has been to the detriment of urban areas through which the vehicles often have to pass at the beginning and end of their journey.

There are urban areas, as we have noted already, where heavy lorries are prohibited; the haulier is then faced with the choice of breaking his load into smaller truck loads at the edge of the urban area or using smaller vehicles throughout. The extra haulage costs should be weighed against the gain to the urban community.

An alternative arrangement is to permit the heavy vehicles to use only specific routes in the urban area, *lorry routes*, where their concentrated impact is calculated to do less harm than under any other arrangement. Where urban motorways exist, they are natural routes for heavy lorries; where they do not exist, other main roads, preferably through industrial areas, have to be selected.

Many lorries, of course, must eventually leave the lorry route in order to reach their destination, and this makes enforcement difficult. Lorry parks, with suitable amenities for drivers, have to be located on or near the routes. To encourage use of the routes, measures should be taken to restrict access by other vehicles. The biggest difficulty in establishing lorry routes, however, is the inevitable opposition from residents and shopkeepers along the routes.

Licensing of Drivers

With modern teenage affluence, the age at which driving licences are allowed has a significant influence on the number of cars owned and the volume of traffic, and conversely on the fortunes of public transport. The age at which boys are permitted to ride motorcycles must have a perceptible influence on the levels of road noise and accidents. It is not suggested that these age limits should be determined by the desire to limit traffic but it could be taken into account if there was some doubt as to the most appropriate age limit.

More relevant perhaps is the effect of a high degree of motorization on the demand for driving licences. In motorized American cities, where public transport has declined and the land use pattern has developed in a way that could not be well served by public transport anyway, the uppermost desire of every youngster on his sixteenth birthday is to get his own "wheels," in order to gain some independence from his parents. In a traditional city, teenagers already have this independence because the places they wish to go to are readily reached by public transport or are within easy walking or cycling distance.

The licensing of unfit drivers is also relevant. In Los Angeles, for example, alcoholics and drug addicts are known to form a considerable proportion of the adult population. It is said, however, that magistrates are loathe to disqualify people from driving, except in extreme cases, because the loss of one's licence in such a car-dependent city can be tantamount to the loss of one's job and the ruin of one's private life. For the same reason there is intense pressure on old people to go on driving as long as they possibly can. In the words of a leading authority, L. P. Spitz:

> The contemporary license administrator is faced with a dilemma, often involving the delicate balancing of competing interests. He must implement legislative enactments and policies designed to protect the general public from incompetent drivers. But he also must give recognition to the fact that in denying, suspending, or revoking a licence he is depriving an individual of the use of a motor vehicle—without doubt the most important mode of individual movement in contemporary American society.[4]

Spitz points out that there are estimated to be some 40 million Americans with hidden medical defects relevant to motor driving, including 2 million diabetics and 2 million epileptics.

A city with a traffic limitation policy would want to avoid granting licences liberally to persons who were really too young, too old, or otherwise unfit to drive. A firm control over driving capability is not likely to be achieved, however, unless it is reasonably possible for people to share in the normal life of the city without the aid of an automobile.

11 TRAFFIC RESTRAINT

Every driver in every car has made a choice. He has chosen to travel from A to B, by car, at a particular time. Some drivers may feel that they have no realistic alternative; they have to make that journey at that time and there is really no other reasonable way of doing it. Other drivers may have feasible alternatives; they could have gone somewhere else, possibly somewhere nearer, they could have gone at another time, they could have gone by train or bus, or they might have chosen a nearer destination and walked. Even in the very short run there are choices; an occurrence of fog, a warning of traffic delays due to roadworks or an increase in parking charges, all have an immediate effect on the volume of traffic, proving that the choices of some drivers can be changed even at short notice. In the longer run the number of feasible choices must be greater because, given sufficient time to react to new travel conditions, people can make locational and modal changes that may not be feasible in the short run; they can choose new homes and jobs, taking account of how they intend to travel between the two; they can decide to buy or not to buy a second car. All their locational decisions affecting schooling, shopping, sport and entertainment, social and recreational activity, must take account of the mode, time and cost of the journeys upon which these various activities depend. If journeys are everywhere easier and faster and not much more expensive by car than by other modes, and people can afford to own and maintain cars, decisions will be largely based on the assumption that journeys will be by car, and the acceptable limits on journey distances will be set by the average car speeds realised on the road system. But if car travel is *relatively* difficult, slow or expensive, people will not allow themselves to become so dependent on cars.

In principle the choices open to car owners are also open to industry and commerce in relation to goods traffic. There is sometimes a feasible choice between road, rail or occasionally water, there are locational choices (especially in the long run)

11.1 Urban network speed-flow curve.

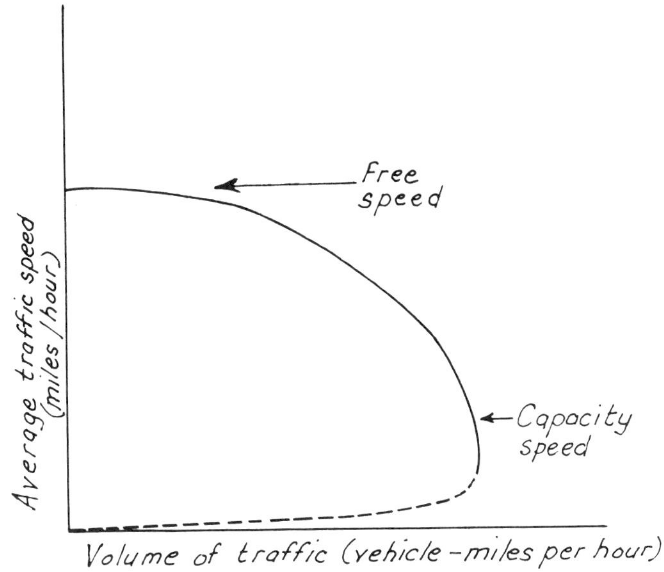

and, much more than with the movement of people, there are choices between different times of day. In practice, however, urban goods traffic has far fewer realistic possibilities of using other modes or of shortening journey distances; the main ways in which the volume of goods traffic in a town can be reduced are by higher load factors and by relocation to the urban periphery of commercial premises which generate a lot of traffic *external* rather than *internal* to the town.

Traffic restraint is intended to influence, without directing, people's choices so that collectively they choose to produce volumes and patterns of traffic on the existing infrastructure, by all modes, that are less costly and damaging to the community. In practice the main objective is to persuade people to make less use of private motor vehicles, i.e. to produce fewer motor vehicle-miles, in towns, especially in the inner and other congested areas. The methods of restraint may be *direct* or *indirect*, and they may be *physical* or *fiscal*.

DIRECT RESTRAINTS—PHYSICAL

CONGESTION

Traffic congestion is defined as the average delay imposed by vehicles on each other. It is therefore measured by the difference between the actual traffic speeds and the speeds achievable when the volume of traffic is low enough to occasion no delay.

In view of the importance of speed and congestion it is desirable to define some relevant concepts. Figure 11.1 shows the typical relationship between the volume of traffic on an urban network and the average traffic speed. When there is no delay due to traffic, vehicles move at their *free speed* (or *design speed*) as determined by the road characteristics; in Central London, for instance, the free speed is only over 20 m.p.h. mainly on account of signalised intersections. With increases in the volume of traffic, measured in vehicle-miles per hour, the speed comes down. Each successive increment in traffic is gained at the expense of the "existing" traffic; as the process continues and the slope of the curve becomes steeper, longer delays are imposed on larger volumes of traffic. Depending on the traffic demand, there may come a point before the maximum capacity of the network is reached at which further increments in traffic are not worth the losses to the rest of the traffic. This is the economic optimum sought by road pricing theorists. It defines the economic capacity of the network and, with it, the *economic speed*. The economic speed will vary between different times of day, depending on traffic demand. With further increases in traffic the network eventually reaches maximum capacity, i.e. the state where through-put is maximised. The highest speed at which maximum capacity can be sustained is the *capacity speed*, which might be as low as 5 or 6 m.p.h.

In practice, the *actual speed* is usually none of these. The difference between free and actual speed gives the amount of congestion; the difference between free and economic speed gives the amount of *economic congestion*, and the difference between economic and actual speed (assuming the latter to be lower than the former) gives the amount of *excessive congestion*.

At present, traffic congestion is probably the most effective restraint on traffic in many cities. The level of congestion is usually a major factor in maintaining a stable, if uncomfortable, equilibrium in the transport pattern of the city. But it is seldom treated as such. A degree of congestion—that which is described as economic congestion—is desirable in the sense that a free flow system would often be highly wasteful of resources. A larger degree of congestion—excessive congestion—is by definition not economically desirable but might nevertheless be preferred to either (1) expansion of the road system, or (2) other available methods of keeping down the volume of traffic.

This is becoming increasingly recognised. In 1967 the British Ministry of Transport stated, "Traffic congestion acts to some extent as a restraint on road use. Consequently, the growth of vehicle ownership does not seem likely to cause traffic to 'grind to a halt.'"[1] And in 1972 the Greater London Council stated, "The present general restraint of traffic to the Central Area is by congestion and parking control."[2] The same could be said of many other cities, and there is no doubt that congestion will continue to exercise an important restraining influence on the volume of traffic.

When this fact is recognised and congestion is no longer regarded as temporary, accidental and unexpected, a new concept emerges: planned congestion.

Planned Congestion

If congestion is expected to be a permanent factor helping to determine the volume and pattern of traffic in the city, there are advantages in organizing the road system so that the congestion occurs in the most suitable places where it can be properly controlled. Congestion is a form of queuing. Planned congestion is a form of crowd control, designed to minimize the danger and nuisance caused by haphazard congestion. This approach has already been adopted on congested motorways, by means of *traffic metering*, whereby signal controls at the access points aim to concentrate congestion at the access ramps off the motorways rather than on them.

With an ordinary road network the key to planned congestion is to identify where the controlling bottlenecks should be, and to manage the roads so that the queues are concentrated at those places rather than in other streets where they may have more adverse effects on road safety, pedestrian movement and the environment. In this way congestion is systematically organised as a necessary control mechanism, and the bottlenecks are regarded not as problem areas, but as control points. This approach is specially suitable for the protection of city centres from excessive inflows of traffic which tend to become inextricably jammed in the centre itself.

The Bitterne traffic scheme in Southampton is an example of planned congestion. The objective was to organise the roads so that congestion, if it could not be avoided, was concentrated on roads that were *not* bus routes. The Bitterne Road is a major bus route into Southampton and was regularly congested in 1970 when the scheme was proposed. The scheme, illustrated in figure 11.2, officially described as follows:

> During morning peak hours it is proposed that vehicles will be able to join the route at a rate which permits the maximum number of vehicles to pass through the overloaded junctions without causing excessive queues at these junctions. On occasions when the route is seriously overloaded and excessive queues on the main route threaten to develop, it is proposed

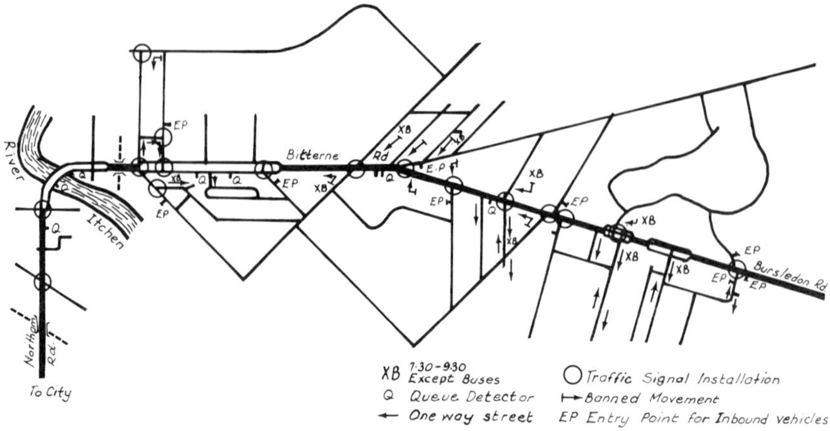

11.2 Bitterne traffic scheme, showing traffic metering on arterial road.

> that the rate of entry of vehicles to the route at the City boundary and from most side roads, should match the capacity of the route. Major queues of vehicles which at present form on the route would not occur; instead vehicles wanting to join the route might form queues in the side roads rather than on the route itself. It is proposed that buses should be able to join the route from separate side roads to gain priority over other vehicles. . . . At present 46 percent of passengers crossing Northan Bridge during the morning peak hours use public transport.[3]

The control of entry from side roads was to be effected by linked traffic signals with three alternative linking plans, calculated for the morning peak period, the off peak period and the evening peak period respectively.

Low Design Speeds

Although driving in congested traffic is unpleasant in several ways, the restraining effect of congestion on traffic volumes is probably due mainly to the low speeds involved. Low speeds can also be engineered without congestion. Speed limitation has hitherto been regarded almost entirely as a matter of safety. But speed is a prime factor determining both the volume and the routeing of traffic. Speed controls are therefore a potential way of influencing these things. At the same time traffic speed is highly relevant to the safety and ease of movement of pedestrians and cyclists, and also to the quality of the environment in residential and shopping streets. Speed controls may therefore serve several purposes simultaneously.

One of the main consequences of low travel speeds, as we shall see later, is that people travel shorter distances, thus reducing the volume of traffic. In this respect the difference between 10 and 20 m.p.h. is the same as between 20 and 40 m.p.h. Arterial roads designed to carry traffic at an average of 20 m.p.h. will generate less traffic than motorways designed to carry traffic at 40 m.p.h. A speed of 20 m.p.h. caters well for short-distance traffic without encouraging much long-distance traffic. The transport policy proposed by the amenity movement in London adopts this principle of designing roads in the inner urban area for short-distance traffic, with the intention of encouraging long-distance travellers to use the metropolitan railway network.[4] Similarly, at the 1972 inquiry into the West Cross Route in Inner London, the amenity organisations proposed that the road should be built to normal arterial standards with a design speed of 20 m.p.h. (allowing for grade intersections) rather than to motorway standards, partly in order to lessen traffic generation.

Especially in large cities, the design speeds of main roads are of strategic importance in influencing locational decisions and hence the volume of traffic. Even if the roads are not able to operate at their design speeds during peak hours the influence during off-peak hours may still be important. Lower design speeds will obviously add to the travel times of some activities, and this has to be taken into account. But, as will be discussed later, higher speeds have been found to lead to longer distances being travelled, almost in proportion; there is no evidence that people spend a smaller part of their lives travelling, or indulge in more activities, in cities where travel speeds are high than in cities where they are low. Nor is it simply a matter of higher speeds leading to lower building densities; higher speeds can be expected to lead to longer journey distances even if there is no change at all in density. The real, long-term effects of higher travel speeds on people's lives in large cities have never been investigated in depth. The presumption that shorter travel times between fixed points are *wholly* beneficial, after allowing for differences in travel cost and safety, has never been seriously examined. Hence the real disadvantages of low (versus high) design speeds on main roads are not known. But their effectiveness as a general measure of traffic restraint has been amply demonstrated.

At a more local level, speed controls can be extremely effective in limiting the use of particular roads to local traffic. Low *speed limits*—of 10 m.p.h. or thereabouts—are not effective, but the use of *speed ramps* (humps or bumps) has been successful and cheap, and requires no enforcement. British examples, where traffic speeds have been reduced from 30–40 m.p.h. to about 15 m.p.h. by this method, can be seen at Broadmoor Hospital, the Royal Military Academy, Sandhurst, and Guildford University. Elsewhere speed ramps have been erected at the entry to residential streets and in pedestrian areas where traffic has to cross pedestrian streets.

The speed of traffic on new local roads can be greatly reduced by the design of the roads themselves, incorporating minimum lane widths, rough surfaces and numerous sharp bends. An example in the Brow neighbourhood of Runcorn New Town was described as follows: "Somehow, by nimble and ingenious bending of rules and standards, Runcorn's planners avoided heavy by-law roads and kerbs and pavements. In the end they produced a twisty, informal layout of narrow access roads which gives the nieghbourhood its sense of visual and social cohesion and keeps car speeds down to 4 or 5 m.p.h."[5]

Speed control on local roads is not likely to affect significantly the total volume of traffic but it can be effective in restraining the use of local roads by through-traffic and thus helping to concentrate traffic on main roads. Even if it has no effect on the distribution of traffic, the reduction in speed helps to achieve some of the objectives of traffic limitation.

Allocation of Road Space

The width of a road extends from the property frontaging one side to that on the other. The space thus available can be used for (1) transport purposes, including parking, by vehicles of many kinds, and by pedestrians; (2) non-transport purposes, such as trading, public gatherings, townscaping and recreation (which includes standing, sitting, strolling and other pedestrian activities, including crossing the road, which are not really acts of transportation). The way in which city streets are allocated to these often incompatible activities varies greatly. There are no rules, conventions or agreed priorities. But for half a century there has been steady pressure everywhere to obtain a larger allocation for motor vehicles. Street traders have been forced out, trees felled, verges concreted over, footpaths narrowed, traffic islands removed, pedestrians regulated, bus stops moved, tramways closed down and every odd corner claimed for car parking. When the objective is to increase traffic capacity, these measures constitute a logical process. But when the objective is to limit traffic, either locally or generally, it is logical to consider setting the process into reverse. There are many ways in which road space can be reallocated in favour of non-transport activities or forms of transport other than private motor vehicles. Every step in this direction represents a gain for the activities that receive more space and a restraint upon those that lose it.

The most common measures of this kind involve the reallocation of road space in favour of buses, pedestrians and cyclists. Often the twin aims are to help those modes to operate more effectively and to restrict the competitive power of the private car, with the objective of shifting passengers from cars to other modes.

Bus Priorities

The first important bus priorities were *bus lanes*, which are now commonplace and rapidly spreading. In certain circumstances, depending on the number of buses, the number of other vehicles and the width of the road, a bus lane can improve the speed of both the buses and the other vehicles; and at first traffic engineers tended to adopt this as the necessary criterion for a bus lane. But the argument was put (by Smeed, Webster and others) that even though the speed of other traffic was adversely affected at first, if the improved bus service attracted a proportion of motorists to use buses instead of cars the consequent reduction in traffic could enable speeds to rise again to their previous level or higher. The objective of influencing modal choice in order to maintain or improve travel conditions for everybody on the affected bus routes thus became accepted; but it is not easy to assess the effect of a bus lane on modal choice. Today a stronger view tends to be taken of the need to persuade commuters, in particular, to use buses instead of cars, for a variety of reasons in addition to congestion; and this leads to a more radical criterion for bus lanes, namely that the *losses* to other traffic should not exceed the gains to bus users. Bus lanes thus become accepted as a positive restraint on other traffic. In Oxford, at least, the recent proposals for bus lanes are intended as much to restrain cars as to speed up buses. The Camden bus lane study proposed a *network* of bus lanes within the borough of Camden, which covers a large area of inner and central London.[6] The bus lanes would affect about 8 miles of main road, reducing their effective width for other traffic by an average of 27 percent. This was estimated, after allowing for restraint and modal transfer, to increase bus speeds by 40–50 percent but to reduce the speed of other traffic by about 30 percent.

Bus lanes are only one kind of bus priority. Other methods such as bus priority signals,[7] bus exemption from restricted turnings, busways and restricted access to bus routes, all entail some restraint on other traffic. The most comprehensive concept of bus priorities is probably that proposed by the writer in Bogota and London, namely the *continuous bus priority network*. The following description brings out the element of traffic restraint involved in the proposal.

What is ultimately required is a continuous network of roads on which the movement of buses takes first priority. This network need not include all bus routes, but should include most of the important ones. The principle of the bus priority network is that the volume of other traffic using the network should be limited to a level which does not impede the buses from giving an efficient, reliable service. Some parts of the network may need no special management at all; others may have to be turned into busways, i.e. reserved exclusively for buses; but most parts of the network would fall between these two extremes and would require a degree of management varying according to local circumstances. Bus lanes, bus actuated signals and exemption from certain general traffic restrictions would all play a part, as would parking restrictions and control of access; but an important additional feature would be prohibition of through-movements at certain intersections except by buses and, capacity permitting, by taxis, cycles and lorries. Such controls could be applied during the hours when they were needed, and relaxed at other times.[8]

Cycle Lanes

Evidence from Dutch towns and from Stevenage and Cambridge in England, where cycle paths and lanes have been generously provided, suggests that substantial numbers of people would use bicycles, especially for work journeys, if cycling conditions were good. The virtues of cycling are generally recognised and are known to find favour, in Manhattan and London, of course, most potential car drivers go to work by bus and train. It is not at all clear, therefore, to what extent cycle lanes would attract people away from their cars or away from public transport.

Cycle lanes are generally approved of but rarely provided. The reason is invariably that road space cannot be spared. There are also difficulties caused by introducing a flow of cyclists between the footpath and the main carriageway when vehicles, including buses, have to load and unload goods and passengers. As a method of traffic restraint, however, cycle lanes have the advantage of taking road space away from motor vehicles in order to promote a generally approved activity which itself should reduce the use of motor vehicles.

Where cyclists have to mix with motor traffic it is important that the nearside lane should be 2 or 3 feet wider than normal.[9] The unusually generous width of nearside lanes in

Manhattan possibly helps to explain the relative popularity of cycling there. Sometimes traffic engineers are faced with the choice of four generous lanes, which give room for cyclists and pedestrian islands, or six sub-standard lanes which make no allowance for either.

Pedestrian Facilities

When a road is wholly reserved for pedestrians, this is a clear case of traffic restriction. When pedestrians share roads with motor traffic, the pedestrian facilities are a restraint upon the traffic. The wider the footpaths, the greater the restraint on traffic. The critical points are at intersections where traffic capacity is most scarce and where most pedestrians want to cross. The design of the corners is then important: a high kerb radius (i.e. a well rounded corner) speeds up turning traffic but makes pedestrian crossing movements longer and more dangerous; a tight kerb radius has the opposite effect and helps to restrain the dominating tendency of motor traffic in the urban environment. Channelising islands at junctions are also useful in keeping traffic movement orderly and predictable and in allocating to pedestrian use road space which is not necessary for correct traffic movement.

Between junctions, pedestrian crossings and refuges are other ways of allocating road space to pedestrians, which inevitably have a restraining effect on traffic.

The overall effect of allocating space to pedestrians by all these means may be to discourage the use of cars and encourage patterns of movement that can be served by walking.

Traffic Management

A number of traffic management techniques can be employed to discourage, without preventing, traffic from using certain streets. One-way systems can make it difficult and tedious, without making it impossible, to drive through residential or other environmental areas. There are numerous examples of this technique. Figure 11.3 shows the planned system of primary circulation in the centre of Bologna. A carefully devised combination of one-way streets and contra-flow bus lanes make it impossible for vehicles other than buses to drive through the city centre. Traffic with a destination inside the city centre is encouraged to go round the ring road and enter from the point nearest to its destination.

Another traffic management technique is to design side street junctions in ways which discourage through-traffic from using the side streets as short cuts or in order to avoid congestion on the main roads. Turning movements across the main

11.3 Bologna, primary circulation system, showing traffic management measures to exclude through-traffic except buses.

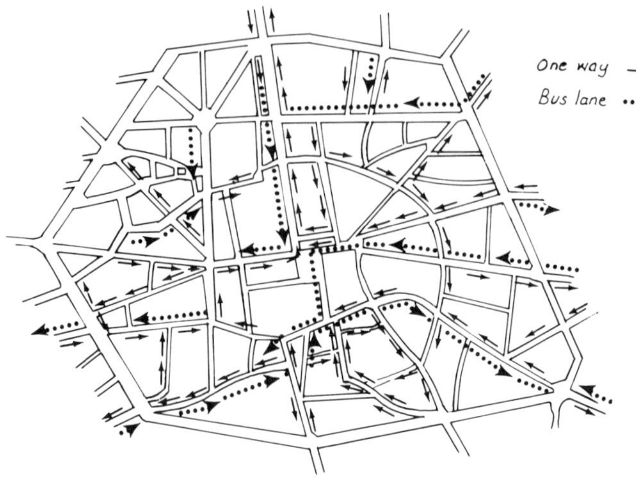

road may be prevented by regulation or by a centre kerb, or the footpaths may be reshaped so as to create a *funnel* or *neck* which acts as a psychological barrier to traffic that has no business in the street. This method is successfully used in the Pimlico traffic restraint scheme, illustrated in figure 11.4.

An important, though more conventional, use of traffic management for restraint purposes was proposed for Midtown Manhattan.[10] The proposal included a variety of methods:

1. pedestrian areas with minibus service in 48th and 49 Streets, Madison, Lexington and Broadway;
2. widening of footpaths in Fifth Avenue;
3. Central Park Drives except for the transverse drives to be closed to traffic;
4. two minute cycle progressive light timing on First, Second, Tenth and Eleventh Avenues to *attract* traffic; and lights on Third, Fifth, Eighth and Ninth Avenues to be timed against the traffic to *discourage* through-movement;
5. reduction of parking space;
6. changes in one-way street system;
7. construction of by-pass on land fill in East River.

The strategy was to create more road capacity on some routes and use various restraints to force existing traffic on to these routes without generating new traffic on the old routes.

Another aspect of traffic management is the adverse effect of one-way gyratory systems on the volume of traffic. Whatever the effect may be on speed and capacity there is no escaping the fact that gyratory systems increase the length of journeys and hence increase the volume of traffic. This is well il-

11.4 Pimlico traffic restraint scheme.

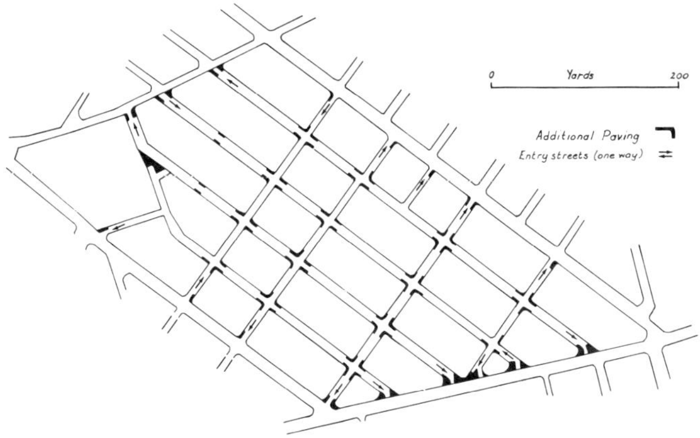

lustrated in figure 11.5, extracted from the report of a technical committee on traffic engineering,[11] who point out that "the volume of traffic on each arm of a triangular road pattern made one-way increases by 50 percent." To compensate for the added distance the traffic speed on the affected roads must increase by 50 percent. Consequently there is a 50 percent increase in both the volume and speed of traffic before the scheme begins to produce any traffic benefit. With four-sided systems the proportionate traffic increase can be still greater. In central London it has been estimated that traffic management has raised the total volume of traffic by about 5 percent on account of lengthening journey distances.[12] These aspects of traffic management may conflict with the objectives of a traffic limitation policy.

DIRECT RESTRAINTS—FISCAL

The physical restraints described above are mostly open to the objection that they reduce traffic by reducing traffic capacity, and they are therefore not likely to reduce congestion. They may reduce noise, air pollution and accidents, they may improve conditions for pedestrians, cyclists, bus passengers, shoppers and residents, they may remove congestion from some areas to others, but they will not make the movement of private motor vehicles easier or faster, because the nature of physical restraint is to do just the opposite. In themselves, they may make congestion worse, although they should always be regarded in conjunction with other policies.

If the objective of traffic limitation is solely to speed up private motor traffic, physical restraints are not therefore the way to do it. *Fiscal* restraints aim to reduce the volume of traffic by financial inducements without physically modifying the road system in any way. They do not produce the beneficial

11.5 One-way gyratory scheme, showing effect on traffic volumes.

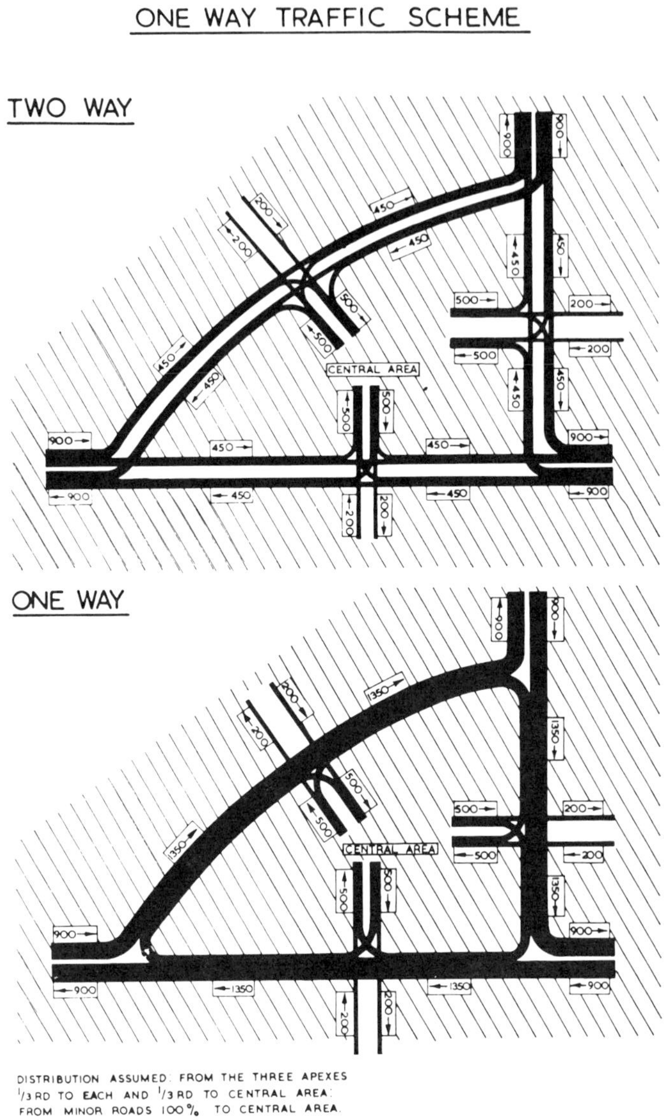

side effects of the physical modifications but insofar as they reduce traffic volumes they do speed up traffic movement.

Direct fiscal restraints are charges or taxes on the *use of vehicles*, with the object of reducing the volume of use, or the sort of vehicles used, in specific places and at specific times. The broader political and economic issues raised by these methods will not be discussed here; our concern is with the available methods and their implementation.

Fuel Tax

Taxes on petrol and diesel oil for motor vehicles have been introduced in most countries either in order to raise general revenue or to finance roadbuilding or, in oil importing countries, to limit the expenditure of foreign exchange for this purpose. The tax automatically enters into the 'price' of using vehicles and, if it is high, may be expected to influence (1) the annual mileage of vehicles, (2) the number of vehicles owned, and (3) the type of vehicles owned.

The effect of fuel tax on (3) is much easier to detect than on (1) and (2). A high tax gives a strong incentive for commercial operators to use diesel instead of petrol vehicles and for motorists to use small, low powered cars; national differences in the *composition* of the vehicle populations are readily analyzed and related to fuel tax. It is difficult to attribute national differences in vehicle *use* and *ownership* to fuel tax because there are other more important geographical and economic factors present and, in any case, annual mileages are usually not known. In high tax countries the tax may amount to over 20 percent of the annual cost of owning and running a cheap car; hence, although most car owners could and would, if necessary, economise by running a cheaper car rather than by running no car, the tax is bound to affect ownership among lower income people who are only just able to run any sort of car at all; and there are a lot of people in this category.

Fuel tax may also affect annual mileage, although in Britain, where fairly good estimates are available, increases in fuel tax have had no detectable effect.[13] But the increases themselves have never been large. If there is any effect, it is likely to be on long distance journeys and of little significance in towns. Car trips in towns are usually short and slow whereas between towns they tend to be much longer and faster; hence the fuel tax on urban trips is smaller absolutely, and even smaller relative to total time and money costs, than on inter-urban trips. To be an effective restraint on traffic in towns the fuel tax would need to be much higher, and although this might be justified in towns it would not be so outside towns. The main argument against fuel tax as a means of restraint is that it cannot be increased on town roads without being similarly increased on rural roads, and in towns it cannot be increased on congested roads at congested times without being increased on all roads at all times. As put by the Smeed Report,

It fails completely to discriminate between those situations in which congestion costs of road use are high and those in which they are low, and we have already mentioned how very large this difference can be. A charge of 1d per mile is higher than the cost of using uncongested roads and far lower than the cost of using congested roads and thus discourages the use of lightly-trafficked roads and encourages the use of crowded roads, though it is not known by how much.[14]

The Smeed Report investigated the possibility of a *differential fuel tax*, charging a higher tax in towns than elsewhere, but rejected it as impracticable on an effective scale. As put by Lyle Fitch as long ago as 1957, "The situation calls for special charges for the use of high cost urban roadways, but there is no way of putting special taxes on gasoline consumed in metropolitan areas."[15]

In many countries a tax is also charged on *car ownership*, i.e. an annual registration fee. In the view of the Smeed Report, "the universal use of a fixed annual charge appears 'a priori' to have even less effect than the fuel tax in restraining congestion" (p. 10). If the fixed charge were abolished on cars (and reduced on commercial vehicles) and fuel tax raised correspondingly, the average annual mileage recorded by all motorists would probably be reduced but there would be an incentive for more people to own cars, especially second cars. It is not clear whether the change would reduce the volume of traffic in towns.

The main impact of fuel tax as a traffic restraint in towns, therefore, is not likely to be on the volume of traffic, but on the use of large cars with high petrol consumption and on petrol commercial vehicles.

Other Mileage Taxes

Taxes on tyres, ton-miles, axle-miles and weight-miles have been used, notably in the United States or Australia, with the aim of relating revenue to distance travelled and size of vehicle. These taxes, of course, are designed mainly for commercial vehicles. It is difficult to see how they could be applied usefully to the problems of passenger vehicles in urban areas.

Area Licences

A system of area permits was discussed earlier, and great difficulty was anticipated in handling the task of selection. The difference between area licences and area permits is that the problem of selection is avoided by making the licences freely purchasable. G. J. Roth has advocated a *transferable annual licence* and expounded a theory whereby a rental market would be formed for these licences. Administratively the scheme is attractive: the highway authority would decide each year how many licences it wished to issue and would sell them for whatever price it could get. The buyers would mostly be dealers (probably garage proprietors) who would make a business in renting out the licences. Thus the authorities would achieve their objective of traffic limitation, making a profit in the process, and the administrative task would be handled by private enterprise. The main practical objection to this ingenious proposal is that an *annual* licence would be very valuable—a price of $100 was suggested for central London—and would therefore cause considerable security problems since it would have to be displayed inside the windscreen.

A less imaginative proposal, which would avoid the security difficulty but impose more work on the authorities, is the *daily licence* whereby vehicles would be banned from moving in specified areas unless they had bought a licence valid for one day only. Licences would be sold to all comers at a price determined by the authorities and obtainable at garages (gasolene stations), tobacconists, post offices and possibly from coin operated machines. Licences would be sold in monthly books, not at a discount but simply for convenience, with the right of refund for licences not detached at the end of the month. The licence would be affixed to the windscreen and would be designed so as to indicate clearly the valid date. Li cences could be issued free, or at reduced prices, to special classes of road user such as residents and disabled drivers. Different prices could be charged for commercial vehicles by requiring them to display a different colour of licence.

Area licensing requires that a suitable area be defined. In smaller cities the natural area would be the city centre, in large larger cities it might include some of the inner suburbs. One difficulty is that the creation of a boundary would place business premises on one side at a disadvantage against competitors on the other side. This problem could be alleviated by drawing the boundary along natural breaks in the land use structure, such as rivers, railways and parks.

A more sophisticated version of this method is *differential area licensing* where, say, three areas are defined, a central "red" area surrounded by an inner "blue" area and an outer "yellow" area. Licences would then be sold in the three colours, the most expensive red licence giving access to all areas, the less expensive blue licence giving access to the blue and yellow areas, and the cheapest yellow licence applying only to the yellow area. This system is probably suitable only for very large cities. The stepped increase in price as one approaches the city centre would help to alleviate the boundary problem.

The process of enforcement for area licensing would be the same as for ordinary vehicle licensing except that vehicles parked on the road would not necessarily be obliged to carry a valid licence since the licence would only be required by moving vehicles. However, vehicles arriving at car parks or parked at meters could be checked. It might be possible for this routine inspection work to be performed by wardens with the power to issue licences at a penalty price to non-complying motorists. A study by the British Ministry of Transport in 1967, however, was doubtful about the prospects of enforcing a daily licence system, for the following reason:

> Normal licensing control at present results in some 120,000 cases of infringement of vehicle excise licensing being reported annually in Greater London. The number of undetected offences is, of course, not known. The effectiveness of this type of enforcement is limited. Its adequacy probably depends on the nature of the licensing system. For example, it might be adequate for a differential rate of excise duty (especially if the differential were comparatively small, when it would, of course, act only as a weak restraint), but would be less likely to be adequate for supplementary licensing for single days or short periods. In general, the shorter the period for which a licence could be obtained, the less do we consider that normal licensing control would suffice as a method of restraint. Motorists would soon find that for a period of, say, a week or less the risk of detection would be relatively small. It would not then be a sufficient deterrent to evasion unless there were a high fixed penalty, which would raise problems of public acceptability in relation to the treatment of other motoring offences.[16]

It is clear that a positive effort would be needed to enforce the system. If it were possible to reduce the chances of escaping detection below 10 to 1, a penalty of ten times the licence fee should then be effective and would still be quite small compared with other fines for motoring.

A licensing scheme might be expected to have greater effect on through-traffic in the area than on terminating traffic but the extent of this effect would greatly depend on the size of the area, because the larger the area, the smaller would be the proportion of through-traffic and the longer the distance necessary to by-pass the area. Schemes designed solely to cover small city centres would induce a relatively large volume of through-traffic to circumvent the licence area, causing congestion around the periphery unless there was a suitable ring road. Given a large area, it is more difficult to generalize about the reaction of through-traffic; some vehicles might "disappear," others might divert some way from the licence area, others would pay in order to go through. There should be less peripheral traffic but there might still be considerable pressure on the roads immediately by-passing the area.

On the other hand there should be relief of congestion on the radial roads into the licence area. It is not necessarily correct to assume that motorists restrained from entering the area would drive to the area, park just outside and complete their journey on foot or by bus. Some would do this but others, if they were going to change modes, would do so at the cheapest and most convenient place, which may be farther out, for example at a suburban station. Others might come the whole way by public transport or not come at all. Thus there would be a reduction in traffic outside the area as well as within.

A daily licence scheme was studied in London in 1965.[17] The area covered about 8 sq. miles of the city centre and was bounded by the river for nearly 4 miles and by Hyde Park, Regent's Park and Buckingham Palace for another 2 miles. The scheme required cars and motorcycles to display licences when moving in the area between 8 A.M. and 7 P.M. except on Sundays and national holidays. The study recommended a price of 6 shillings a day for cars, 3 shillings for motor cycles. These prices were estimated to maximize the net traffic benefit, i.e. the saving to the remaining traffic less the imputed losses of utility to the restrained traffic.

The idea of area (or supplementary) licensing has been revived at different times by the Greater London Council.[18] They stress the need for a "buffer zone" around the licence area to prevent over-parking around the boundary, and also for the definition of by-pass routes for through-traffic.

Tolls

The earlier studies of road pricing paid scant attention to the possibilities of tolls. The Smeed Report simply stated that, "for ordinary roads in urban areas they (tollgates) are costly and inefficient and impede the flow of traffic, and even with modern refinements we do not regard them as practicable" (p. 17). The "Better Use of Town Roads" study commented:

"For a controlled area in a city, toll collection areas would be needed on at least all the main access routes. On most routes into city centres this would entail considerable and expensive road-widening schemes at the point of entry, involving some demolition of buildings. It is mainly for this reason that the Smeed Panel appear to have found the classic toll system unsuitable for urban use, and so do we" (p. 27).

The presumption that large toll collection areas would be needed is, however, not convincing. On motorways, with a practical capacity of 1500 pcu's (passenger car units) per hour per lane, it is necessary to provide two to three toll booths per lane in order to cope with the volume of traffic. On urban arterials the lane capacity is 600–900 pcu's per hour, depending on the conditions regarding intersections and parking, but the capacity of a toll booth is the same as on a motorway. In other words the hourly flow through a toll booth is similar to that through one lane of a typical urban intersection. Consequently, although some road widening would probably be necessary in order to accommodate the booths themselves, the additional area required would not necessarily be so large as to rule out the possibility of tollgates altogether. For the same reasons there is no justification in assuming that urban tollgates would add significantly to congestion.

The principal drawback of urban tollgates is the number of gates that might be needed, and the consequent cost of operation. In Central London, for example, about one half million vehicles enter the area between 7 A.M. and 7 P.M. If the maximum capacity per booth were 500 vehicles per hour, a minimum of 80 booths would be required, assuming that minor access roads were blocked and all traffic concentrated on larger roads. In practice the number of booths required would be somewhat more than this and, assuming two shifts per day, at least 200 toll keepers would be needed. But the order of expenditure involved, though considerable, is not more than 10 percent of the expected benefits of such a scheme; and financially, the annual cost of the system would be recovered from toll revenues in about one week.

A toll system may *appear* to be more attractive in cities like New York and San Francisco where there are few access roads to the centre requiring correspondingly few toll points. Tolls are charged on the bridges into San Francisco and have often been proposed for the bridges into Manhattan, e.g. by the New York State—New York City Fiscal Relations Committee as long ago as 1956. An explicit purpose of the proposed tolls was to moderate the traffic flow into Manhattan, and it was suggested that the toll revenues should be used to subsidise other transport facilities. Although the number of toll *points* may be small, however, the number of toll *booths* required is related to the volume of traffic. Key bridges and their approach roads are designed for very high capacities. Not only would a large number of toll booths be required but also a large area to accommodate them. The toll collection area on the Oakland Bay Bridge is a striking example of this.

The two British studies which rejected urban tollgates were made when urban motorways in Britain were practically non-existent. The case for tolls on urban motorways was therefore not considered. The argument against tollgates on ordinary street systems clearly does not apply to urban motorways, as shown by their effective use in Tokyo and Boston, for example. In these cities tolls are collected, so far as is known, in order to raise revenue, not to restrain traffic, although of course they may have a restraining effect. It is necessary to consider whether tolls on motorways alone could be used with advantage as a means of traffic restraint.

An economic argument against tolls on *rural* motorways is that, if the road is uncongested, a toll stops people benefitting from a facility the use of which costs the community very little and certainly much less than the typical toll. The argument is put most forcibly by A. A. Walters, who concludes, "It is of the utmost importance . . . that the method (of charging) should avoid, as far as possible, imposing any price on the *use* of uncongested roads. As long as there is neither wear nor congestion, the services of the road should be free since they are pure social goods. If we *do* impose a price for a free good, then it must be justified by arguing that the best alternative would be inferior."[19]

In urban areas, however, there are two reasons why this argument may not be relevant. First, many urban motorways or their access points are regularly congested; the case for a toll is then straightforward. But secondly, even if they are not themselves congested, they may be discharging large volumes of traffic that cause congestion on the surrounding streets. A distinguishing feature of urban, as against rural, motorways is that most of the vehicles are on relatively short journeys of which an important part must be travelled on ordinary secondary and local roads. With a fairly extensive motorway network for every mile travelled on the motorways there is on average approximately half a mile travelled on the secondary network. The secondary road travel naturally tends to be concentrated in the city centre and other major centres.

Motorways both divert traffic from other roads and by generating new or longer vehicle journeys, add new traffic to the other roads. The balance of relief and generation provides an important equation to be solved before the case for tolling uncongested urban motorways can be analysed. Insofar as tolls cause traffic to divert to the untolled secondary roads they increase the congestion on the latter. But insofar as they restrain traffic altogether they reduce congestion on the secondary roads, particularly in the terminal areas where congestion is likely to be worst. In practice motorways giving access to the city centre and major suburban centres are more likely to have an adverse overall effect on traffic congestion, and to become congested themselves, than are motorways in outer suburban areas. There is a case, therefore, for considering motorway tolls on the inner parts as an alternative to tolling the whole network within the urban area. Congestion, of course, is only one of the possible reasons for wishing to restrain the volume of traffic. There may be other reasons for a comprehensive restraint on motorway traffic.

Vehicle Metering

The term "road pricing" has become closely associated with the metering of vehicle use in congested areas. The Smeed Report concluded that an economically efficient method of charging (or taxing) road users should be flexible enough to enable a wide range of charges to be made in different areas at different *times of day* in accordance with variations in congestion. The only method that came close to this requirement was metering. In England the Transport and Road Research Laboratory have made a major effort to develop meters that are reliable, cheap and easy to administer.

Meters may be distinguished as either on- or off-vehicle. On-vehicle meters are located on the vehicle and are activated by electrical loops buried in the road surface. Meters may be designed for either *continuous* or *point* pricing. The continuous meter is switched on automatically when the vehicle crosses a loop entering a pricing zone and continues to run until it is switched off when the vehicle passes out of the zone or parks. The point meter simply registers one unit every time the vehicle passes over a loop, i.e. a pricing point; a large number of pricing points are needed but no definition of zones. Development work has concentrated on the point pricing meters.

The problem of enforcement makes it desirable that the source of power for the operation of the meter should be either external to the vehicle or wholly internal to the meter itself and secure against interference. Ideally the meter would either be energised from the road surface or contain an encapsulated battery with a life as long as that of the meter itself. Various possibilities have been investigated, including clockwork mechanisms, but a satisfactory solution, at reasonable cost, has not yet been discovered. Further development of long-life batteries might possibly provide an answer.

Off-vehicle metering also works on a point pricing basis. An identification unit fitted to the vehicle enables the identity of the vehicle to be recorded by roadside equipment when the vehicle passes over a pricing point. This information is then stored and processed by computer and an account is presented to the vehicle owner every 3 months, say, in much the same way as for telephones. The identification unit requires no power source on the vehicle. It may therefore be fully encapsulated and cannot be removed or broken into without being irreparably damaged. Vehicles without a unit in working order can be immediately spotted when they pass a pricing point by their failure to transmit an identity number to the roadside apparatus.

The off-vehicle metering system has now been developed by the TRRL to a high standard of performance at relatively low cost. It is hoped that identification units could be mass produced at about $5 each, and pricing points would cost about $2000 each. Computing costs would not be excessive. The equipment has been under test on 50 laboratory staff cars for two years, with results which technically justify further tests in a "real life" situation. There are, of course, political and practical difficulties in doing this. The simplest way of advancing from laboratory conditions to a real situation might be to install pricing points at a toll bridge or tunnel and to fit units to the cars of regular users.

In the United States there has been discussion of the possible adaptation for pricing purposes of automatic vehicle monitoring equipment, developed for the operational control of goods vehicles and buses.

Given its successful technical performance, vehicle metering offers a powerful and sophisticated means of traffic restraint. Pricing points would normally be laid in intersecting cordons so that motorists could not avoid points by deviating from the normal route. Cordons would be laid closer together in higher

11.6 Cordon schemes for pricing points.

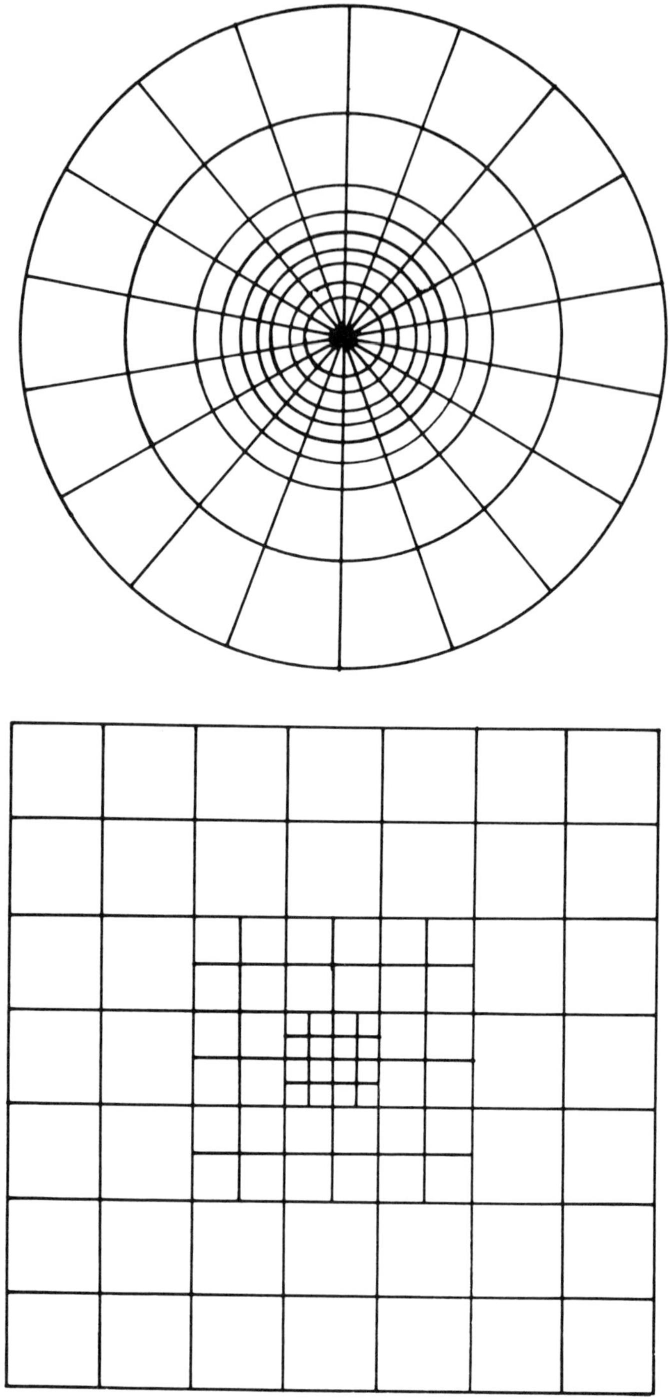

priced areas. Two schematic arrangements for a typical concentric city are shown in figure 11.6, but in practice these would have to be carefully "fitted" to the road network to take account of geometric irregularities in the main roads. Additional pricing points could be located at strategic places to discourage traffic from entering environmental areas or needlessly using shopping streets. Different classes of vehicle could be charged at different rates per unit.

Vehicle metering has been developed as a means of charging road users with the congestion costs they cause. It could be used to achieve other objectives. The level of prices and the location of pricing points offer an opportunity to influence the volume, composition and distribution of traffic throughout the urban area. The "Better Use of Town Roads" report considered that pricing would be much the most efficient way of restraining goods traffic, encouraging higher load factors and better organisation of delivery schedules (p. 16).

The main technical problem foreseen with vehicle metering is fitting identification units to all vehicles entering the area. Some vehicles such as buses, emergency vehicles and possibly taxis would doubtless not be required to have them, and there would be no difficulty in issuing and fitting them to regular users. But a large number of vehicles belonging to visitors and tourists may enter the area only once or very occasionally, and it would not be practicable to fit them and invoice their owners just for one or two trips in the area.

This is a major difficulty. It has been estimated in central London that although "infrequent users"—i.e. cars that visit the area less than once a month—account for only 5 percent of the cars visiting the area on any one day, nevertheless over the course of a year they may account for roughly 70 percent.[20] It can be deduced that if 170,000 cars visit the area on a typical weekday, approximately 1,200,000 different cars visit it during a year, of which about 850,000 do so not more than six times. The task of fitting all these infrequent visitors with identification units, would present a real problem.

There are two possible solutions. First, if vehicle metering was introduced on a national scale and was adopted by many towns and cities it would become practicable to fit all or most cars in the country, and only visiting cars from other countries would need special treatment. Even if this were done, though, it could not be done overnight, and there would be a period of years during which the system was being gradually extended. At least during this interim period some way of dealing with occasional visitors would be needed. Secondly and more realistically, it might be necessary to operate permanently an alternative pricing system for vehicles not fitted with an identification unit. The most suitable alternative would be a special or daily licensing system. Cars entering from abroad or from other parts of the country, and not entitled or willing to be fitted for vehicle metering, would be required to display a special short-term visitor's licence instead.

A major conclusion about road pricing, therefore, is that a daily licence system is not just a simple alternative to metering but a necessary precondition for a metering system. Metering may be regarded as a mechanized and refined form of daily licensing which would only be applied to regular users.

INDIRECT RESTRAINTS—PHYSICAL

The practical difficulties of charging directly for the *use* of roads, or otherwise restraining excessive *use* of them, has led to the adoption of indirect measures affecting (a) vehicle ownership and (b) parking, on the principle that people restrained from owning or parking vehicles will not contribute to the traffic problem. Again one can distinguish between physical and fiscal restraints.

Restriction of Parking Space

In most, if not all, cities the number of motorists driving into the city centre has been limited by the effects of the motor vehicle itself, in three ways: (a) traffic congestion, (b) scarcity of parking space, and (c) dispersal of employment, shopping and other activities to places more accessible by car. The third effect is found mainly in some American cities. Of Detroit, for instance, it has been written that, "it is comparatively easy to get around by automobile in the core area. . . . There is always plenty of parking space available in Detroit, for the simple reason that there is no longer sufficient urban activity to fully utilize it."[21] But in most cities the limitation of traffic has been a direct consequence of either congestion or parking scarcity or both. In central Bologna it is reported that traffic moves fairly freely because the mediaeval city contains only 10,000 parking spaces. In central London in 1966 it was found that the available parking space was never more than 76 percent occupied, the city centre being "sheltered" by congestion on the approach roads, in particular by a "gluepot ring."[22] But empty parking space is not necessarily available to everyone nor in the places where it is wanted; in London, as in most cities, traffic is limited by both congestion and parking scarcity. Many cities have concluded that they can relieve the congestion by reducing, or at least not increasing, the amount of parking space. The policy adopted in Liverpool and other British cities is to limit parking capacity so as to keep the volume of traffic within the economic capacity of the road network. As recently described by Lyle Fitch, "A still more sophisticated notion involves a planned balance between capacities of arterials leading into cities, local street systems which distribute traffic, and parking capacity. This has not been systematically attempted in any United States city, to my knowledge."[23]

There are several major difficulties with a policy of parking restriction. The first concerns the concept itself, namely that there is a direct relationship between parking capacity and the volume of traffic. In any urban area the traffic includes buses, taxis, bicycles and commercial vehicles, which are not reduced by parking restriction; in fact they are likely to be increased by a diversion from cars. Then there is through-traffic which does not park in the area. The smaller the area under consideration, the larger is the proportion of through-traffic likely to be. Parking restriction can only encourage the growth of non-parking traffic, including cars that would have parked in the area but now drive through and park just outside.

The relationship between parking capacity and the volume of traffic is further blurred by the fact that parking capacity and road capacity are not comparable parameters. City parking space is largely empty at daybreak and steadily fills up until the early afternoon when the number of departures begins to exceed new arrivals. There cannot be a scarcity of parking space until the available space has been filled. If this happens early in the day there will be nowhere for later arrivals to park. The simple fact is that commuters arrive first and will naturally try to occupy space before the arrival of visitors, shoppers and business people, whose need is usually regarded as greater. Also, of course, it is the commuters who contribute to the peak traffic.

Consequently a simple policy of parking restriction has little to recommend it. To achieve its objective, a policy must be contrived to restrict the availability of parking during the morning traffic peak without making it totally unavailable after the peak. This can be done in two ways: by keeping some car parks closed until after the peak, and by placing time limits on some parking space in order to exclude the commuter.

The second major difficulty with parking restriction is the practical problem of withdrawing space from parking use and of preventing new space being provided. If a large proportion of the available space is on the streets, there is no technical problem in taking some of it out of use by means of parking bans. But if, as is often the case, there is a great deal of street parking space in private ownership there may be no way in which the authorities can control its use.

In most cities developers of new buildings for offices, shops and hotels are permitted to provide as much parking space as they wish and are often forced to provide a minimum. New multistorey car parks are permitted, if not encouraged. Possibly the first city to reverse this situation was London in the early 1960s, when *maximum parking allowances* replaced *minimum parking requirements* in new buildings, and the new maximum was often less than the old minimum. In New York the City Planning Commission has attempted to limit the construction of new parking garages in midtown Manhattan in order to relieve street congestion. In Germany, Kühnemann and Witherspoon report that:

> In Bavaria, the State Ministry of the Interior is considering an alteration to existing building ordinances to enable communities to restrict the construction of garages within central areas. At Munich the municipal council has decided that builders of commercial buildings in the city centre can either provide parking in accordance with building ordinances or, if the council requires it, contribute a sum about 10 percent less than prevailing construction costs for the required parking to financing car parks near S- and U-Bahn stations.[24]

In Chicago, driveways are not permitted into streets with pedestrian flows exceeding 25,000 per day. This effectively limits the construction of car parks in the city centre and forces them to the fringe of the area. In Melbourne too the problem of excessive driveways has led to an arrangement whereby developers, instead of providing parking, pay towards municipal car parks.

Despite the practical difficulties, some degree of parking restriction is nevertheless possible in all towns. To be effective the restriction must prevent some motorists from parking either at their destinations or near enough for them still to drive into the area by car. A policy of pure restriction is a first-come-first-served policy. It has the inherent disadvantage of transferring the delay, uncertainty and frustration of traffic congestion from the highway to the parking place. Many towns find themselves unintentionally in this predicament anyway, having seen the demand for parking space grow greatly in excess of supply, at least in some areas of the town. Intentional or not, if the supply of parking space falls well below demand some method of allocation other than first-come-first-served is badly needed. This can be provided by controls or by fiscal means.

Parking Controls

The term "parking controls" is used to mean non-fiscal methods of allocating parking space. In most cities a great deal of parking space is both privately owned and privately used; it is not available to the public but is privately controlled by its owners for the use of management, senior employees and officials, customers, guests and so on. Great trouble is taken to control the parking space so that those who are privileged to use it may do so without difficulty and those who are not do not try to. Rota systems, permits, individual reservations and methods of collective discrimination are all used and are enforced by gatekeepers or commissionaires. If *public* space is to be controlled equally effectively, without the use of the price mechanism, careful administration and enforcement are equally necessary.

The methods used to control private parking space are all forms of *individual* or *collective privilege*. Spaces are explicitly reserved for the managing director, the Minister, official visitors, customers, and so on. With public parking space the equivalent of individual privilege is a system of *parking permits* whereby space is reserved for permit holders, e.g. residents.

Public parking controls, however, are almost invariably based on collective, not individual, privilege. By far the most common is that provided by *time limits*, which are a thinly disguised discrimination against commuters in favour of shoppers and callers. As a method of traffic restraint their effect is ambivalent. By restraining commuters they reduce peak hour traffic, but by catering for short-term parkers they can accommodate a larger turnover of cars and hence encourage more traffic, at least in the off-peak hours. The turnover of 2-hour meters (at nominal charge) in London has been found to be 5.5 cars per day as compared with only 2.25 on streets without time limits.[25] Thus although time limits are an important way of allocating parking space, they may not reduce the volume of traffic.

Open discrimination against commuters is effectively achieved by closing car parks until after most commuters have started work. The simplest method is to lock the gates, but if the park is used by overnight parkers wishing to leave before the gates are opened it may be preferable simply to prohibit cars from entering during peak hours.

Other controls may be needed to overcome the problems caused by parking restriction to residents of the area. The larger the restricted area the more likely is it to include residential streets, and inner city residents often have nowhere to keep their cars except the street. *Residents' permit schemes* then become a practical necessity as a means of reserving on-street parking space for holders of permits granted on the basis of residence.

INDIRECT RESTRAINTS—FISCAL

Parking Charges

The price of parking can be a powerful influence on the volume and distribution of traffic. A parking tax or surcharge, i.e. an addition to the normal charge, if any, is often advocated as a means of traffic restraint. The organisation of parking facilities, however, is a complex matter, presenting considerable practical difficulties in the implementation of a parking charge policy. It is necessary to distinghish five types of parking space.

1. *On-street parking.* In principle, since the streets are controlled by public authorities, it is possible to charge for parking on them, except in some countries like Turkey and Italy, for example, where constitutional laws prevent this. In other countries two methods are in use: meters and licences.

Parking meters are now used in the majority of large cities. When a street is converted from uncontrolled to metered parking its parking capacity is usually reduced by at least one-third, which in itself is an important traffic restraint. In Central London between 1961 and 1971 the number of on-street spaces has been reduced from 63,500 to 30,000. The traffic effects are complicated by the combination of price and time limits. If, as often happens, the meters are fully occupied and cars are cruising the streets looking for a space, the price restraint adds little to the physical restraint. If, on the other hand, the price is so high that plenty of meters are unoccupied, the physical restraint is superfluous because a larger supply of spaces would not attract more traffic. The effect of time limits, as noted above, is to reduce long-term parking and, if this makes short-term parking easier, to increase the number of short-term parkers. But again, if the price is high enough and is related to time, it too will restrain long-term parking and will to some extent usurp the role of the time limit.

Parking meters vary greatly. The time limits vary from 10 minutes to 10 hours and prices vary from a nominal charge to an effective market price. The prices are usually time-related, either by the hour, half-hour or quarter-hour, but are sometimes a fixed charge. Enforcement varies in effectiveness from city to city. The financial penalties for excess parking and the probability of being made to pay also vary and are reckoned by some motorists as part of the price structure. Because of these variations it is impossible to generalize about the current role of parking meters as a method of traffic restraint. But where the prices are high they are probably effective in either reducing long-term parkers or causing them to park elsewhere, depending on the size of the metered area and the policies adopted towards other types of parking space.

2. *Municipal car parks*. There is no technical difficulty in raising charges at car parks owned by the city authorities. In some cities there are no municipal car parks, reflecting the view—at least in the past—that parking is properly a service to be supplied by private enterprise. In other cities the opposite view is taken; in Bremen, for instance, the State-owned Bremer Parkplatz GmbH owns most of the parking space in the city and is actively engaged in constructing multistorey car parks.

3. *Private commercial car parks*, i.e. privately owned but open to the public on a commercial basis. In London the Greater London Council have obtained licensing powers to control the prices charged by private parking companies. The objectives are to prevent the prices from being too low and to control the structure of charges in order to discourage long-term parkers and other peak hour traffic. (Price structure is discussed below.) In most cities these powers do not exist at present. An indirect way of achieving the same object is to withhold planning permission for new private car parks until the scarcity of space has forced up parking charges.

4. *Semi-private car parks*, i.e. privately owned but available on a non-commercial basis to restricted sections of the public, e.g. customers, visitors, employees. Occasionally a charge is made to employees on a monthly or annual basis, but the great majority of these car parks have no charge and no organisation for making a charge. They are often small and tend to be numerous, being attached to shops, offices, hotels, restaurants, libraries, swimming pools and all manner of places. To extend parking charges to these car parks raises formidable problems. No city has attempted to do so.

5. *Private parking sapce*. Even in the very centre of a city there is invariably some parking space attached to residential property. Yards or lock-up garages may also be owned and used by small traders, manufacturers and other people working in the area. Cars can be, and often are, parked in odd corners and spaces on private property. The application of a charging

policy to these parking areas is virtually impossible. It is no small task just to discover where they are.

Clearly a policy of restraint by parking charges cannot be fully implemented. A proportion of parking space will remain freely or cheaply available to privileged motorists, and in some cities it might well be large enough to make the restraint policy ineffective. But if it is not too large and can be prevented by planning controls from growing larger, the charges on remaining parking space can still be of some effect. The fact that some motorists can avoid paying for parking does not prevent the use of charges to restrain the less fortunately placed; but it does mean, of course, that to achieve the same effect, the charges must be higher than if they applied to everyone, and the policy is inequitable.

Parking charges or taxes have the advantage over parking restrictions of providing an easier and economically superior method of allocation. They share the drawback of not applying to non-parking traffic. Largely for this reason the writer, in a quantitative study of parking taxes and daily licences, concluded, "Little advantage is seen in the proposal for a parking tax. The maximum benefit of such a tax, assuming it to be imposed on all parking space, is only about half that of the daily licence. In practice it is unlikely that the tax would touch more than 20 percent of the traffic."[26]

The Smeed Report came to a similar view: "We believe that the parking tax could bring significant benefits, in spite of its inequitable results and its undesirable effects in encouraging non-parking traffic and penalising local traffic. Preferable to a parking tax might be a system of daily licences, which would embrace all traffic in the areas concerned but would in turn give rise to difficult boundary problems" (p. 41). The "Better Use of Town Roads" report recognised the theoretical superiority of the daily licence over parking charges but had serious doubts about the enforcement of licensing. It therefore preferred a combination of parking charges and parking controls.

Many cities are beginning to look to parking charges as a principal means of traffic limitation. Meters are to be found in most large cities; they have at last been introduced in Paris and are expected soon in Vienna. In Stockholm and Götenborg the policy is to charge 2 Kr./hour for on-street parking in the central area and 1 Kr./ hour outside; these are relatively high prices.[27] In Germany a Federal law sets a very low maximum of 20 pfennigs/hour to parking meters; but for off-street parking the policy of the Bremer Parkplatz GmbH, for example, is to charge 1DM for the first 2 hours, then 50 pf./hour with a discount of 50 pf. if the motorist makes a purchase in the city centre shops.

In Glasgow the role of parking charges has been well described by J. Armour, as summarised in the following sentences:

When completed, the Highway Plan (for Glasgow) will cater for a large increase in motor car usage throughout the city but it has not been designed to cater for unlimited access of motor cars to the central area. . . . The Highway Plan, now being implemented, involves a restraint on entry to the central area. In terms of parking spaces, the measure of this restraint is that it is aimed to provide (by 1990) 27,000 spaces against an estimated demand of 65,000 spaces. . . . It is not sufficient merely to place a limit on the number of parking spaces. It is necessary also to ensure that the spaces are used in such a way as to produce the approximate balance between long- and short-term parking. If the commuter parker can be restricted by controlling the number of spaces available for long-term parking, substantial reductions could be achieved in peak hour travel. Control may be exercised by setting a limit to the number of spaces to be provided, the charges to be made for parking and the times when certain spaces may be opened and closed. . . . It is expected that the number of spaces for which parking charges must be paid will increase from 8,000 (out of 18,000 in 1971) to 18,000 (out of 21,000) by 1980.[28]

An almost identical degree of restraint is proposed by the Leicester Plan, which was claimed in 1964 to be, "the first city plan to say NO to the motor car scientifically." An upper ceiling of parking accommodation was established at 22,500 car spaces, as against an estimated demand for 67,000 spaces if unlimited use of cars was provided for.[29]

In some cities, especially in the United States, parking charges are regarded as a restraint that arises "naturally" if off-street parking is provided by private enterprise and if on-street parking is priced commercially. Although, theoretically, free market prices for parking do not reflect the incremental system costs caused by traffic, there is no doubt that parking charges in American cities are among the highest in the world, both on- and off-street. A similar view seems to be taken in Athens where off-street parking is left entirely to private enterprise and the prices are quite high; but as yet there are not many meters in Athens.

There are few city centres where new parking space can be provided at less than £1,000 per space if a normal market price is paid for land,[30] and in others (e.g. Toronto) costs of over £2,000 per space have been reported. If parking prices are high enough to cover interest, maintenance and running costs, they cannot be much less than £1 per day and could well be more.

If this is accepted as the market level and similar prices are charged for on-street parking, a formidable instrument for traffic restraint is created. Curiously, in some cities like London where the principle of the parking tax is generally accepted, parking charges tend to be far below a true market level and much lower than in American cities where the principle is not generally accepted.

Structure of Parking Charges

The effectiveness of parking charges as a method of traffic restraint depends not simply on the level of the charges but also on their structure. There are several possibilities, as follows:

1. *Fixed charges.* A lump sum charged for the act of parking may seem logical if it is intended to be a surcharge for the use of roadspace for movement rather than of parking space, because there is no way of knowing how far a vehicle has travelled before it parks or will travel after it leaves the parking space. But fixed charges would weigh heavily on local motorists and others who need to make numerous short stops, e.g. salesmen and other businessmen.

2. *Constant hourly charges.* The hourly charge may appear unfair to the long-term parker. He does not necessarily cause more congestion and nuisance than does the short-term parker, just because he remains parked longer. *If* there is a strong correlation between long-term parking and travelling in the peak hours, then there is a rough logic in a charging structure that weights heavily on the long-term parker. Certainly it is effective, as will be shown later.

3. *Two part tariff.* It is possible to charge a fixed sum on arrival plus an hourly charge for the duration of parking. This can be designed to make long-term parking by commuters (i.e. 8–9 hours) as expensive as with a straight hourly charge, without making short-term parking too cheap.

4. *Variable hourly charges.* The most effective anti-congestion charging structure consists of a basic charge, either fixed or hourly, plus surcharges on vehicles arriving or leaving during the peak hours. Thus a typical commuter might pay a basic 4 shillings for a day's parking plus 3 shillings for arriving between 8 and 10 A.M. plus 3 shillings for leaving between 4:30 and 6:30 P.M.

Vehicle Licence Fees

The annual tax on the registration or licensing of motor vehicles is a restraint on vehicle ownership and hence on the volume of traffic. The tax is normally imposed, of course, by central or state government, not by the city. So far as is known, the tax on cars has nowhere been imposed or increased with the explicit intention of restraining car ownership. The political and social implications of a tax that makes a minority of households give up their cars altogether are quite different from those of a tax that makes a majority of car owning households use their cars less. With the general growth of motorization, however, *two car households* tend to become numerous. For *most* households situated in large towns with effective public transport, the second car is a luxury whereas the first car is regarded as a necessity. The second car is therefore likely to be more price sensitive than the first car and a more acceptable target for restraint, but there seems to be no practicable way of charging a higher tax on the second car (in the household) than on the first. It could be argued that, if first cars are price inelastic and second cars price elastic, the tax could be set high enough to discourage a lot of second cars without being so high as to discourage many first cars. This is an argument for a moderately high tax. But there are no factual studies to support it.

Licence fees for goods vehicles are probably more important because they tend to be much higher than for cars. In some countries these taxes have been deliberately used to swing goods traffic from road to rail. This does not dispose of the need for road collection and delivery at rail terminals in urban areas, but it may eliminate some through-traffic.

Concessionary Vehicle Licence Fees

It is possible to charge lower fees for the registration of vehicles that conform to certain desirable standards as regards noise, power-weight ratio, size, safety and air pollution. As discussed earlier, all these features can be regulated by law. Vehicle regulations are limited, however, by their universality; since they apply to everyone they must not be unreasonably severe on anyone. Concessionary fees are a way of obtaining higher standards from those who can comply cheaply, without forcing them from those who cannot. Therefore they can achieve higher standards than can regulations.

The vehicle characteristics amenable to tax concessions are probably limited to those that can be permanently certified in the vehicle registration papers, having been approved for the particular model of vehicle. But these include some characteristics that are currently causing much concern, such as noise suppression and clean exhaust.

Hire Purchase Restrictions

Limits on the maximum period and minimum deposit allowed under hire purchase contracts relating to cars have proved effective in reducing the purchases of cars. These restrictions are normally imposed or increased as a temporary measure only, for reasons of national economic policy. It is difficult to know to what extent their impact is purely a deferment of car replacement rather than a real reduction in car ownership. Clearly, though, if marginal car owners obtain cars on hire purchase as soon as they can afford to, restrictions on hire purchase will delay the day when they become car owners and will to that extent reduce the volume of traffic.

Import Tax and Purchase Tax

Taxes are imposed on the retail sale of motor vehicles in most, if not all, countries. Again, the declared purpose is seldom if ever to restrain the volume of traffic in spite of the growing awareness that rising purchases of cars and lorries spell traffic congestion and demands for costly road construction. In countries that manufacture motor vehicles, purchase tax is imposed as a revenue raiser, and in countries which do not, import taxes are usually imposed in order to relieve the balance of payments. In some vehicle importing countries the taxes are very high and the effects on vehicle ownership and use are marked. In Colombia, for instance, the tax on cars reaches 200–300 percent depending on the model. The capital city, Bogota, is perhaps the only city in the world with a population of over 2 million and no serious traffic problem. In Burma the import tax is far higher still—a modest car retails at £5000—and Rangoon, a city of 1.8 million, has no traffic problem. Both these cities have low income levels, of course; but nowadays it is in poor cities like these that traffic congestion is usually worst.

The argument has long been put, for example by Sir William Glanville, that traffic congestion is the result of an imbalance between public investment in roads and private investment in vehicles. One way to remedy the imbalance is to reduce the latter, and import or purchase taxes are an effective way of doing it.

Subsidised Fares

Traffic restraint, as defined at the beginning of the chapter, includes measures intended to influence, without directing, people's choices of journey and mode of transport on the existing infrastructure. This includes subsidised fares to attract travellers from private to public transport.

Most, perhaps all, developed countries have accepted the case for subsidies to help maintain a division of passenger traffic between public and private transport that can be handled without chaotic congestion or intolerable costs, especially during the peak hours. Subsidies are also provided for *social* reasons, to help the poor, the aged and the young. But our concern is with the *economic* objective of checking the drift of passengers from public to private transport; clearly the passengers in question are not poor, nor aged, nor young; they are potential motorists, and the economic purpose of the subsidy is to persuade them to stay on public transport. Those that do so avoid adding to road congestion and at the same time help to shoulder the overhead costs of the public transport system, thus benefitting other travellers on both modes and the general community as well.

Subsidies may be used to produce lower fares or higher quality. The latter is properly a matter of traffic avoidance and will be discussed later. The poorer passengers usually want lower fares; the potential motorists want higher quality. Of course the poorer passengers would also like higher quality but are not able to afford it, and the potential motorists would also like lower fares but are not willing to accept a low quality service. Consequently there may be confusion over the real purpose of the subsidy and conflict over the way in which it should be used.

In this section we are concerned only with subsidised fares. Evidence from about 20 German and American cities collected recently by Baum points to two conclusions:

1. fare reductions lead to a substantial increase in passenger demand; but
2. the extra demand arises mainly for non-work journeys by non-motorists.[31]

These conclusions are confirmed by the recent Rome experiment in free fares. Baum concludes that, "a transfer from private to public transport can be achieved more efficiently by improvements in the quality of public transport; the isolated introduction of free transit seems unable to relieve towns or traffic congestion." It would be wrong to interpret these results as meaning that fares are irrelevant to the volume of private traffic. Most commuters have fairly short journeys and stand to gain little from reduced fares. It is the minority of long-distance commuters, found mainly in large cities, who stand to gain most; and they are the ones who contribute most vehicle-mileage if they drive to work. Cheap fares aimed specifically at outer suburban commuters might be an effective traffic restraint. Secondly, there is a relationship between fares and quality. If high fares drive away non-car-owning passengers, as shown by Baum, this tends to lead in practice to cuts in service frequency, which opinion surveys have shown to be the most important aspect of quality in the eyes of car-owners.

Subsidized fares should therefore be regarded as a significant means of traffic restraint in large cities, but by no means sufficient to solve the whole problem.

Two Part Tariff

One of the factors encouraging the use of private rather than public transport is the pricing structure. Motorists pay many of their expenses in lump sums which are unrelated, or not perceived to be related, to the amount of use they make of the car. The annual licence fee, insurance, interest, driving licence, garaging and annual inspection are costs entirely unrelated to the amount of use; the capital cost (or depreciation) and maintenance cost, being paid at infrequent intervals, are usually not perceived as running costs. Hence, to the user, the actual running cost of the car for individual journeys is low and is usually perceived to be lower still, amounting to no more than petrol and oil. The public transport passenger, on the other hand, except when travelling on a season ticket, pays his full share of the total outlay, including all overheads, every time he travels and in proportion to the distance he travels.

Once a person has acquired a car, therefore, he has a financial inducement to use it in preference to public transport. He is on a two part tariff that strongly encourages consumption. Proposals have therefore been made to put public transport on an equal footing by offering two part tariffs there too. There are numerous ways of doing this. Sherman has proposed *club subscriptions* for commuter buses:

> Consider an institution which we shall call the Passengers' Club. Membership extends over a long time, perhaps three years. In addition to a fixed membership fee (payable in convenient monthly installments), members would pay a direct charge (for which they would be billed monthly) based on their usage of each different public transport mode. Membership commitments over shorter periods could also be available.[32]

The ideal, from the passenger's point of view, is a *pass* which authorises free use of the entire public transport system, but it is impossible to find a price for such a pass that is neither so expensive that few people would buy it nor so cheap that the operating authority would lose revenue as a result. There are many other possibilities, and the best arrangement will undoubtedly vary according to the size of the city and the proportion of car owners.

Community Transport Centres

The alternative solution to the problem of pricing structure is to change the way in which private transport is paid for, in other words to eliminate or modify the two part tariff structure. A radical proposal to do this has been made by Fishman and Wabe, who suggest the creation of community transport centres, preferably attached to large council estates and other closely knit communities.[33] The main purpose would be to provide car rentals, at average cost rates, to members of the community. It would give the benefits of private transport more cheaply than private ownership, and with numerous other advantages, but the *average cost pricing* would encourage people to travel by public transport unless they really needed private transport, and to share private transport where possible. This proposal offers big advantages to the city, from traffic limitation, and to the individual, but it has received little attention.

12 TRAFFIC AVOIDANCE

The third branch of traffic limitation is the most fundamental and least understood. Industrial empires are familiar with the problem of optimising transport requirements, i.e. locating their widespread operations in a way that maximises the gains from spatial separation *less* the transport costs involved. Transport is treated as a cost to be avoided unless there is a net gain from doing otherwise; operational planning seeks to avoid (a) journeys by employees or goods unless they are necessary and worth the cost, and (b) journeys of greater length than is necessary unless the additional distance is worth the cost.

These basic, elementary principles have never been systematically applied to the planning of cities, and are not successfully incorporated in present transport planning methods. Indeed current methods often treat transport as a benefit, not as a cost. Given two alternative plans, the one that produces more vehicle trips and more vehicle-miles is reckoned to produce correspondingly greater benefit. If a short walking trip in one plan is replaced by a longer car trip in the other—as a result of different land use and transport strategies—the latter is assumed to be superior to the former in direct proportion to the time and cost of the car trip. If fast new roads widen the area of competition for jobs among workers and lead to more and longer commuting, the extra vehicle-mileage is costed and counted as a benefit. These anomalies result from the application of static analytical concepts to dynamic situations; they are incompatible with the principles of traffic avoidance.

The underlying assumption of traffic avoidance is that most journeys are costs, to the individual and society, and are therefore undesirable in themselves. They are means to an end. People make journeys because they are the easiest means of achieving their ends, given the urban complex of land uses and transport facilities confronting them. The aim of traffic avoidance is to enable people to achieve similar ends with fewer means. This can be done in four ways:

1. by reducing the *number* of necessary journeys;

2. by reducing the necessary *length* of journeys;
3. by organising land uses and transport facilities so that more people *prefer* to use other modes than private motor vehicles;
4. by organising the life of the town so that more people *prefer* to travel at off-peak times.

In short, the aim of traffic avoidance is to organise the town in such a way that its inhabitants have no need or desire to use private motor vehicles to such an extent that they create a major social problem. It may not be possible, of course, to achieve this aim fully.

The principal methods of traffic avoidance are strategic land use planning, local land use planning and transport structure planning.

STRATEGIC LAND USE PLANNING

The largest single factor determining the volume of traffic in a town is the geographical relationship between people's homes and their jobs. Until quite recently the location of houses and jobs was decided by private enterprise in most countries and still is in some. Private enterprise takes account of transport facilities from the point of view of individual users but not from that of the city as a whole. The main task of strategic land use planning is to decide on the pattern of locations for new houses, jobs, shops, services, recreation facilities and other land uses. Many objectives have to be satisfied, and much depends on the geographical situation and the inheritance from past generations, but one desirable objective is traffic avoidance. Houses of the right types must be located near employment areas if work journeys are to be kept short. This does not ensure that work journeys *will* be kept short, because the houses will not necessarily be occupied by the people for whom they are intended; but at least the opportunity is created.

More important than the proximity of homes to jobs are the densities and geometrical arrangement, because these determine the ease with which work journeys can be served by public transport. For traffic avoidance, it is more important that work journeys should be made by public transport than that they should be short. Public transport can only provide a fast, cheap, frequent service if there are large numbers of passengers wishing to travel in the same direction. Therefore housing and employment must be arranged so that the journeys between the two can be concentrated into a few corridors well served by public transport. High density housing and high density employment (e.g. offices) should therefore be located in these corridors, and low density housing and other land uses should have good access to the transport corridors.

The next most important locational decision concerns shopping. Shopping centres should be sufficiently compact to avoid the need for vehicle journeys between one shop and another. Traffic avoidance requires shops to be located so that the journeys between homes and shops are either short or well served by public transport. It follows that the principal shopping centres in a town should be situated along the main trnasport corridors.

High density old cities with compact centres and radial road systems, as found throughout Europe, necessarily developed along lines of traffic avoidance, because they developed at a time when most passenger movements had to be on foot or by public transport. If cities were developed today without private transport, they would presumably follow a similar pattern, but there are two important differences: first, more efficient methods of public transport are now available; secondly, cities are richer and better able to afford lower density living and the transport that necessarily goes with it. Railways have pushed city suburbs out farther than have cars. Cars have enabled people to fill in the spaces between the railway lines and between the stations. Low density living developed along railway lines (in England at least) before the car was invented, but it was not for the poor.

Traffic avoidance need not therefore imply a return to very high living densities. But if it is adopted as a planning objective, the form of the city is more likely to resemble that of a traditional European city than of a mid-West American city.

Given the density of the city, the *form* that minimises transport requirements is derived from two principles. First, the *principle of central location* states that the location of an enterprise, which serves customers evenly distributed over a circular or rectangular area, will minimise total travel distance if it is at the centre of the area. Secondly, the *principle of a hierarchy of centres* states that enterprises in a city can be placed in centres that form a hierarchy of four scales defined by the size of area that they need to serve and to draw labour from. The four scales are city, sector, suburb and neighbourhood, and they are identified as follows:

1. *City scale.* Some enterprises, like government, major law courts, financial houses, specialist shops and services, head offices and headquarters of all kinds, theatres, museums, cathedral, university, etc. serve the whole city (and often the whole

region or nation beyond) and cannot feasibly or desirably be broken up into units serving only parts of the city; they have a strong claim to a location that is highly accessible from all parts of the city; geography permitting, the city will tend to develop so as to make this centre the geometric centre. The essence of these enterprises is that they must, on account of uniqueness, specialisation or economies of scale, serve a city-wide market and/or employ rare skills not easily obtained in any one corner of the city.

2. *Sector scale*. As cities become large the market becomes large enough in terms of both sales turnover and travel distances, to support major shopping centres outside the city centre, each serving a sector of the city with a population in the range 500,000 to 1,000,000 people. As the city grows, some city centre functions can be hived off into these sector centres; this need not damage the city centre itself which should be receiving growth in its other functions. London is a classic example; first, manufacturing functions started to leave the city centre, then routine shopping, then routine office work, but the employment and prosperity of the centre did not decline, because off-setting growth continued in specialised functions such as government, business headquarters, tourism, entertainment, higher education, etc. The essence of a sector enterprise is that most of its necessary transport linkages, with employees, customers and other enterprises, should be within the sector. They will naturally tend to gravitate towards the most accessible place within the sector, e.g. a major sub-centre. In Athens six sector centres have been designated; in London there are seven.

3. *Suburban scale*. A major sub-centre is equivalent to the centre of a large town and is unlikely to achieve sufficient accessibility to attract all the routine weekly shoppers who can easily be supplied more locally. This suburban scale relates to populations between 50,000 and 200,000. As with the higher scales, the essence of the suburban enterprise is that its transport linkages should be largely confined to its own suburb, in which it will choose the most accessible location. In addition to shops there will be banks, garages, cafes, cinemas, local government offices, police station, fire station, small hotels and numerous "small town" activities.

4. *Neighbourhood scale*. Of great importance socially and psychologically is the identity and character of the small area in which people "belong." It is also important from the standpoint of transport because every little journey made within

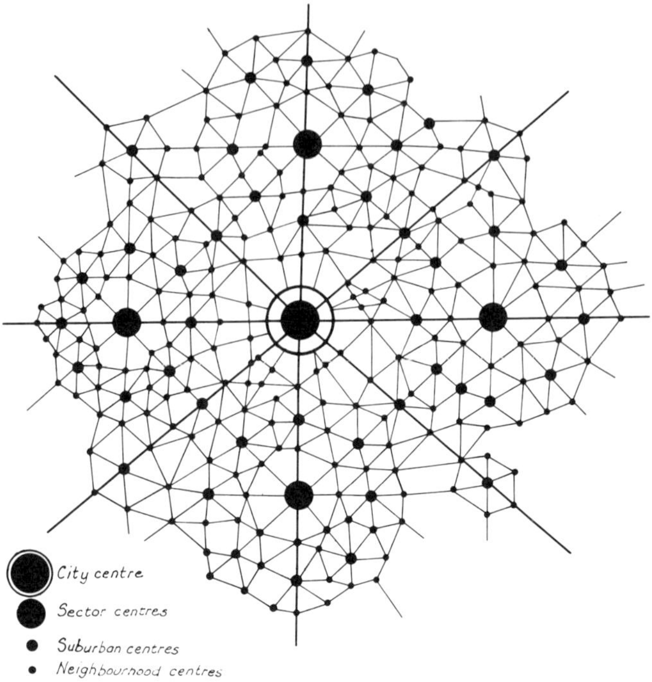

12.1 Diagrammatic representation of a "minimum travel" city (assumed population of 2 million).

this neighbourhood of a few thousand people can be regarded as a substitute for a longer journey; and since neighbourhood journeys are made on foot, by bicycle or, if by car, mainly along local streets, the implications are obvious. Corner shops, local pubs, primary schools, ordinary secondary schools, general practitioners, churches, village halls, social clubs, belong definitely to neighbourhoods and account for many journeys which collectively produce only a modicum of traffic. It is arguable that some other functions should be encouraged to operate at neighbourhood rather than suburban level, for social and transport reasons.

Figure 12.1 gives a diagrammatic picture of a city arranged on the four-tier structure described above. Without going into mathematical niceties, one might expect a land use structure of this general sort to give the opportunity to minimise the volume of travel for a given quantum of activities.[1] Whether the opportunity is fully taken depends upon the quality and management of the transport system. Of course there is no reason to suppose that travel *minimisation* is a desirable objective, even within the constraint of a given volume of activities. Other land use patterns are possible with very different travel implications; but the pattern described gives an indication of the way in which land use planning can help to reduce the demand for transport capacity (of one kind or another) by avoiding needlessly long distances. The fact that old European cities have tended to develop in this kind of form is no accident; it reflects the very real need until modern times to organise urban activities on the assumption that most trips would have to be made on foot supplemented by a few railway lines.

Traffic avoidance, on principles similar to those described above, has played an important part in the strategic land use planning of some cities. An early example was the Copenhagen "finger plan" of 1945, with five fingers stretching away from the city centre along railway lines, although this was probably motivated more by the assumption that most people would not have cars than that they should be encouraged not to use them. Current planning in Copenhagen seems to be departing from the strongly radial pattern of the finger plan by including orbital motorways.

The development plan for Hamburg adopts a similar approach with seven axes, each with a freeway, a regional railway and bus feeder services to the railway. At the end of each axis a major suburban centre is situated, to attract counterflow traffic. Subsidies will be available to force through this development. One planning criterion is that all journeys to work should be possible by public transport in not more than 45 minutes.

Near Amsterdam the satellite town of Bijlmermeer is being built for a population of 110,000, four-fifths of whom will live within 500 metres of Metro stations on two lines going to the city centre and to the major sub-centre of Amstel. The future growth of Stockholm is similarly planned along two suburban railway extensions. In Tokyo, where a vast growth of population is expected, 25 major suburban centres are planned (called New Towns but located within the metropolitan area). It is intended that some of these centres should become specialist centres in some industry or occupation. This should help to attract people to live and work in the same area. Again the centres are planned along railway lines from the city centre, to be connected by an outer ring railway.

LOCAL LAND USE PLANNING

At the local level it is possible to avoid the need for some vehicle journeys by locating some activities within walking distance of as many people as possible. Clearly this is easier in high density residential areas. A large number of trips are made to sundry places like the tobacconist, newspaper shop, local bar, laundrette, doctor and dentist. In some towns activities of this sort are zoned out of residential areas and are forced into locations uncomfortably far from many people's homes. Good location planning can convert many car trips of this sort into walking trips and at the same time make life more convenient for local residents.

The *location of schools* is also of importance. If the distances are rather far for children to walk, a large number of cars tend to appear on the roads during the morning rush hour, with children being taken to school at the inconvenience of their parents. In a detailed study at St. Albans and Stevenage, Levin and Bruce concluded: "If the walking distance from home to school is less than 0.4 miles, it is highly unlikely that under normal circumstances a parent will consider the use of a car to be necessary. Thus the greater the proportion of distance below 0.4 miles, the less a primary school will act as a generator of two-way car traffic—in particular the smaller will be the number of cars parked outside the school at 3:30 P.M."[2] Clearly the traffic generating capacity of a school is critically dependent on the size of its catchment area, which is directly related to the size of the school.

Careful planning of commercial premises can help to reduce goods traffic. Barnstead has noted in Toronto that the generation of many small consignments in the city centre arises from the lack of *storage space* in office buildings and restaurants.[3] He suggests that planning controls might be used to ensure that such premises do contain adequate storage space.

A common suggestion for reducing goods traffic, especially heavy goods traffic is the creation of *road freight terminals* in the suburbs close to motorways, where goods consignments can be broken down or consolidated. In New York, for instance, it has been reported: "Distribution of truck freight in the Central Business District is notoriously inefficient. Over 25,000 trucks ply the 9 square miles of the District each day frequently with overlapping routes and stops. A truck travelling within the area may have a capacity of 8 or 10 tons but only deliver 1 or 1¼ tons of freight each day."[4] The Tri-State Transportation Commission has studied the possibility of setting up freight consolidation terminals and has estimated that 15 percent of truck trips in the central district of Manhattan could thereby be eliminated.[5] A system of this sort is already used by the United Parcels Service and other large organisations. The difficulty arises in getting numerous smaller organisations to form a cooperative system. In Brussels a study of a joint road-rail terminal was carried out by the Société de

Recherche Opérationelle et d'Economie Appliquée, which concluded:

> . . . that if the joint terminal is not necessarily a universal solution to congested urban roads, it can constitute a valid one, and need not necessarily be located outside of the urban areas. . . . These studies convinced us that the solution of maintaining the central location was by far the best. In effect, the peripheral site had the disadvantage of replacing one heavy vehicle by five to ten light vehicles on the roads into the centre, effectively concentrating and increasing traffic. With the central installation, the dispersion in a radial pattern was immediate. . . . It goes without saying that the central solution is not always possible, nor a universal remedy for congestion.[6]

The most important aspect of local land use planning, however, concerns the public transport system. The need to change from one vehicle to another is a great deterrent to the use of public transport. The more people living within walking distance of suburban stations, therefore, the better for traffic avoidance. This is obviously relevant to the siting of new stations and to the location of high rise apartments, as strikingly demonstrated along new underground lines in Toronto where it has been stimulated by a system of bonuses for development within 1500 feet of stations or bus feeder routes. It is also a reason for having entrances at both ends of a station because this can make a considerable difference to walking distances.

The city centre stations of the public transport system affect much larger numbers of people than do suburban stations. In London it has been found that of commuters arriving at rail terminals and with jobs within two-thirds of a mile, 80 percent walk. Massive office development above or adjacent to stations reflects the belief that this will attract, ultimately if not immediately, people to commute by train. Major examples are at Grand Central Station, New York, and Montparnasse Station, Paris. In London similar office development is planned at several main line terminals. In San Francisco it is being encouraged along the new BART lines by bonuses on floor area near to, or with direct access to, stations.

TRANSPORT STRUCTURE DEVELOPMENT

The earlier chapters on traffic restriction and restraint dealt with methods of *managing* a city, with only minor modifications to its physical structure, in order to limit the volume of traffic. In this chapter we are considering changing the structure itself, and we have already discussed the land use structure. This brings us finally to the transport structure which holds together the land uses. How can transport facilities be restructured or extended in the interests of traffic limitation?

There are two complementary approaches to this question: (1) to build new or better facilities which will encourage people to use public transport and to walk or cycle; and (2) *not* to provide facilities that will have the opposite effect. Both these approaches can obviously be carried too far. Cities must decide how far they wish to limit traffic. But in recent years a number of major cities have taken big steps to curtail their roadbuilding plans and to increase investment in public transport. They include San Francisco, Boston, New York, Chicago, Toronto, London, Edinburgh and Munich; and similar policies appear to be emerging in Copenhagen, Stockholm, Zurich and Vienna. In every case the proposition that certain projected roads should *not* be built rests on the belief that, if they are not built, much of the expected traffic growth will not materialize, or to put the matter the other way round, that the new roads would generate traffic growth.

Restricted Roadbuilding

Without doubt the most effective way of limiting the volume of traffic is not to provide roads for it. The London Transportation Study predicted that Plan 3, with 337 miles of motorway, would produce 59 million vehicle-miles of traffic per day, while Plan 1, with 142 miles of motorway, would produce 42 million v-m/day.[7] A decision in favour of Plan 1 rather than Plan 3 would be the most effective possible limitation on traffic, because there is no other feasible way of removing 17 million v-m/day from the roads of the city.

In order to assess a policy of restricted roadbuilding it is important to understand the exact process by which traffic generation occurs. This has been described by the writer elsewhere, as follows:

Only quite recently have many people really begun to appreciate that the reaction to more roads is yet more traffic; and the full implications of this interaction between the amount of roadspace and the amount of traffic are still not adequately realised. New roads alter the pattern of accessibility provided by the network on which they are imposed; that is to say, travel by road becomes faster between certain origins and destinations within a given disposition of activities. The effects of this change may be grouped into short-term, medium-term and long-term effects.

Short-term effects. There are six identifiable effects in the short term.

1. The new roads will relieve those congested streets that lie closely in parallel with them; from the day they open, there will be a straightforward diversion of traffic from the old to the new.

2. Before long other motorists (and other vehicle users) already making trips within the area will find that alternative destinations have become more accessible to them on account of the new roads and they, too, will be encouraged to divert.
3. There will then follow a general redistribution of traffic over a relatively wide area of the street system as road users adjust to the new pattern of accessibility.
4. In practice it is often, though not necessarily, found that the introduction of fast new roads induces road users to make longer journeys, through diversions of route or changes of destination. Longer journeys mean more traffic, of course, for the same number of trips.
5. The improved journey times and reduced journey costs of road transport relative to rail may lead to some diversion from rail to road.
6. The performance of private road transport will sometimes be improved relative to that of buses and this may lead to some diversion from public to private road transport.

Medium-term effects. After a while three further effects will begin to be felt:
1. People will gradually learn that certain destinations are more accessible than before and that various trips formerly not contemplated are worth making, including some by bus. Thus new trips (as distinct from 'diverted' old trips) are generated by the new roads.
2. The enhanced advantage of having a car will accelerate the growth of car-ownership and hence the use of cars instead of public transport.
3. The increased car-ownership itself will stimulate more trips.

In practice the magnitude of all these short- and medium-term effects (together with the growth in demand which would have occurred anyway) is often so great that not long after new roads are opened they become congested and traffic conditions on the other streets are little better than before. For example, the American traffic engineer, K. Moskowitz, stated in 1959 that the worst congestion to be found in California was on the Bay Shore motorway, built only five years previously (Highway Research Board, *Special Report* 56, 1959). It is easy at this point to assume that the rapid growth of traffic is solely the reflection of greater activity and affluence, and that the new roads were built just in time to avert chaos, and that further road construction is necessary to meet the needs of similar growth in the future.

Long-term effects. There are also a number of gradual, but ultimately important, changes set in motion by the improved accessibility provided by new roads:
1. Accessibility to raw materials, labour and markets is an important influence on the location of industry and other sources of employment. Accessibility to employment and other personal activities is an important influence on the individual's choice of home. Improved accessibility leads gradually to greater dispersal of associated activities, involving longer journeys and hence more traffic for a given volume of activities. The most important example of this is the journey to work. New roads induce people (over a long period of time) to live farther from their work or to work farther from their homes.

2. This inducement towards greater dispersal, created by new roads, tends to take people to areas less easily served by public transport and is therefore an added inducement towards higher car-ownership and a further shift away from public transport.
3. The steady erosion of demand for public transport commonly leads to higher fares and a decline in the quality of public transport, manifested by reduced services and financial stringency, which in turn lead to a further shift from public to private transport, thus exacerbating the difficulties on both sides.
4. Since individuals do not make major changes in location (of home or job) frequently, the increasing road congestion and deteriorating condition of public transport gradually "trap" some people in locations which they would not have chosen had they foreseen these unfortunate developments. This leads to discontent and to demands both for better toads, to fulfill the expectations of motorists, *and* for better public transport, to restore the quality of service expected by public transport users.

These are the processes in the short, medium and long term, by which new roads lead to traffic generation and demands for more roads.[8]

Nearly all the urban road projects that have been abandoned recently have been motorway projects. The high speeds afforded by motorways are responsible for much of the traffic generation. Lower speed roads would generate less traffic.

The particular ways in which the restriction of roadbuilding limits traffic growth will obviously vary from one city to another. It is clearly desirable to try to analyse the process and discover what the underlying implications are. If people travel fewer vehicle-miles by car, they must make either fewer trips, shorter trips or more trips by other modes. In practice they are likely to do all these things. Do these differences reflect an inferior use of land and labour, an inferior pattern of activities, and inferior travelling conditions? If so, to what extent is the life of the town adversely affected and is it an acceptable price to pay for not having additional roads and traffic?

Subsidised Investment in Public Transport

In the earlier section on subsidised fares the point was made that whereas lower income groups tend to want lower fares, potential motorists are more concerned about higher quality. A subsidy to raise the quality of service is therefore likely to be a more effective means of attracting the motorist than a subsidy to reduce fares.

These alternative uses of a subsidy are, however, not comparable from an economic point of view. A fare subsidy in effect transfers money from taxpayers to passengers, who may be largely the same people anyway; it is a *transfer payment*, not a real cost to the community except insofar as the operating costs of the transport system may be increased, e.g. if more buses are needed to meet the demand generated by the reduction in fares. An investment subsidy is a *real* cost, involving the consumption of scarce resources. From a social point of view the two subsidies are totally different and should not be regarded as alternatives.

These considerations underline the fact that investment subsidies are a costly means of avoiding traffic. A higher quality of service has to be provided for all the existing passengers, who as a group are not willing to pay more for it, in order to attract new passengers, who are not willing to accept the existing quality; and the latter are usually a small fraction of the former.

One answer to this dilemma is to provide two classes, although this is not always popular. Buses with first, second, third and fourth class compartments may be acceptable in India but not in Europe or North America. On the other hand, Europeans and Americans seem willing to accept *separate* buses of different classes: the standard bus with one scale of charges and the luxury, bookable, air-conditioned bus with another. On the railways, first and second class survive on the Paris Metro and on the longer London suburban services. Public attitudes towards class differentiation are mixed and confused; yet this lies at the heart of the subsidy problem. Higher income people want better service, are prepared to pay for it and will move over to private cars if necessary in order to get it; lower income people obviously like higher quality too but are not able to afford it. In principle the most effective solution should be to provide two classes of service, subsidizing quality in one and fares in the other.

This is only possible up to a point. Fares are divisible between classes, but quality in some respects is not. The most important aspects of quality are reliability and frequency, which are common to all passengers. If potential motorists are to be *attracted* away from their cars, public transport must be frequent and reliable, and stations must be clean and pleasant; and these qualities must necessarily be provided for all passengers, not just those with the option of using a car. The additional comforts of first class travel—mainly space—can be provided for those who are willing to pay more.

Hence subsidies for traffic restraint are inextricably bound up with subsidies for social welfare. New York is a striking example. Only 8 percent of commuters to Manhattan come by car, yet this is enough to cause severe congestion. The majority of commuters come by transit, the quality of which leaves much to be desired. In order to attract a quarter of car commuters, i.e. 2 percent of all commuters, from automobile to transit, thus solving the traffic problem, the quality of the whole transit system would need to be improved. This would mean new rolling stock, modernisation of stations, and new lines to relieve overcrowding. The expense would be enormous in relation to the 2 percent of commuters expected to be attracted to the transit system, or even to the 6 percent of commuters remaining in their cars. The great majority of beneficiaries would obviously be those who were already using transit.

The construction of superior systems of public transport, then, is a costly method of traffic limitation, yet this is the most widely practised method at present. In at least 40 cities throughout the world, urban railway systems are being built and in practically every case they are expected to make a financial loss which governments have agreed to bear. The reason for this universal willingness to subsidise urban railways is, of course, traffic avoidance, which is sometimes seen by governments as a way of avoiding even more costly and unpopular road construction. As long ago as 1956 the local authorities in San Francisco formally opposed the construction of the proposed Western Freeway and expressed their belief that "mass transit held a promise of solving the problems of crowded thoroughfares,"[9] a belief which took 16 years to be put to the test. In 1962 the first serious attempt to prove that the subsidy was justified in terms of traffic avoidance was made by Foster and Beesley in their study of the Victoria Line in London; nearly half the predicted net operating benefits of the line were attributed to reductions in congestion costs.[10]

The importance of railways as a means of avoiding road congestion has been described elsewhere by the writer as follows:

> It is difficult to compare the severity of congestion between one city and another, but we believe that the worst conditions have tended to arise in cities which have no rail services, i.e. no alternative to roads. The acute difficulty in cities without railway systems arises not because of the lack of capacity of the roads to carry the required movement of *people* and goods, but because the lack of market prices for costly road and parking space removes the one big advantage that buses have over cars, namely their economical use of space. If the only alternative to the car is the bus, then no matter how slow the car becomes through congestion, the bus will always be slower, as long as they use the same roads. The effectiveness of new roads in relieving the congestion is strictly limited by the enormous power of traffic generation which exists in any large

dense city with even quite low car ownership. The only ways of really relieving the situation in such cities are to make the use of cars much more expensive than buses, to reserve parts of the road system for the buses, and to build a rail system. We know of no cities where the first two methods have been adopted on any scale, but the building of a rail system has appealed to many cities including American cities with very high levels of car ownership.

It is essentially the rail system that provides the relief valve for congestion on the roads. The worse the quality of the rail system, with its bus linkages, the worse will be the level of road congestion. The basic reason is the almost limitless potentiality of traffic generation in a large, dense city of high car ownership.[11]

What then of buses? If bus services become sufficiently bad, some people will buy cars and use them. But people who already own a car will not normally use a bus, no matter how good, unless (1) there is a penalty on using the car, such as a parking fee or a special tax, or (2) the bus is enabled to give a quality of service competitive with that of a car. The possibilities of bus priorities were discussed earlier. A more radical approach is the construction of *busways*, or roads for the use of buses—and possibly taxis—only.

There are two kinds of busway. The first is demonstrated by the Shirley Highway in the Washington, D.C., area. These busways are really bus lanes on motorways. They *may* enable buses to overtake cars at the access point, if there is a queue there, and they may have special provision for stopping along the motorway. They make use of the motorway in order to provide a rapid transit service that falls not far short of the service given by a car.

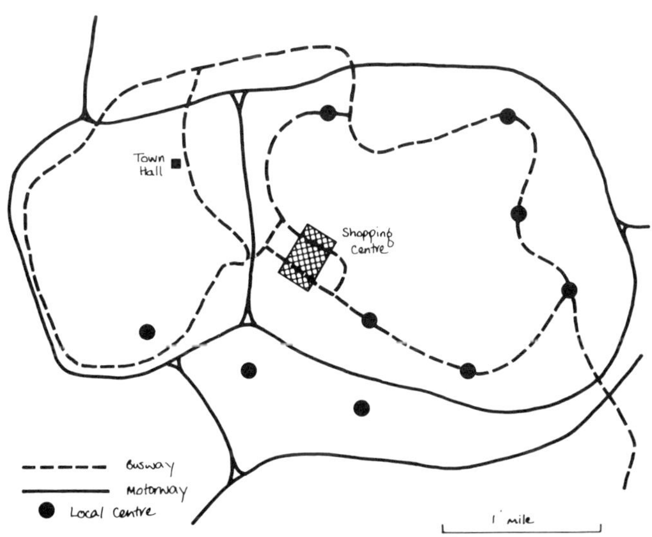

12.2 Runcorn busway.

The other kind of busway aims to give buses a distinct advantage over the car; it by-passes congestion and takes passengers closer to their destination than is possible by car. The pioneering example of such a busway is at Runcorn, England. The full Runcorn busway pattern forms a figure-eight, about 10 or 11 miles long, passing through the centre of residential areas and taking passengers right into the new town centre which is not directly accessible by car (figure 12.2). The busway, of which 75 percent of the cost is paid by the central government, is an integral part of a transport plan which aims to achieve a 50:50 split between public and private transport. The residential, industrial and shopping areas have been planned together with the busway and car parks so that the average walking distance will be no greater for bus passengers than for motorists. Runcorn busway is a bold example of traffic avoidance.

13 EFFECTIVENESS

In the previous chapters, in order to avoid repetition, little mention was made of the effectiveness of the various methods of traffic limitation. Up to a point they are all pushing against the same force of self-interest in the use of private motor vehicles. How strong is this force and how easily can it be resisted in the public interest? In answer to this question we shall simply point to a certain amount of scattered evidence.

EFFECT OF CONGESTION

Although congestion is an inefficient way of restraining traffic, it is instructive, because if one can restrain traffic effectively by congestion, one should be able to do so by more efficient means. The effectiveness of congestion as a restraint was noticed long ago in city centres. It was observed that in most large cities congestion tended to settle at a similar level, irrespective of big differences in population, car ownership or road capacity. Predictions that traffic would "grind to a halt" were not borne out, despite rapid growth of the vehicle population unmatched in most cases by any great expansion of the road network. Traffic speeds tended to fall to about 10 miles per hour at peak periods, as indicated in table 13.1. In London, where good records exist, the volume of traffic entering the city centre in peak hours reached a ceiling in 1962 and has remained fairly stable since. The same trend must have occurred in other cities, otherwise speeds would have fallen lower. There is no way of knowing to what extent this restraint of traffic growth was assisted by other factors such as shortage of parking space. But the magnitude of the restraint is indicated by Professor Smeed's calculations in London, where he concluded: "that to meet the potential demand for traffic capacity in Central London with a short peak traffic period might well necessitate an increase of road capacity to at least 20 times its present value."[1]

Similar conclusions could be reached in other large cities. Table 13.2 shows the percentage of car commuting reported in some major city centres. It is evident that in all these cities a

Table 13.1 Average Traffic Speeds Recorded in City Centres during Peak Hours (mi/hr)

City	Average speed
London	10
Paris	10
Calcutta	7–10
Brisbane	10
Salt Lake City	18
New York	10
Athens	9–12
Manila	7
Leeds	11
Glasgow	9
Birmingham	13
Leicester	11

Note: The above figures are neither reliable nor strictly comparable, owing to data deficiencies and different definitions of "city centre" and "peak hours," but they give a general picture.

Table 13.2 Percentages of Car Commuters in City Centres

City	Percentage of car commuters
Tokyo	5
New York	8
London	10
Sydney	13
Chicago	15
Paris	20
Manchester	22
Toronto	37

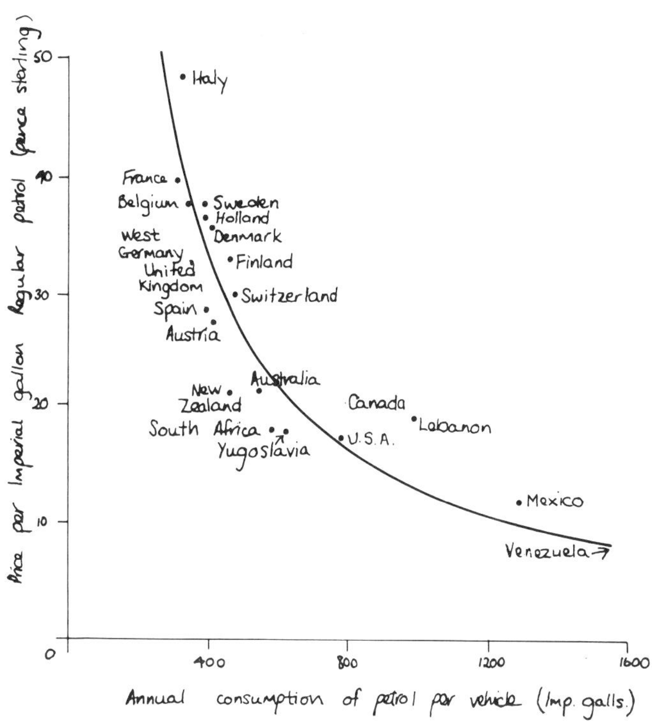

13.1 Relationship between price of petrol and annual vehicle consumption. Note: Prices are for regular grade as at February 1971 expressed in sterling at July 1972 exchange rates. The Venezuelan price is unreliable. Consumption is for 1971 and includes consumption by motorcycles and other petrol uses in addition to cars and commercial vehicles. Vehicle registrations used are for 1970 and include cars plus two-thirds the number of commercial vehicles (including buses); the reason for this procedure is that no breakdown between diesel and petrol vehicles was available except for the United Kingdom, from which the factor of two-thirds was derived. Japan was omitted because an exceptionally high proportion of vehicles are registered as commercial, with no indication of how many are petrol driven.

very much larger number of people could and would use cars if fast roads and cheap parking space were available. All these centres are prosperous and increasingly in demand as centres of employment, trade, tourism and other activities. It appears therefore that, in certain circumstances, *heavy limitation* of traffic is possible without having obviously harmful effects. It should be noted, however, that in none of the cities mentioned has the volume of traffic been *reduced*; it has only been restrained from growing. Whether it is practicable to stop people from using cars after they have become accustomed to doing so is another matter. In Toronto, for instance, there are plans to reduce the commuting percentage from 37 percent to 15 percent, and in San Francisco and Boston there are also hopes of persuading established car commuters to transfer to much improved public transport.

EFFECT OF FUEL TAX

There is evidence that the price of motor fuel has a considerable effect upon consumption. In figure 13.1 the average consumption of petrol per vehicle is plotted for a number of countries in which the price of petrol varies greatly on account of tax. These figures do not reveal the effect, if any, of fuel prices on car ownership. Nor do they reveal to what extent lower consumption reflects smaller use of cars rather than use of

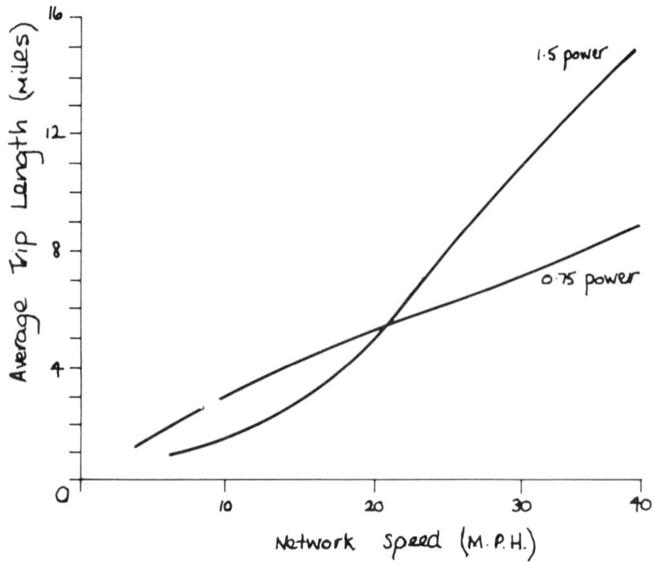

13.2 Relationship between network speed and trip length. Note: The position of the curves is taken arbitrarily from a point at 5 miles and 20 m.p.h. In a particular case the position would be determined mainly by the population and form of the city.

smaller cars. But they do suggest a strong relationship between price and consumption. The curve fitted to the data has a price elasticity of unity, i.e. for a given percentage change in price there is an equal and opposite percentage change in consumption.

EFFECT OF SPEED

Numerous studies have shown that higher traffic speeds lead to longer trip distances ahd hence larger volumes of traffic. According to Voorhees and Bellomo, "Various simulation studies of the work trip have demonstrated that trip length in miles for a constant urban form is proportional to about the 0.75 power of the change in network speed."[2] But in a study of 5 cities they show that historically changes in speed appear to have caused trip length increases to the 1.5 power. "The difference between the 0.75 power and the 1.5 power must reflect the amount of change that was due to the change in the urban structure, where the cities tended to spread out." These relationships between network speed and work trip lengths are shown in figures 13.2. The position of the curves, as distinct from their shape, will depend mainly on the size and form of the urban area.

In another study of 8 European and American cities of similar population, Bieber and Jorry found that the network speed in the European cities was about half that in the American cities and the average distance to the city centre was also about half.[3]

Another example of the same phenomenon is taken from the Los Angeles Regional Transportation Study. Putting forward proposals for new freeways, and predicting their traffic effects, the Study says: "It was also implied that people would be willing to travel, in a given length of time, a greater distance in 1980 than in 1960 because of higher speeds made possible by more miles of improved facilities. The results showed that the overall average auto-driver trip length in 1960 was 6.0 miles . . . versus 7.9 miles in 1980" (p. 28).

EFFECT OF STRATEGIC LAND USE PLANNING

The work by Voorhees (*et al.*) throws light on the relationship between urban form and trip length.[4] Naturally there is a strong tendency for trip lengths to increase with the size of city: the data collected by Voorhees from 23 American cities showed that the average length of work trips by car ranges from 2.0 miles in Hutchinson, with a population of only 38,000 to 8.7 miles in Los Angeles with 6.5 million. More interesting are the comparisons between old and new cities of similar size. Pittsburgh (1.8mn) and Baltimore (1.4mn), for instance, have averages of 4.2 miles and 7.0 miles respectively; and Voorhees makes the following comment about New Orleans (845,000):

> New Orleans, one of the oldest and most compact of the cities listed, has an average trip length and duration typical of that normally found in newer cities only one-tenth its size. The length and duration in the spread-out city of Fort Worth (503,000) is more than twice that of New Orleans—8.1 miles and 15.7 minutes against 2.5 miles and 7.4 minutes—while the latter is somewhat larger in terms of population (p. 2).

An even more striking comparison can be made between London (8.8mn), where the average work trip by car is 4.5 miles long, and the American city in Voorhees's list which has the same average trip length: it is Pensacola, a town of 128,000 people.

EFFECT OF LOCAL LAND USE PLANNING

A proposal for road freight terminals in the New York metropolitan region was put forward by the Tri-State Transportation Commission in 1969 and summarised by Wilfred Owen as follows:

> Some 430,000 trucks perform local trucking services and generate 11 percent of all vehicle-miles driven in the (New York Metropolitan) area. One-third of these trucks do not move at all on any given weekday, and of those moving about half carry no load. The proposed solution is to divide the region into centers of truck activity averaging about 3 square miles each, where off-street sorting terminals would combine small

Table 13.3 Effects of Price Changes on Usage of South Bank Car Park

	Number of parkers		
	Before	After	Change (%)
Arrivals			
07.00–10.00	560	177	-68
10.00–18.00	172	348	+102
Durations			
less than 3 hours	73	217	+198
3–6 hours	55	95	+73
more than 6 hours	523	139	-73
Purpose			
work	526	178	-66
employer's business	38	94	+148
other	64	154	+140

shipments for delivery to other terminals. By consolidating shipments into 12-ton loads, it is believed that transport between centres could be reduced by 90 percent.[5]

EFFECT OF PARKING CHARGES

In spite of the importance of parking charges as a traffic restraint, little systematic study of their effectiveness seems to have been made. In a survey by the writer the effect of different charges at parking meters in Central London was investigated.[6] At that time, in 1966, most meters in the area had a 2-hour limit, and the charge was either 6 pence per hour, 6 pence per half-hour or 6 pence per quarter-hour. The survey covered about 5500 motorists divided almost equally between the three varieties of meter. The average length of stay was as follows:

6 pence per hour: 1.6 hours

6 pence per half-hour: 1.2 hours

6 pence per quarter-hour: 1.1 hours

An experiment by the Greater London Council in 1971 produced striking evidence of the reaction of motorists to off-street parking charges.[7] The charges were altered from a flat 15 pence per day to an hourly charge of 5 pence per hour at South Bank car park, which holds 662 cars and was mainly used by commuters. Similar changes were made at three smaller car parks. Before and after surveys were made at these car parks and at 10 others nearby which might be used as alternatives. The results given in table 13.3 show that the number of commuters using South Bank park fell by about 70 percent. Peak hour arrivals at the four affected car parks fell from 628 to 121; at the same time the other 10 car parks showed an increase from 331 to 352.

In a sociological study in Newcastle-under-Lyme, C. S. Riley found motorists insensitive to the cost of their annual licence or depreciation. But, "it is extremely surprising to find frequent mention of parking fees as a disincentive to car ownership. This suggests an unexpected evaluation of car-owning costs by the car-owner in which the owner is more sensitive to small costs imposed subsequent to ownership than to the major and unavoidable costs of motoring."[8]

EFFECT OF ROAD PRICING

A number of attempts have been made to predict the effects of road pricing schemes. They all encountered the difficulty that, in the absence of practical experiments, there is no direct evidence of motorists reactions to these new kinds of taxation. The first studies, by Thomson and Tanner, simply considered a range of demand elasticities between -½ and -2. Studies in South America for Bogota and Caracas have had still less information. The Bogota study estimated that road pricing (in 1980) would reduce the number of car trips by about 23 percent, if prices were set equal to marginal congestion costs.[9] These South American studies are rather hypothetical and do not reflect any intention to introduce road pricing in the two cities mentioned.

The only road pricing study to investigate the magnitude of demand elasticity appears to be that published by the writer in 1967.[10] Four methods were used to obtain an approximate estimate of motorists' demand elasticity. These were based on statistics describing four observations where there was clear evidence of traffic restraint from congestion and/or parking charges. Briefly, the four observations were:

1. In Central London the volume of car traffic between 1947 and 1964 increased at only half the rate experienced in other parts of London and other urban areas.
2. It was discovered by Tanner that households of similar income in city, town and rural areas incurred a similar outlay of time and money on travel, if one valued their time at 5 shillings an hour, despite big differences in situation and transport facilities.[11]
3. The London Traffic Survey showed that car-owning households used private transport for 71.3 percent of their journeys to the non-central area but for only 38.9 percent of their journeys to the central area.
4. The LTS also showed that there was a 45 percent greater probability that a car-owner working in Central London would

Table 13.4 Comparative Effects of Daily Licence and Parking Tax (average effects on main survey roads of Central London)

	Peak hours			Off-peak hours		
	Without restraint	With 6s. daily licence	With 9d. hourly parking tax	Without restraint	With 6 s. daily licence	With 9d. hourly parking tax
Flow (veh/hr)						
Cars	1,201	633	854	863	426	761
Taxis	208	298	283	295	295	245
Cycles	432	261	267	70	42	57
Commercial vehicles	369	428	398	610	676	626
Buses	139	193	175	100	100	100
Total	2,349	1,803	1,977	1,866	1,539	1,789
Total p.c.u's/hr	2,522	2,225	2,348	2,325	2,049	2,264
Average speed (miles/hr)	9.1	11.4	10.4	10.4	12.7	11.0
Net benefits per mile-hr (d.)	–	11,434	6,766	–	3,534	1,739

go by car if he travelled there before the morning peak than if he travelled there during the peak.

Analysis of these situations produced elasticity estimates (with respect to time plus money costs) between -1.25 and -1.49, suggesting that in London motorists could be restrained fairly readily by means of additional taxes. An elasticity of -0.8 was estimated for goods traffic, suggesting that it would not be so easily restrained. With the aid of these estimates, the effects of a parking tax and a daily licence system (for cars and motor-cycles only) were preducted as shown in table 13.4. (The parking tax was assumed to be paid by *all* parked cars and motor cyclists.)

EFFECT OF LOW FARES

Baum (op cit) quotes a study by V. Kindt, who uses regression analysis to estimate the price elasticity of public transport in 7 German towns. These estimates, with others from Holland, United Kingdom and the United States, are reproduced in table 13.5. They suggest elasticities between -0.3 and -0.4 in Europe, and between -0.15 and 0.30 in the United States. Unfortunately this potential response to cheaper fares does not necessarily signify a transfer of passengers from private to public transport but is more likely to represent increased travel by non-car owners. Moses and Williamson estimated in 1963 that free fares in Chicago would attract only 13 percent of car commuters from their cars.[12]

Table 13.5 Price Elasticity of Public Transport

Germany		U.S.A.	
Aachen	-0.32	San Francisco	-0.18
Bremen	-0.30	New York	-0.18
Duisburg	-0.32	Boston	-0.19
Essen	-0.27	Portland	-0.28
Hamburg	-0.41	Salt Lake City	-0.12
Oldenburg	-0.33	Connecticut	-0.28
Wiesbaden	-0.30	Atlanta	-0.28
		Cincinnati	-0.24
Holland			
		Philadelphia	-0.14
The Hague	-0.4		
		Baltimore	-0.08
Utrecht	-0.6		
		Chicago	-0.30
U.K.			
London	-0.4		

Sources:
Germany: V. Kindt, University of Hamburg, 1968
U.S.A.: Charles River Associates, Inc., "An Evaluation of Free Transit Service," DOT, Washington, 1968.
Holland: J. Meyer, "Theorie und Praxis der Stadtverkehrswirtschaft." Internationalen Kongress des Internationalen Verins fuer oeffentliches Verkehrswesen, UITP, Barcelona 1967.
U.K.: London Transport Executive.

EFFECT OF BUS PRIORITIES

The Camden Bus Lane Study (referred to in chapter 11) estimated that:

> If the bus segregation schemes were restricted to the Camden study area the effect would be to reduce other vehicle traffic by 6.7 percent from 1673 p.c.u./hr. to 1560 p.c.u./hour. If such schemes were introduced to the whole of the Central Area, other traffic would be reduced by 10.2 percent. Some of this lost traffic would represent . . . trips no longer made or possibly made to an alternative destination outside the bus segregation area. However, the majority of the lost traffic will be the short distance traffic which in terms of percentage journey times is most affected by the increased traffic congestion. Much of this traffic will undoubtedly divert to the improved bus service in the area, although other trips will be made on foot instead of by car (p. 37).

EFFECT OF PARKING RESTRICTION

In the 1966 survey carried out by the writer in Central London the following question was put to 4,600 motorists parking in the area: "If it was impossible to find a parking place in Central London would you (a) use public transport, (b) use a taxi, (c) walk or bicycle, or (d) not make the journey?."[13] Over 95 percent of the replies were usable but some motorists replied that their journey was essential and could only be

Table 13.6 Reaction to Restraint

	Percentage
Would use public transport	62
Would use taxi	16
Would walk or bicycle	5
Would not make the journey	13
Car essential	4
Total	100

Table 13.7 Primrose Hill Environmental Area: Traffic before and after Restrictions

Morning peak hour	Before	After	Change (%)
Primrose Hill (A) (veh/hour)	1,000	1,200	+20
Primrose Hill (B)	1,180	1,200	+ 2
Regents Park Road (C)	1,450	1,400	- 3
Regents Park Road (D)	1,550	1,620	+ 5
Regents Park Road (E)	1,150	1,150	–
Gloucester Avenue (F)	500	170	-66
Gloucester Avenue (G)	1,660	1,320	-20
Regents Park Road (H)	570	180	-68
Regents Park Road (I)	570	250	-56
Regents Park Road (J)	570	130	-77
King Henry's Road (K)	320	100	-69
Adelaide Road (M)	1,250	1,150	- 8
People experiencing an increase in traffic in their road		332	
People experiencing a decrease in traffic in their road		1,593	

13.3 Primrose Hill environmental area.

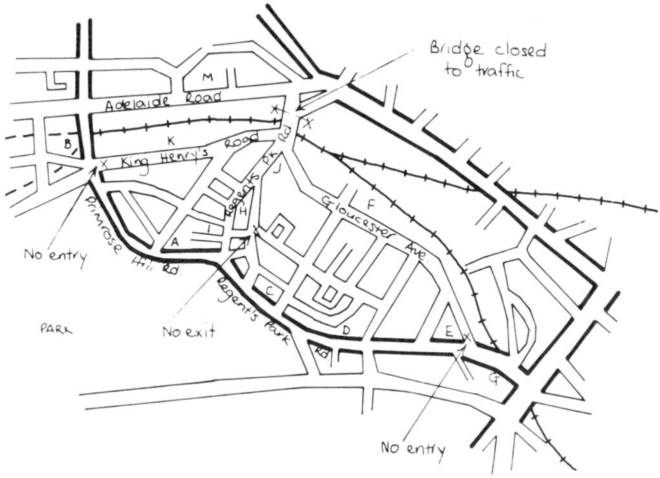

made by car; they were classed as "car essential." Others gave a combination of modes (a), (b) and (c) and were allocated between the modes mentioned. The adjusted results, given in table 13.6 showed that 62 percent said that they would use public transport.

EFFECT OF TRAFFIC RESTRICTION

A scheme to reduce traffic passing through a small residential and shopping area in North London, called the Primrose Hill Environmental Area, was introduced in April 1971 and its effects were surveyed in the summer of 1972. The purpose of the scheme was to prohibit traffic from entering certain streets (figure 13.3), thus forcing it to use the main road along the boundary of the environmental area (Regents Park Road—Primrose Hill Road). It was recognised that initially the scheme would cause more congestion on the main road, and it was hoped that this might discourage a few commuters from using their cars; but obviously one small scheme by itself could not achieve much restraint of that kind. The real purpose was to shift traffic from minor roads to the major road. It was argued that the *relative* reduction of traffic on the minor roads would be large and would create a significant environmental improvement for a lot of people, whereas the *relative* increase on the main road would be small and would create an insignificant deterioration for a smaller number of people there. The results in the morning peak hour are given in table 13.7.

EFFECT OF RESTRICTED ROADBUILDING

All large cities in Britain and the United States have undertaken land use-transportation studies and have drawn up long-term highway plans (which in some cases are still under dis-

Table 13.8 Comparisons between British and American Cities

City	Forecast year	Population (million)	Density (persons per hectare)	Car ownership (cars per person)	Utilisation (trips per day per car)
London	1991	8.1	33.2	0.36	3.47
Birmingham	1981	2.7	27.4	0.32	4.09
Manchester	1981	2.7	25.8	0.31	4.88
Liverpool	1991	2.3	24.3	0.40	3.06
Glasgow	1990	2.2	18.5	0.34	4.08
Average in survey year		3.5	28.5	0.13	4.77
Average in forecast year		3.6	27.00	0.35	3.75
New York	1985	20.8	22.0	0.34	4.27
Chicago	1980	7.8	24.3	0.33	5.48
Baltimore	1980	2.2	9.7	0.34	5.06
Seattle	1985	2.4	9.3	0.47	4.90
Average in survey year		6.1	14.0	0.28	4.76
Average in forecast year		8.3	18.9	0.35	4.65

cussion). Martin and Neffendorf have contrasted the planning approaches in the two countries, as represented by the five largest British cities and four large American cities.[14] The British cities are all planning for restraint; i.e. they do not intend to provide sufficient road capacity to accommodate the full potential use of cars. The American cities are not planning for restraint except inevitably in the city centres of New York and Chicago. Table 13.8, extracted from the paper by Martin and Neffendorf, shows that in the survey years (mostly between 1961 and 1966) the average utilisation of cars was almost exactly the same in the British and American cities, at 4.77 and 4.76 trips per day respectively. But the forecasts for the future are significantly different. The American cities are forecasting a utilisation of 4.65 trips per day, slightly less than before. The British cities, with exactly the same expected rate of car ownership, are planning for a utilisation of only 3.75 trips per day, a reduction of over 21 percent due to restraint. The lowest planned utilisation is in Liverpool, at 3.06 trips per day, which compares with reported rates of 5.80 in neighbouring Manchester (1966) and 5.81 in Los Angeles (1960). The differences between these figures implies big differences in the volume of traffic. Martin and Neffendorf do not show the forecast trip lengths, but they would undoubtedly be much longer in the American than in the British cities. Hence the expected volume of traffic caused by an average car in Liverpool would be far less than—probably only a quarter of—that in Los Angeles.

Martin and Neffendorf make the following comment:

There is no definitive way of assessing how low car usage can become before it is both socially and economically undesirable. What is clear, however, is that if car ownership reaches the level forecast, and if car usage is to be held at about 80% of the level experienced today, then public transport services must be maintained and improved if the planned road systems are to provide more than congested levels of service.

It has been similarly found in Sweden that in Stockholm cars make only about 3 trips per day on average whereas in smaller towns they make about 6 trips per day.

14 CONCLUSION

This paper has described many ways of limiting traffic. Some are more effective than others, all have drawbacks. Nevertheless they comprise a formidable armoury for protecting urban areas from excessive traffic. A city that wishes to reduce the volume of traffic, or its rate of growth, is certainly not without means of doing so. But first, as was pointed out at the beginning, it should decide exactly why it wishes to do so, and to what extent. A clear and precise specification of objectives will help the transport planner to select the most appropriate combination of methods. No single method is likely to suffice; a combination of several methods will be necessary.

In short, a city that is unwilling to allow and provide for free, unrestrained motorisation needs a clear alternative in a traffic limitation policy. This should form part of the wider transport policy of the city and should be complementary with its roadbuilding, traffic management and public transport policies.

It would be wrong to suppose that a traffic limitation policy was opposed to or incompatible with roadbuilding. On the contrary, much of the value of new roads depends on being able to limit the amount of traffic that they generate and, in the case of relief roads, ensuring that the relief they give to other roads is permanent.

Thirty-five methods are discussed in this paper. In chapter 9 the five *functions* of traffic limitation were distinguished: to limit traffic *generally*, by *area*, by *streets*, by *vehicle type* and by *vehicle purpose*. In table 14.1 the various methods are listed and classified according to the functions they can serve.

Most towns and cities are already using some of these methods in a deliberate attempt to limit traffic. Some are merely installing a few parking meters, others are pouring enormous sums into the construction of railways. But few cities as yet have produced a comprehensive policy of traffic limitation. All cities have predicted future trends in vehicle ownership and have drawn up plans to build more roads, but

Table 14.1 Methods of Traffic Limitation and Their Functions

Method	General	Area	Street	Vehicle type	Vehicle purpose
Traffic restriction					
Road closure			X		
Partial road closure		X	X		
Prohibition		X	X	X	
Area permits		X			X
Vehicle regulations	X			X	
Route restriction		X		X	
Driver licensing	X			X	
Traffic restraint					
Congestion		X			
Planned congestion		X	X		
Low design speeds	X	X	X		
Bus priorities	X	X	X		
Cycle lanes	X	X	X		
Pedestrian facilities		X	X		
Environmental management	X	X	X		
Fuel tax	X				
Mileage taxes	X			X	
Area licensing		X		X	
Tolls	X	X		X	
Vehicle metering	X	X	X	X	
Restriction of parking space		X		X	
Parking controls		X		X	X
Parking charges		X		X	
Structure of parking charges		X			X
Vehicle registration fees	X				
Concessionary registration fees	X			X	
Import tax on vehicles	X				
Purchase tax on vehicles	X				
Hire purchase restrictions	X				
Subsidised fares	X				
Two part tariffs	X				
Community transport centres	X				
Traffic avoidance					
Strategic land use planning	X	X			
Local land use planning		X	X		
Restricted roadbuilding	X				
Subsidised investment in public transport	X				

most of the larger cities—and many smaller ones too—know that these plans are not enough; without additional measures to control or accommodate the use of vehicles there will be no escape from the congestion, noise, pollution, accidents and sheer unpleasantness of traffic dominated streets, nor from the deeper effects of excessive motorisation on the physical and social structure of the city. The threat of excessive motorisation has to be met at the local and strategic levels. At the local level much has been done to protect small town centres from traffic, but the rest of the town has usually received little attention. In large cities where the problems are more complex, little effective improvement has been achieved anywhere; the most that can be said of even forward thinking cities like London, New York, Toronto, San Francisco and Hamburg is that in some places they may have prevented conditions from becoming worse. At the strategic level many cities are modernising and expanding their public transport systems, but only a few cities are trying to influence the development of trip lengths and trip patterns.

There is therefore a gap, which most cities would admit, between what they would like to achieve in the way of better traffic and transport conditions, and what their current plans are likely to achieve. In principle, this gap could be closed by a comprehensive traffic limitation policy. Whether this is politically and socially acceptable is obviously a matter for each city to decide. But the basic conclusion of this paper is that, for those who wish to limit traffic and who know where, when, what sort of traffic and by how much, there are many methods that could be used; and although no single method is likely to suffice, there is reason to believe that a careful combination of several methods might prove effective.

NOTES TO PART III

Notes to chapter 9

1. A more detailed analysis is given in an article by the author, "Halfway to a Motorized Society," Lloyds Bank Review, October 1971.

2. M. Beckmann, C. B. McGuire and C. B. Winsten, "Studies in the Economics of Transportation," Yale University Press, 1956, Ch. 4. A. A. Walters, "Truck Costs and Motor Taxation," Journal of Industrial Economics, April, 1954. G. J. Roth, "A Self-financing Road System," Institute of Economic Affairs, 1966. J. M. W. Stewart, "A Pricing System for Roads," Oliver and Boyd, 1965. W. Vickey, "Statement to the Joint Committee on Washington Metropolitan Problems," November 1959. C. D. Foster, "The Transport Problem," Blackie & Son, 1963. Other writers include Beesley, Fitch, Reynolds, Neutze, Tanner, Mohring and Thomson.

3. E.g. see J. M. Thomson, "Speeds and Flows of Traffic in Central London," Traffic Engineering and Control, April, 1967.

4. F. V. Webster and R. H. Oldfield, "A Theoretical Study of Bus and Car Travel in Central London," Transport and Road Research Laboratory, LR451, 1971.

Notes to chapter 10

1. "Windsor Town Bridge; proposed permanent closure to vehicular traffic," Report on Public Inquiry, Royal Borough of New Windsor, January 1972.

2. C. K. Orski, "Vehicle-free Zones in City Centers," OECD, 1971.

3. J. Kühnemann and R. Witherspoon, "Traffic-free Zones in German Cities," OECD, 1972.

4. L. P. Spitz, Executive Director of the American Association of Motor Vehicle Administrators, "The Driver Licence Administrator," Traffic Quarterly, October 1971, p. 577.

Notes to chapter 11

1. Ministry of Transport, "Better Use of Town Roads," H.M.S.O. London 1967, p. 3.

2. Greater London Council, "Traffic and the Environment," July 1972, p. 28.

3. "Bursledon Road, Bitterne Road, Northam Road A3024 Traffic Management with Bus Priority Scheme," Project Description, Southampton City Council, 31st July 1970.

4. "Transport Strategy in London," London Amenity and Transport Association, 1970, pp. 211, 212, and 218.

5. T. Aldous, "The Times," 7th August 1972.

6. "Buses in Camden," London Borough of Camden, 1971.

7. For review see B. van Bilderbeck, "Priority Treatment of Buses at Traffic Signals," Technological University, Delft, February 1969 (for OECD).

8. "Transport Strategy in London," LATA, 1970, p. 220.

9. A nearside lane width of 14 feet is recommended by the British Ministry of Transport, "Roads in Urban Areas," 1966, p. 18.

10. Van Ginkel Associates, "Movement in Midtown: New York City," Ekistics, January 1972, pp. 52-58.

11. Urban Traffic Engineering Techniques, Ministry of Transport, London 1965, p. 45.

12. J. M. Thomson, "The Value of Traffic Management," Journal of Transport Economics and Policy, January, 1968.

13. According to unpublished statistical analysis carried out by the writer in 1965.

14. "Road Pricing: the Economic and Technical Possibilities," Report of a Panel under Charimanship of R. J. Smeed, Ministry of Transport, London 1964, p. 9.

15. Lyle, Fitch, Chapter 12 of "Financing Highways," U.S. Tax Institute, 1957, p. 141.

16. "Better Use of Town Roads," Ministry of Transport, London, 1967, p. 23.

17. J. M. Thomson, "An Evaluation of Two Proposals for Traffic Restraint in Central London," Journal of the Royal Statistical Society, September 1967. Summarised in "Better Use of Town Roads."

18. Greater London Council, "Traffic and the Environment," July 1972, p. 16.

19. A. A. Walters, "The Economics of Road User Charges," I.B.R.D., 1968, p. 20.

20. J. M. Thomson, "Some Characteristics of Motorists in Central London," Greater London Paper No. 13, L.S.E., 1968, p. 37.

21. V. Gruen, "The Heart of Our Cities," 1964 p. 91.

22. J. M. Thomson, "Some Characteristics," p. 19.

23. 39th American Assembly, "The Future of American Transportation," 1971, p. 179.

24. J. Kühnemann and R. Witherspoon, "Traffic-free Zones in German Cities," O.E.C.D., 1972, p. 15.

25. J. M. Thomson, "Some Characteristics," p. 29.

26. J. M. Thomson, "An Evaluation of Two Proposals for Traffic Restraint in Central London," Journal of Royal Statistical Society, September 1967, p. 366.

27. Prices as of 1972.

28. J. Armour, "Parking Objectives in Glasgow," British Parking Association Seminar, May 1971.

29. "Leicester Today and Tomorrow," Leicester City Planning Department, 1968, p. 67.

30. See G. J. Roth, "Paying for Parking," Institute of Economic Affairs, 1965, pp. 57-60.

31. H. J. Baum, "Some Aspects of Fares-free Public Transport," Journal of Transport Economics and Policy, January 1973.

32. R. Sherman, "Club Subscriptions for Public Transport Passengers," Journal of Transport Economics and Policy, September 1967.

33. L. Fishman and J. S. Wabe, "Restructuring the Form of Car Ownership," Warwick Economic Research Paper No. 5, University of Warwick, 1968.

Notes to chapter 12

1. For a theoretical summary of this subject, see K. Robinson, "Central Place Theory: a Review," CES WP2, 1968.

2. P. H. Levin and A. J. Bruce, "The Location of Primary Schools," Journal of the Town Planning Institute, February 1968, p. 60.

3. R. C. Barnstead, "Truck Activities in the City Centre," in "The Urban Movement of Goods," OECD 1970, p. 109.

4. A. Schwartz and Associates, "The Role of Trucks in the Region and Manhattan." Manhattan Auto Study.

5. Tri-State Transportation Commission, "Economics of a Rational Urban Pick-up and Delivery System," February, 1970.

6. A. de Ville de Goyet, "Rail/road Terminals, Road Freight Terminals," The Urban Movement of Goods, OECD 1970, pp. 141-154.

7. London Transportation Study, Freeman, Fox, Wilbur Smith and Partners., 1969 Phase III, p. 22/100.

8. J. M. Thomson, "Motorways in London," Duckworths, 1969, p. 33-35.

9. As summarised by W. H. Lathrop, "San Francisco Freeway Revolt," Transportation Engineering Journal, February, 1971, p. 136.

10. C. D. Foster and M. E. Beesley, "Estimating the Social Benefit of Constructing an Underground Railway in London," Journal of the Royal Statistical Society, 1963, p. 46.

11. "Transport Strategy in London," (op. cit.), pp. 58-59.

Notes to chapter 13

1. R. J. Smeed, "Traffic Studies and Urban Congestion," Journal of Transport Economics and Policy, January 1968, p. 59.

2. A. M. Voorhees and S. J. Bellomo, "Urban Travel and City Structure," 49th Annual Meeting of the Highway Research Board, Washington 1970, p. 6.

3. A. Bieber and J. Jorry, "Formes Urbaines et Systems de Transport dans Huit Metropolis de Moyenne Importance," SETEC-Economic, Paris.

4. A. M. Voorhees, S. J. Bellomo, J. L. Schofer and D. E. Cleveland, "Factors in Work Trip Length," 45th Annual Meeting of the Highway Research Board, Washington, 1966.

5. W. Owen, "The Accessible City," Brookings Institution, 1972, p. 61, referring to the T.S.T.C.'s "Truck Freight in the Tri-State Region," 1969, Vol. 1, p. 2; Vol. 2, p. 14.

6. J. M. Thomson, "Some Characteristics," p. 28.

7. "The Effect of Price on Parking Demand at Four Car Parks on the South Bank," Research paper, Greater London Council, 1972.

8. C. S. Riley, "Motivational Aspects of Car Use," project supported by Ministry of Transport, London.

9. "Bogota Transport and Urban Development Study," Freeman, Fox, Wilbur Smith and Associates, 1970.

10. J. M. Thomson, "An Evaulation of Two Proposals for Traffic Restraint in Central London," Journal of the Royal Statistical Society, September, 1967.

11. J. C. Tanner, "Factors Affecting the Amount of Travel," Road Research Laboratory, TP51, H.M.S.O., London, 1961.

12. L. N. Moses and H. F. Williamson, "Value of Time, Choice of Mode and Subsidy Issue in Urban Transportation," Journal of Political Economy, Vol. 71, 1963.

13. J. M. Thomson, "Some Characteristics," pp. 45–48.

14. B. V. Martin and H. Neffendorf, "Traffic Satisfaction in the Conurbations," Transportation Planning and Technology, April 1972, pp. 75–82.

IV

THE AUTOMOBILE AND THE ENVIRONMENT: Implications for the Planning Process

Marvin L. Manheim and John H. Suhrbier

ACKNOWLEDGMENTS

The findings, conclusions, and views contained in this paper are based on an extensive program of research conducted over a 6-year period by the Transportation and Community Values Project of the Massachusetts Institute of Technology, Cambridge, Massachusetts, U.S.A. The findings reported in the paper are based on research done in the United States on American institutions and planning activities, in hopes that the underlying concepts and issues will be of general interest. Sponsors of this research have been the American Association of State Highway and Transportation Officials, through the National Cooperative Highway Research Program's Project 8-8(3); the California Department of Transportation; the U.S. Federal Highway Administration; and the Massachusetts Department of Public Works. Portions of this paper are based on material to be contained in the forthcoming National Cooperative Highway Research Program Report 156, "Transportation Decision-Making: A Guide to Social and Environmental Considerations."

The authors of this paper, Marvin L. Manheim and John H. Suhrbier, are both affiliated with the Massachusetts Institute of Technology, and also with Cambridge Systematics, Inc.

Dr. Manheim is a Professor in the Transportation Systems Division of the Department of Civil Engineering. He is also Director of the Transportation and Community Values Project and a Principal of Cambridge Systematics, Inc.

Mr. Suhrbier is a Research Associate in the Department of Civil Engineering of M.I.T. and served as Executive Officer of the Transportation and Community Values Project from 1969 to 1974. He is presently affiliated with Cambridge Systematics, Inc. and The M.I.T. Center for Transportation Studies.

This paper draws upon the work of many present and former members of the M.I.T. research staff. Special acknowledgement for their contributions is expressed to Elizabeth D. Bennett, Arlee T. Reno, Frank C. Colcord, Lance A. Neumann, Wayne M. Pecknold, Harry Cohen, Robert Giel, Cathy Buckley, Michael Petersilia, Arthur Hall, Stuart Hill, and Lowell K. Bridwell.

While gratefully acknowledging the contributions of these individuals, and the support of the research sponsors and of OECD, the authors are solely responsible for all statements and views either expressed or implied herein.

SUMMARY

Transportation projects in recent years have become the subject of intense public examination and controversy. In many countries, the construction of new or expanded airports, highways, bridges, and even transit systems is being questioned as a result of changing awareness and attitudes concerning such social and environmental considerations as air quality, noise, water quality, natural ecology, open space, and community structure.

This paper examines the implications of this increased environmental awareness on the process through which transportation agencies plan, design, construct, and operate the multi-modal transportation systems for which they are assigned responsibility. Topics examined include an overall planning objective and strategy, institutional arrangements, decision-making responsibilities, community interaction, evaluation of alternatives, and the interrelation of system and project planning.

Basic guidelines are presented for an integrated approach to systematically incorporating social, economic, and environmental factors into transportation decision-making. Because the paper is directed at transportation planning conducted in a variety of cultural, institutional, and environmental settings, the discussion emphasizes overall procedures, strategies, work styles, and ways of viewing problems which generally should be applicable to transportation agencies in all OECD member countries. At the same time, the paper also provides a limited number of specific techniques and examples so as to illustrate the means through which the guidelines can be implemented.

Three key findings form the basis of the proposed approach:

1. The overall process through which social, economic, and environmental considerations are brought into transportation planning and decision-making is as important as the particular techniques used for predicting impacts. The issues of "the environment" cannot be handled simply by hiring a few sociologists or ecologists; nor can opposition be eliminated simply by a program of citizen participation. Decision processes must be structured so that the inevitable conflicts among competing interests can be resolved with a full understanding of the choices made.

2. Issues of social equity must be explicitly recognized and taken into account. The planning, design, implementation and operation of a transportation system is as much a political process as it is a technical one: every decision involves the need to balance the gains to some interests against the losses to others. This requires that the total set of effects, on all groups, be considered using procedures that pay particular attention to the differential effects—how individual groups gain and lose.

3. Different groups of people can be expected to have different interests and differential priorities; the importance people attach to a particular factor depends on the context. Consensus objectives, if they exist, are generally at too abstract a level to provide guidance to those responsible for specific choices among alternative actions.

The planning and design of transportation projects should adhere to the basic principles of:

- early, effective, knowledgeable involvement of local officials and the public to clarify issues and aid in the development and evaluation of proposals, with each group or sector of the public that may be potentially affected having full access to all relevant information;
- consideration of a range of transportation improvements that are sufficiently different so as to represent real choices. Options should involve various types of modal facilities, relationships with other modes, regulations, controls and constraints, and the no-build choice.
- timely identification of the nature, magnitude, and incidence of potential social, economic, and environmental effects so that in all phases of transportation planning, alternatives may be developed that avoid or minimize adverse effects and that take full advantage of opportunities to increase benefits. Effects considered should include those considered to be important by any particular segment of the public, whether or not those effects are readily quantifiable.
- provision of fair and just compensation. Where adverse impacts result for some groups of interests in order that benefits can be provided for other groups, those adversely impacted should be adequately compensated.

It is proposed that the objectives of all phases of a transportation planning and design process should be:

- to clarify issues of choice;
- to fully inform decision-makers;
- to achieve substantial, effective community agreement on a course of action that is feasible, equitable, and desirable.

It is particularly important that increased consideration be given to environmental issues during system planning, especially in decisions on programming. At the individual project level, many options desirable from an environmental viewpoint already may have been foreclosed, and many effects already may have been established by system planning decisions. Most transportation agencies have found it difficult to deal with regionwide effects during project studies.

To help alleviate these difficulties, system planning should be viewed, not as a phase of planning preceding project studies, but as a framework within which project decisions can be made, serving to coordinate ongoing project studies. System plans should be defined using a strategy of implementation decisions being made over a period of time, thereby enabling system-level options to be kept open longer. Plans formulated according to such staged strategic commitments can more readily respond to changes in conditions as portions of a system are brought into operation or as new information becomes available, than can target year master plans.

15 INTRODUCTION

THE PROBLEM: A DESIRE FOR CHANGE

The context of transportation decisions has been changing rapidly in recent years. Many transportation projects in OECD countries have become the subjects of public controversy; airports, highways, bridge, and even in a few instances transit systems are being subjected to intense examination and criticism.

These debates are symptoms of significant and far-reaching changes in public awareness and attitudes:

- greater public concern for air quality, noise, water quality, and the natural ecology;
- greater public concern for more efficient use of limited natural resources;
- greater public concern for the quality of life.

As a consequence of these public concerns, fundamental changes also are occurring in the nature and scope of transportation decisions.

In the past, the dominant considerations in transportation decisions were usually transport efficiency—how to get the greatest mobility for the lowest cost. Now, however, because of the shifts in public values, many significant social, economic and environmental factors must be considered explicitly. People have become much more aware of the wide-spread effects of transportation system changes, and to the differences in opinions among various groups as to which effects should be achieved; and a wider variety of interest groups than in the past have become effective politically at making their views heard.

To illustrate the phenomenon of increased awareness of effects, consider the construction of a new transit system or even an extension of a previously existing one. Such a transportation project may provide major direct benefits: efficient movement of people and goods and increased mobility and access, particularly for the elderly, handicapped, and other groups who may not have direct use of an automobile. There may also be positive effects which are more indirect, such as provision of jobs, increase in the trade and industry of a region, and other general multiplier effects on the economy due to the direct benefit accruing to users and to those involved in construction of the facility.

On the other hand, the transit project may also impinge in an adverse manner upon matters that individuals and communities value highly, such as preservation and enhancement of neighborhoods, noise levels, takings of land, residences, and businesses, increased traffic on local streets; and increased requirements for local public facilities such as fringe area parking lots, improved local roads or bus service, and sewage disposal. The construction of a large scale urban transport system may channel the development of a region in ways which are desirable, such as serving as a structuring device to preserve irretrievable open spaces, or in ways which go counter to regional growth objectives.

As this example illustrates, the result is a situation in which there exist adverse as well as beneficial effects. Furthermore, the incidence of these effects on particular groups or interests is differential: some interests will benefit greatly, others only slightly, while other interests will be only adversely affected. When such a differential incidence of effects occurs, there is always the potential for conflict.

This potential for conflict has become reality in many situations, as more and more interest groups have recognized these effects and become more active politically to defend what they perceive as their interests. In recent years, individuals in general have gained increased political effectiveness. In many countries, transportation conflicts have arisen involving the public in conflict with elected officials and with bureaucratic transportation agencies. In some cases these conflicts have become increasingly vocal and emotional, often resulting in a highly polarized impasse. One consequence has been a "crisis of confidence," in which many citizens have begun to feel that transportation professionals are no longer doing their jobs well, and do not reflect the values and objectives which those citizens feel transportation officials should reflect.

THE RESPONSE

As concern for the natural and social environment has moved to the forefront, those agencies and officials having transportation-related responsibilities at various levels of government have responded in various ways. The general thrust of these responses has been first, to increase the understanding of the effects of transportation systems in such areas as land use, air quality, and noise, and second, to provide mechanisms whereby transportation facilities and services can be provided so as to minimize, if not avoid completely, any adverse consequences.

The formation and work of the Environment Committee itself represents one response at the international level. The Committee's reports and working papers have contributed to an improved understanding of the problems, issues, and public policy implications in such areas as automotive air pollution and noise, energy resources, materials resources, traffic limitation, transport in developing countries and urban public transport.

In many countries, new laws have been passed or regulations promulgated affecting such things as public involvement, coordinated comprehensive regional planning, special treatment of sensitive lands, and detailed reports of the probable effects of proposed actions. One example in the United Kingdom is the Noise and Property Value Compensation Act of 1973. This Act is directed towards increased compensation for those suffering adverse social or economic effects from public works programs; achieving equity through compensation of adverse effects is a primary objective.

The United States has taken a different approach with emphasis on performance standards and a documented, open, participatory planning process. Examples are the National Environmental Policy Act of 1969 establishing the requirements for environmental impact statements, Section 4(f) of the Department of Transportation Act of 1966 specifying that certain publicly owned land cannot be used for highways unless there are no "feasible and prudent alternatives," the Federal-Aid Highway Act of 1962 establishing 3-C ("continuing, comprehensive and cooperative") urban area highway planning, and the Clean Air Act of 1970 and its amendments, establishing national ambient air quality standards which have resulted in transportation control plans for various air quality control regions.

Similar environmental-related laws are in effect or under consideration in other nations.

Within transportation agencies, in particular, the need for increased consideration of social, economic and environmental factors has been recognized and efforts are being devoted to improving procedures for the identification and measurement of such potential impacts. However, while significant advances are being made in the development and refinement of these methods, there is increasing recognition that improved analysis tools by themselves cannot insure that transport proposals will fully reflect environmental considerations and community interests. In particular, the importance of early information on potential impacts and community viewpoints has become apparent; attempts to respond to community wishes and to mitigate adverse effects by modifying substantially completed proposals have proven both costly and relatively ineffective. Moreover, there is a clear-cut need to tie together social and environmental data and the decisions that are made in defining basic transportation systems, so that the choices made reflect the knowledge gained through technical studies.

In short, the overall *process* through which social, economic, and environmental considerations are brought into transportation planning and decision-making is as important as the particular techniques used for predicting impacts.

What is needed is a framework for the planning and design of transportation projects in which environmental considerations and public attitudes are integral components of the planning and design process.

This report presents basic guidelines for an integrated approach to systematically incorporating social, economic, and environmental factors into transportation planning and decision-making. The guidelines are based on the broad principles of:

- early, effective, knowledgeable involvement of local officials and the public to clarify issues and aid in the development and evaluation of proposals, with each group or sector of the public that may be potentially affected having full access to all relevant information.
- consideration of a range of transportation improvements that are sufficiently different so as to represent real choices. Options should involve various types of modal facilities, relationships with other modes, regulations, controls, and constraints, and the no-build choice.
- timely identification of the nature, magnitude, and incidence of potential social, economic, and environmental effects so that in all phases of transportation planning, alternatives may be developed that avoid or minimize adverse effects and that take full advantage of opportunities to increase benefits. Effects considered should include those considered to be important by any particular segment of the public, whether or not those effects are readily quantifiable.

- provision of fair and just compensation. Where adverse impacts result for some groups of interests in order that benefits can be provided for other groups, those adversely impacted should be adequately compensated.

The objectives are to develop a planning process that results in achieving:

- decisions that are technically and politically sound; and
- confidence on the part of the public and governmental officials in a nation's transportation agencies.

Because this report is directed at transportation planning conducted in a variety of cultural, institutional, and environmental settings and is intended to serve a range of needs, the discussion emphasizes overall procedures, strategies, work styles, and ways of viewing problems which generally should be applicable to transportation agencies in all OECD member countries—at the national, provincial or state and local levels of government; in rural as well as urban settings; and for highway, transit, airport, rail, and other transport projects. At the same time, the report also provides a limited number of specific techniques and examples for community interaction, evaluation of alternatives, and the interrelationship of system and project decisions so as to illustrate the means through which the approach can be implemented.

16 A PLANNING PROCESS

A number of specific guidelines can be proposed for the transport planning process, based upon the broad principles laid out in chapter 15.

PROCEDURAL GUIDELINES

Differential Effects

Any decision in transportation planning affects many groups. What transportation mode is utilized, what kind of service is provided, which of several possible areas or corridors is served, what decisions are made about location and design treatment of a facility, even the determination of design or vehicle standards: although these are seemingly only technical decisions, they almost always have significant social consequences.

The total set of effects, on all groups, must be considered, with particular attention paid to the differential effects—which groups gain and lose. It should be noted that, while a change to the transportation system, such as introduction of a vehicle exclusion zone, may bring benefits to certain people in an area, some individuals or groups may bear a particularly high cost, and some may receive little or no benefit.

Every decision about transport involves the need to balance gains to some against losses to others. It is essential that the process of planning, designing, implementing, and operating transportation systems explicitly recognize and take into account such issues of social equity. *The planning, design, and operation of transportation systems is as much a socio-political problem as a technical and economic one.*

Community Values

There is no single set of values underlying any community. Different groups of people can be expected to have different interests and different priorities, and the importance people attach to a particular factor depends on the context. For example, one group might generally prefer the development of recreation facilities to the preservation of wilderness areas; another group might generally prefer just the opposite. But in a specific instance, people who are usually supporters of wilderness preservation might opt for building a park.

Table 16.1 United States Environmental Protection Agency Transportation Control Strategies

I. Reduce motor vehicle-miles of travel
A. Transit operations
1. Bus lanes on city streets
2. Bus lanes on freeways
3. One-way streets with two-way buses
4. Park-ride, kiss-ride*
5. Service improvements and cost reductions
B. Regulation
1. Parking bans
2. Automobile-free zones
3. Gasoline rationing
4. Four-day, 40-hour week
5. Congestion passes
C. Pricing policy
1. Parking tax
2. Road-user tax
3. Gasoline tax
4. Car pool incentives
II. Increase efficiency of traffic flow
A. Freeways
1. Reverse-lane operations
2. Driver advisory displays
3. Ramp control
4. Interchange design
B. Arterials
1. Alignment
2. Intersection widening
3. Parking restrictions
4. Signal progression
5. Reversible lanes
6. Reversible one-way streets
7. Helicopter reports
C. Traffic improvements
1. Traffic responsive control
2. One-way street operations
3. Loading regulations
4. Pedestrian control
5. Traffic Operations Program to Increase Capacity and Safety (TOPICS)
D. Staggered work hours
III. Apply supplementary motor vehicle emission controls
A. Inspection and maintenance
B. Idling restrictions
C. Retrofit of emission control devices
D. Conversion to gaseous fuels

* A situation in which a passenger is driven to a public transportation terminal and dropped off has come to be called a kiss-ride.

It is clear, then, that trying to determine "public preferences" in the abstract (by certain kinds of opinion polls, for instance) is a futile exercise. Similarly, techniques such as linear scoring functions and cost/benefit analysis which assign different weights (i.e., importance) to factors, cannot be assumed to represent "public preferences." Not only do different people have different preferences, but any one person's preferences depend on the choices available and the specific effects of those choices.

The best way to find out about people's transportation preferences is to ask them how they feel about specific alternatives and *why they feel that way.* Some people will stand to gain from a particular alternative; others will lose. By identifying who would gain and who would be harmed, the planner can modify alternatives to reduce negative impacts and increase benefits and can develop compensatory programs.

Consideration of Alternatives

A wide range of both long and short-term courses of action are available in regional as well as facility oriented transportation studies and should be examined as part of any decision-making process. These include: investment in fixed facilities, operating policies, pricing policies, and choice of technology. In addition to direct transportation system options, there are a variety of options available regarding the activity system, including land use controls and staggered or flexible work hours. Table 16.1 illustrates possible "control strategies" defined by the United States Environmental Protection Agency for use in meeting ambient air quality standards required by the Clean Air Amendments of 1970. These options were used in various mixes by local authorities in development of a transportation control strategy for their area. Another typology, illustrating operation, technology, and institutional options available in urban public transport, is presented by Burco.[1] These options include bus rapid transit, demand-responsive vehicles, activity center circulation, central city bus malls, tramways, rail rapid transit and other guideway systems, financing and pricing, continuous and discrete people movers, and area-wide personal rapid transit service.

Transportation options should be considered as only one part of a more comprehensive course of action, being effectively coordinated with such non-transportation components as replacement housing, impact amelioration programs, joint development, and direct compensation. Impact amelioration programs may include modifications of geometric design features, noise barriers, or other such design options that have the effect of minimizing or eliminating adverse impacts. They may also include provision of sound insulation to adjoining residences, such as that provided by the 1973 U.K. Noise and Property Value Compensation Act, or development of adequate replacement housing if it is determined that a suitable inventory of housing stock does not exist. For example, the Uniform Relocation Assistance and Replacement Housing Act (1970) in the United States represented a major step toward equitable compensation by recognizing that fair market value compensation to displacees was not sufficient reimbursement for all incurred disruptions. The 1973 U.K. Act goes significantly further in achieving equity by compensating adjacent residents and businesses for decreases in property value caused by such physical factors as noise, smell, fumes, grit, dust, or artificial lighting.[2]

Alternatives should be developed and presented at all levels of technical studies so as to display the full range of possibilities and to catalyze meaningful and constructive debate in the community of affected interests. Discussion of alternatives can assist in bringing out issues, identifying tradeoffs, and clarifying community objectives. The development of alternatives should be early enough in the process so that this debate can influence further development and refinement of alternatives, with adequate time to explore the issues that different segments of the community think are important.

It is especially crucial that alternatives involving no new construction and low capital investment be fully examined. This examination should be consistent with that given other kinds of more capital intensive alternatives and should provide the basis for use as a reference point in defining potential beneficial and adverse impacts.[3]

Alternatives proposed by affected interests should be studied in sufficient detail to determine their feasibility, desirability, and equity. This investigation should be consistent with the attention given to alternatives developed by professional staff. Communities generally have a unique and valid understanding of their problems and of potential solutions, and should be encouraged to develop alternative transportation-related proposals.[4]

Identification of Impacts and Affected Interests

In analyzing transportation options, a wide variety of beneficial and adverse impacts must be considered, including user-oriented consequences such as travel time, out-of-pocket expense, and level of service; facility-related factors such as capital investment, maintenance costs, and safety; and social, economic, and environmental factors such as land use and activitity effects, neighborhood character and community cohesion, displacement of families and jobs, air and water pollution, effect on tax base, property values, and public services, and disruption of eco-systems. Identification and study of impacts must occur in coordination with technical studies so that alternatives may be modified to take advantage of beneficial opportunities and to avoid or reduce the effect of potentially harmful consequences. Numerous procedures to predict various impacts are available, ranging from quick approximation methods to precise measurement tools, and the choice of prediction methodology should be a function of the needs of the study.[5]

Not all social, environmental, and economic impacts can be expressed in quantitative terms and measured with a high level of accuracy. The process of identifying, analyzing, and evaluating such effects, therefore, must be designed to explicitly account for qualitative information and to recognize uncertainty. Particular care should be taken that important effects are not disregarded solely because they cannot be described in quantitative terms.

It is essential that the incidence of impacts be determined so that effective programs to mitigate adverse effects can be developed. Also, impact prediction should be carried out in association with the public involvement activities, since many impacts can be identified most readily through interaction with the affected community and the significance of many impacts is dependent on personal values and priorities. Other agencies can contribute much useful information and may be able to provide technical assistance and resources.

Involvement of Other Agencies and the Public

Early effective, informed participation of the community—national, provincial, county, and local agencies and officials, interest groups, and individual citizens—is necessary in all phases of transportation planning, starting during regional and urban area system planning and continuing through corridor, location, and design studies and even into construction. Such interaction helps the transportation agency to identify and predict both the incidence and magnitude of social and environmental impacts and to learn what issues various people consider to be important. Also, community groups can serve as a useful source of suggestions for solutions to transportation and related community problems.

Information about the existence, status, and results of studies should be continuously available. Such information should be clear and comprehensible and should include the alternatives being considered, the effects of alternatives whether beneficial or adverse, and the time schedule of major points in the process which would be of public interest.

The public must understand, accept, and be able to provide inputs to the decision-making process. The transportation agency should make deliberate efforts to search out differing viewpoints and to secure participation of a wide variety of groups and individuals, in order to obtain all points of view for consideration by the decision makers.

Different levels of participation should be provided depending on interest. Participation may range from general awareness to periodic attendance to intensive involvement.

A community interaction program will normally utilize several techniques simultaneously as part of a coordinated and carefully managed program to gather information, to distribute information, and to interact with community groups. Emphasis should be placed on small group, face-to-face interaction and on the use of existing institutions and channels of communication. Agency staff must have an attitude of openness and responsiveness and be able to listen and learn from community groups.

System Planning

System plans refer to the sum of the facility, operating, and policy changes proposed over time for the transportation system of a particular geographic region—be it a nation, province, or individual urban area. System planning studies should treat all modes together with emphasis on their complementary and competitive relationships. A proposed project improvement, be it the introduction of a new modal system or an operational improvement to an existing system, must be seen as part of a total multi-modal system; in competition with other modes (existing or new), and with full consideration of the total door-to-door trip, including access modes. The key characteristics of a transport system are the service it offers to prospective users and its costs (direct and indirect, including externalities such as social and environmental impacts). That is, technology hardware is a means and not an end.

The improvement must be implemented as a component of the overall transportation and development strategy of the region. Implementation of a project, set of projects, or an entirely new modal system must be appraised in terms of potential market response, through an analysis of demand in the context of equilibrium in a multi-modal network. Sensitivity analyses should be done to explore the significance of uncertainty on both the supply and demand sides.

A wide range of variations on the generic transport system should be explored—for example, network structure and coverage, vehicle type and size, operating policy, feeder services (if any), pricing policy, financing, etc.

The network flow modelling tools used in past studies have most often been oriented toward the analysis of a single mode without fully evaluating the relationships among all modes. For example, existing techniques in highway network analysis often assume that total demand is independent of the existing level of service or the level of service to be provided in the future. Also, in assigning trips to links, capacity limitations often are ignored. Such techniques are unrealistic and tend to bias evaluation toward favoring more and larger facilities without examining the full range of consequences of such a policy. Further, where more than one mode is concerned, the technologies are generally tailored to provide approximately equivalent level-of-service, and then costs are derived as functions of volume.

Failure to explore the potentially wide range of variations in service, costs and impacts achievable leads to two major difficulties. First, we should expect that different technologies will be appropriate for different service levels, since the choice of technology is sensitive to both operating policies (vehicle sizes, frequencies, fares, etc.) and demand parameters. Setting a particular level of service a priori, therefore, imposes a bias in the comparison of systems. The best way of taking differences in levels of service into account is through use of an explicit demand function which enables a prediction of the market responses to differences in service and cost. The choice of technology then can be explicitly related to specific options about how a system is operated as well as to demand elasticities. Second, by examining service characteristics alone in making preliminary decisions on the types of facilities needed or desired, critical social and environmental issues are ignored.

OBJECTIVES OF THE PLANNING AND DESIGN PROCESS

As the transportation field evolves, the objective of simply planning and building a facility—be it line haul or terminal—no longer reflects the breadth and sophistication of the transportation agency's activities. In fact, when the "plan and build" objective is carried too far, it can work counter to the best interests both of the agency and the public it serves.

These procedural guidelines reflect a particular set of objectives. Specifically, it is suggested that the objectives of the transportation planning and design process should be as follows:

- to clarify issues of choice;
- to fully inform decisions-makers
- to achieve substantial, effective community agreement.

Clarifying Issues of Choice

The transportation agency should work to clarify for the public and the decision-makers the issues of choice—to explore with interested groups and individuals the range of transportation decisions which might be made, to determine the implications of each as fully as possible, and to explitily bring out the tradeoffs among alternatives. Opportunity for meaningful negotiation on what comprises an equitable distribution of gains and losses should be provided.

Informing Decision-Makers

Authority to make transportation decisions may rest solely with transportation personnel, or it may be shared with state or local officials, other agencies, or even a special task force. Regardless of who the decision-makers are, they need full information on the study to date and the potential consequences of each course of action in order to make wise choices.

Many other decision-makers also need to be kept informed on the proceedings of transportation studies, even though their authority may not extend to transportation decisions. For example, those with responsibility for public utility systems need to coordinate their plans with transportation plans. The transportation agency has the responsibility of making sure that *all* decision-makers have sufficient, up-to-date, accurate information on transportation proposals.

Achieving Substantial, Effective Community Agreement

Ultimately agreement is desired on a final choice to be implemented; indeed, that is the main objective of a transportation study. There are other choices in the planning and decision-making process though for which agreement is important, and

16.1 Levels of agreement.

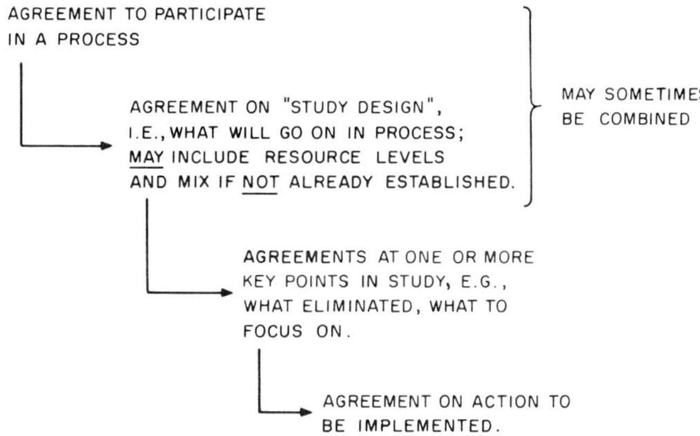

which, if achieved, may help to achieve agreement on a final choice (figure 16.1).

For example, agreement should be sought on:

1. general aspects of the planning and design process, including
• the scope and timing of studies,
• the roles and responsibilities in a process of national or regional staff, of local agencies and officials, and of interest groups;
2. the community interaction program, including
• agreement to participate in a process,
• means for direct participation,
• use, if any, of indirect techniques such as surveys and field work,
• availability of data and draft reports and the public role in reviewing them;
3. interim decisions, including
• what alternatives are deserving of further study, and whether new alternatives are needed,
• what impacts should be given highest priority for study and most importance in decision-making,
• whether the scope of studies or the schedule for decisions should be modified.

Agreement on these interim choices does not guarantee that agreement will be reached on the final choice. Full consensus even on these earlier choices may well be unattainable when the potential gains and losses involved are significant. But by striving for substantial agreement, the transportation agency is likely to unearth the major issues of choice and thus will be better equipped to develop equitable solutions and to inform decision-makers. Furthermore, without general agreement that the study *process* itself was equitable, it will be much more difficult to reach agreement on the final decision.

PROCESS DYNAMICS: A FOUR-PHASE STRATEGY

Although the specific tasks to be carried out must be determined for each study, a four-phase strategy is recommended as providing a workable framework for any stage of a planning process—system planning; corridor, location, or design studies; or even preparation of plans, specifications, and estimates.

The four phases of the basic process strategy are:

I. Study Design
II. Exploration of Alternatives
III. Detailed Analysis
IV. Choice

Phase I: Study Design

The objectives of the study design are to initially define the transportation problems to be addressed; to establish an initial data base; to acquire an initial understanding of the interests, needs, plans, and objectives of potentially affected communities and citizens; and to develop a first work program.

The transportation agency will have some conception of the transportation problems of the area under study at the outset. However, interested parties may have different views of their transportation needs and may be able to point out needs not identified by the agency. It is necessary then to have public input in defining the transportation needs and desires that should be addressed in the study. To do this effectively, the agency should publicize that studies are beginning, make available the data that led to the study initiation, inform the public about prior transportation decisions and agency responsibilities, and provide mechanisms for public input.

The study design phase is the appropriate time to assemble basic data on social, environmental, and economic characteristics of the study area, as well as transportation and other technical data. Coordination with other agencies is vital; they are a valuable source of data and also can provide input on their own plans and objectives. Community interaction activities should be initiated as an additional useful source of data and to provide an initial sense of what the significant technical, social, environmental, economic and political issues are likely to be.

The results of this phase are:

1. an initial definition of the scope of studies, including the types of alternatives to be considered, the roles and responsi-

bilities of transportation agency staff, other agencies and officials, and interest groups, and the community interaction program; and

2. an initial work program, including identification of data needs, scheduling of technical studies, and timing of community interaction activities.

Phase II: Exploration of Alternatives

The objective of this phase is to develop an understanding of the available courses of action by discussing with the public a variety of alternatives, each of which reflects different objectives. The intent is *not* to develop a final solution but to bring out the issues involved in the particular study and to help all concerned to improve their understanding of the advantages and disadvantages of various alternatives. Because a wide range of possible alternatives are to be explored, sketch planning and approximate impact prediction techniques are appropriate.

Extensive interaction with the public is conducted during this phase, with special emphasis on bringing into the process all those who may be affected by or interested in the planning process. Information on alternatives and their impacts is presented, and the public's reactions and suggestions for modifications or for additional alternatives are gathered and fed back to the technical activities. It is also important to coordinate the transportation planning activities with other community and regional plans.

Data collection continues throughout this phase. It is important to maintain careful data files and to document the alternatives considered, their impacts, and the responses of groups and individuals. Draft reports on studies may be produced and reviewed by interested parties.

During this process, the agency and the public will develop a sense of what alternatives are deserving of further study, what issues must be addressed, and what additional data are needed. Modifications should be made as necessary to the scope of studies, the work schedule, and the community interaction program. The result of Phase II is a clarification of the issues of choice, with a preliminary identification of a few alternatives which seem to have the greatest potential for acceptability.

Phase III: Detailed Analysis

The objective of Phase III is to subject the selected alternatives to detailed analysis to develop feasible, equitable, and desirable courses of action.

In-depth prediction of the potential impacts of the alternatives under study is performed. By examining the incidence and magnitude of these impacts with the public, the agency develops ways of modifying alternatives to alleviate negative impacts and to obtain additional beneficial impacts. Associated programs such as land use control plans, public services and utilities plans, joint development programs, and relocation assistance and compensatory programs also are developed in detail, and the means of funding both transportation plans and related programs are determined.

Community interaction activities focus on determining the acceptibility of the alternatives and associated programs and may suggest additional impacts to be addressed and new alternatives or modifications to alternatives to be considered. The agency works with the community to find feasible compromises which would result in more equitable and desirable plans. This often will require bargaining and negotiation with particular interests. Detailed evaluation reports are prepared for each alternative.

The results of this phase are one or two detailed courses of action which represent the agency's best efforts and which best reflect the public's preferences.

Phase IV: Choice

The objective of Phase IV is to make a decision on a particular course of action.

If substantial community agreement has been reached on a particular course of action during the detailed analysis phase, Phase IV merely formalizes that agreement at a public hearing with appropriate reports being prepared and circulated.

In many cases, however, no clear-cut preferred course of action will have been found. In this case, the agency prepares reports on the alternative courses of action, discussing for each alternative its advantages and disadvantages and the views expressed by various groups and individuals and pointing out the tradeoffs among alternatives. Public participation is aimed at checking that these reports are accurate, obtaining further input, and when possible, attempting to find acceptable compromises.

Complete reports on the alternatives, their impacts, and the study and negotiation process are then provided to the designated decision-making authority for selection of a preferred course of action.

INSTITUTIONAL ARRANGEMENTS AND TRANSPORTATION AGENCIES

While the major thrust of this working paper is procedural, the policy framework and organizational structure within which these procedures are used should be viewed as part of a single social system. When one of the dimensions is to be modified (planning procedures in this case), changes and adjustments to the other two (policy framework and organizational structure) should be carefully planned and implemented if the intended objectives of the change are to be fully achieved.[6]

The experience of the environmental movement in the United States has frequently demonstrated that the arrangement and organizational structure of political and technical institutions influences the manner in which social and environmental effects are investigated, and to a large degree determine the effectiveness with which these considerations are incorporated into planning and decision-making. The issues raised in this and in preceding Environment Committee working papers and reports therefore relate directly to the arrangement of governmental institutions and the allocation of decision-making responsibilities to government. As Robert Burco stated:

> . . . technology alone . . . is not enough to ameliorate the transportation difficulties of the major cities of the world. Institutional innovations (in the form of planning, administrative, and financial initiatives . . .) may contribute as much or more to the resolution of both present and future urban transport difficulties. "These dim prospects . . . lie . . . in the gaps of communication and responsibility between those institutions with specific transportation missions and those with differing tasks, such as enhancing the quality of the urban environment . . .[7]

Clearly then, the changes made by transportation agencies in responding to environmental issues cannot be limited to the introduction of new or revised procedures, but must extend to fundamental changes in policy and institutional arrangements if the problems referred to by Burco are to be overcome. For example, transportation agencies have generally been oriented to an individual mode and have had a broad mandate to *build* facilities for that single mode. In the United States, as in many countries, highway engineers first were asked to "get the farmer out of the mud" and then to create a network of fast, safe, and efficient freeways spanning the country. Many public transportation agencies operate with a similarly narrow mission—to build a master network of facilities for a single mode, most frequently some form of rail rapid transit. Only rarely is the rail network well integrated with the area's bus systems, and frequently the bus systems are not even under the same domain as the rail facilities.

Transportation agencies can no longer be *mission-oriented construction* agencies, but must be *service-oriented transportation* agencies. They should work:

- to provide transportation as a service,
- utilizing all modes of a region, and
- utilizing all options.

The emphasis of these agencies can no longer be exclusively or even predominantly on construction of a single mode. Equal attention should be devoted to other modes and in urban areas, especially to various technologies of public transportation. The concern is with the efficient operation of these facilities as a multi-modal system to move both passengers and freight, not primarily with their construction. Relevant options for all modes include not only capital intensive investments in fixed facilities, but also operating, pricing, and regulatory policies; vehicle control strategies; and experiments and demonstration projects.

THE RESPONSIBILITY FOR TRANSPORTATION DECISION-MAKING

Transportation and political institutions in many countries and at all levels of government are presently experiencing changes in the manner in which transportation decisions are reached. For example, three general trends are clear within the United States:

1. responsibility for decisions is being delegated to lower levels of government;
2. interests outside the immediate domain of transportation, for example a state department of natural resources, are being given increased authority over transportation decisions; and
3. elected political officials and even the judicial branch are playing increasingly strong decision-making roles, with a corresponding lessening of the formal powers statutorily assigned to the chief administrative officer of the highway agency. Thus many more officials, agencies, and interests now have some either formal or informal decision-making responsibility. Unfortunately, the result of these changes too frequently is a decision process that is complex, not understood, and indecisive.

All levels of government have a legitimate interest in major transportation decisions. The difficulty is in determining the appropriate mix of local, regional, provincial and national responsibilities in decision-making. For example, while there are practical reasons for placing responsibility for regional and local area transportation planning and programs with institutions at the corresponding levels of government, there are legitimate state and national interests which need to be taken into account in making decisions.

The challenge is to reach an appropriate balance between national or regional concerns and the objectives of the affected local communities. The local communities cannot be viewed merely as parts of the broader region or nation; quite commonly, their interests in fact may be in conflict with those of the region. Presumably, a nation's broad concerns with transport facilities will relate to its objectives to maximize mobility, to encourage economic development, or to provide access to some area of special interest. To a smaller community through which a particular proposed facility may pass, these broad objectives may be of relatively little concern in comparison with more local objectives, such as preservation of the viability of neighborhoods or maintenance of traditional environmental amenities. The institutional structure for making decisions must be responsive to the legitimate, though sometimes inconsistent, interests of all levels. Conflict is inevitable among these interests and institutions and the structure should make it possible for such conflicts to be resolved constructively. To assist in translating these broad objectives of responsiveness and legitimacy of all interests into an operational decision-making process, three needs must be met.

First is the need for *equal access to decision-makers and the decision-making process*. Decision-makers must be as accessible to those groups whose concerns are *not* primarily transportation-oriented as they are to more transportation-concerned groups. The points of access must be made clear to everyone, and must be as readily available to the public as possible. The points of access also must be credible; people must see their points of access as having the same degree of influence on the decision-makers as anyone else's.

Second is the need for *clarity of decision-making authority*. In many cases, authority to make transportation decisions is shared by several institutions. All concerned must understand who has the capacity to make transportation decisions, and who makes what decisions. Recent controversies over transportation facilities have raised serious questions as to whether those traditionally and legally understood to have such authority really do have it. A related question often raised is whether they *should* have it. Most important, the public should have a clear understanding of who has the authority for the final decision.

Third is the need for *clarity of a process of appeal* of transportation decisions. It is traditional in systems of democratic government that the public has the right to question official decisions. Appeals may be made politically to higher authority and legally to the courts. In many instances, however, these needs can be met through procedural arrangements rather than through organization structures. Such procedural arrangements for appeal may minimize the potential for long drawn-out legal battles and their resulting excessive delays of normal transportation decision-making.

Existing institutional arrangements and organizational structures should be examined to determine whether in fact they are consistent with the three identified principles of access, clarity, and appeal, and to discover ways of improving their capacities to effectively deal with social and environmental considerations. In some cases, redefinition of responsibilities, reorganization, or creation of new institutional structures may be desirable.

A desirable pattern of institutions, however, is not something that can be determined in the abstract, but must be tailored to the customs of each nation and to the local situation. A national government, however, can play a strong positive role in providing incentives to institutions at the regional and local level to stimulate the development of capabilities which would allow them to assume increased responsibilities for transportation planning and design.

17 COMMUNITY INTERACTION

INTRODUCTION

One of the basic guidelines described in Chapter Two was the need for timely public involvement. In many countries there already exist legal requirements for official "public hearings" of one form or another. However, a public hearing is just one of many possible ways of obtaining timely, constructive public involvement. In this chapter, we will explore the area of community interaction to describe both general principles and specific techniques.

Community Interaction as a Communications Process

Community interaction can be viewed as a communications process where information is exchanged during all phases of a study by the transportation agency and other agencies, officials, interest groups, and the general public. Thus community interaction involves:

- who sends information and who receives it;
- information content—the kinds of information sent and received;
- communication channels—the means available for sending information, and the means available to receive it;
- sources of interference, distortion, and interruption—the things that may prevent information from being received or understood as intended;
- feedback mechanisms—ways of obtaining responses to messages sent and to check on the accuracy of transmission and reception; and
- efficiency of communication—how well particular mechanisms are transferring information.

Community interaction is also a cooperative relationship. The agency and the community must work together to identify and clarify relevant issues, to develop and analyze a range of transportation alternatives, and to attempt to reach agreement on a variety of decisions. This requires direct, personal contact between transportation agency staff and community members. Community interaction is concerned with the opportunities provided for citizens to get involved; with operating styles and how they enhance or detract from good working relationships; with the timing of participatory efforts; and with the relationships among participation activities, technical work, and decision-making.

In short, community interaction is all the ways in which the agency and the community learn about each other and work together to enhance the transportation planning process. A *community interaction program* is a coordinated set of activities to stimulate community interaction.

A Community of Interests

The community of concern is everyone with a stake in the transportation planning process. Included are the residents, property owners, and employers and employees within a study area; national, provincial, and local agencies; elected and appointed officials at all levels of government; potential users of the transportation facilities; and interested persons from outside the study area. The community of concern, then, is more than the concept of neighborhood: it is a community of interests.

A literal interpretation of this definition could indeed be "society at large." But in practice, the community with which the agency will deal will be those who may be directly affected by the study and those with strong concerns about the potential impacts of the transportation decisions to be made.

Contribution to Other Technical Activities

The community is an important source of information. People may be able to provide data that the agency could obtain indirectly only at great expense, if at all. But perhaps more importantly, the need for and implications of transportation projects depend in large measure on the attitudes and aspirations of citizens. People may have differing goals and priorities. In some cases conflicts may arise, the resolutions of which will require value judgments to be made. Participation thus helps obtain information needed to guide the development of alternatives, to expand the data base, to help in evaluation of both data and alternatives, and to inform decision-makers.

While a good community interaction program can improve the quality of technical studies, community interaction is also dependent on, and affected by, technical studies. Indeed, the public expects the agency to present and discuss their technical studies if they are to make meaningful contributions.

In a sense, there is a partnership between the public and the agency staff. The staff members are experts with responsibilities for performing studies, but they also must be open and sympathetic to the public. Agency staff should carry out their technical work with input from the community, while informing the community of the uncertainties and implications of the work. In response to community requests for analyses, they should either perform the requested analyses or explain in understandable terms why such explorations would be wasteful.

It follows that community interaction activities and technical studies must be coordinated to meet each other's informational needs. But sometimes there is a tendency to insulate technical work from community interaction. The community interaction program is then not planned to meet the informational needs of the technical process. The result is a less useful interaction program and a technical process that may not address community needs: neither the community nor the agency is served.

Measuring the Effectiveness of Interaction

Some argue that interaction should be limited to elected officials since they are directly elected by the public and it is their responsibility to represent the interests of their constituents. While the delegation of power to representatives by voting in elections is indeed a legitimate form of participation, the vote is too occasional and non-directive to act as a powerful force in an individual issue area such as transportation planning. Because it is virtually impossible for the public to indicate its preferences on a single issue by casting ballots for candidates, there is a need to provide opportunities for citizens to participate directly.

It is difficult to measure effective interaction. One method is to count heads at meetings, the thought being that the agency's program is successful if it can stimulate high attendance. But there are too many unknown factors influencing meeting attendance for it to be a reliable indicator. Low meeting attendance may indicate a desire to let others do the planning, or a preference to learn of the plans by other means, or any of a number of things rather than poor publicity. Likewise, large numbers of persons may only indicate dissatisfaction with the agency.

A similar measure is the percentage of people responding to a survey, the reasoning being that the higher the response rate, the greater the effectiveness of the participation. This fails, however, to take into consideration the quality of the survey, the means of distribution, or the nature of the information received and whether or not the information contributes in any way to the study.

Some persons equate effective interaction with a "build" decision. If a community is opposed to construction because of misconceptions and narrow thinking, then a program that "sets right" such thinking and results in building may be viewed as successful. But, it is also true that construction may be opposed because its adverse effects are felt to outweigh potential benefits. "Successful" community interaction in this case could result in a no-build decision.

The more meaningful parameters of successful community interaction are less visible than meeting attendance or a "build" decision. The important concepts are whether or not all affected persons (1) were allowed an equal opportunity to participate, (2) understood the planning process, and (3) understood the issues of choice.

Providing equal opportunity to participate is not as easy as it might seem at first glance. Providing everyone with exactly the same ways of participation—treating all groups alike—does not insure equal opportunity. Different groups may require different agency efforts if equal opportunity is to be achieved. Different efforts also may be required to ensure that the planning process and the issues of choice are understood. The agency can do much to increase understanding of the process by clearly laying out what it can and cannot do and by carefully documenting its actions. Understanding of issues can be increased by talking about alternatives rather than abstract values and by using models, photographs, and other display techniques—by relating planning to those things that concern people and by helping people visualize possible changes.

In sum, the way in which a community interaction program is structured and carried out strongly influences who will participate and how effective the interaction program will be.

Community Interaction during System Planning

Opportunities for participation should be continuously provided throughout all stages of transportation studies, including system planning and programming. This is useful even though during system planning it may be difficult to identify and predict many impacts, and thus difficult to identify who would be affected or how particular interests would be affected. Most people are primarily interested in short term issues that affect them directly, and it will always be difficult to motivate some people to deal with decisions that involve consequences that may be far removed, perhaps twenty or thirty years in the future. Thus special efforts are needed to obtain public input in system level studies.

One strategy is to seek out groups and individuals with an ongoing interest in particular kinds of impacts, for example, ecology or economic development. While such spokesmen will change as time passes, many of the impacts for which they are concerned will remain of interest.

Another strategy is to discuss some of the shorter term issues that interest the average citizen together with the longer range "big" questions; those who come to talk about present problems will find out about the long range problems as well, and how these may relate to their present concerns.

There also should be substantial public involvement in programming decisions. Decisions on the programming of projects for implementation or for further planning have a substantial effect on each project. Therefore, those concerned about each project are likely to have a stake in programming decisions.

Programming, as will be shown in Chapter Five, provides the most general link between system and project planning, and involvement of the public in programming can enable them to better understand the linkages between long term and short term issues. In fact, programming decisions can be one type of shorter term issue to stimulate interest in system-level planning.

TECHNIQUES

Catalog

An effective community interaction program requires not only

• *direct interaction*: two-way, face-to-face contact and communication between agency staff and the community; but also
• *information gathering*: the collection of data about the community, its nature, its needs, and goals, and the different points of view of its members; and
• *information giving*: the dissemination of agency information on the scope and process of studies, the issues of choice, and the data relevant to those issues.

Some available techniques for each of these three basic activities are identified in table 17.1. This collection of techniques is indicative of the range of techniques which might be used; there is no attempt to be comprehensive and an agency may find other techniques or variations of those listed to be equally useful. There is, of course, some overlap among the three categories; working meetings, for example, are listed as an interaction technique, but they also both collect and disseminate information. The categorization merely indicates the *primary* direction of information flow—from the community

Table 17.1 Catalog of Community Interaction Techniques

Information gathering	Information distribution	Interaction	Special purpose
Existing Sources Compiled statistics Descriptive information	Posters, billboards, and signs	Small group meetings	Referenda
Working with local officials	Mail notices, brochures, newsletters, fliers	Working meetings	Technical assistance
Monitoring new developments	Newspapers Legal notices Advertisements News articles Feature columns and articles News releases Letters to the editor	Workshops	Mediation and arbitration
Analyzing plans, programs, and reports	Radio and television Announcements News coverage Talk shows and community oriented programs Documentaries	Hearings and other large public meetings	Ombudsman
Monitoring mass media Newspapers Radio and television Other media	Community organizations	Field offices	Charette
Field work	Displays, maps, and models	Public information centers	
Surveys		Advisory committees, steering committees, and other groups	

to the agency, from the agency to the community, or both ways at once.

In addition, a few special purpose techniques are identified. They are put in a separate category because they would generally be used only in unusual circumstances.

An effective community interaction program will require the use of several techniques from each of the three basic areas. The main consideration in selecting techniques is to use several means for each activity to increase the comprehensiveness of coverage. Timing is important, for activities carried out too early or too late in the process will not yield maximum benefits. Similarly, the agency must consider the resources available for community interaction, and must remember that the type of community being dealt with, and the particular groups in that community, should affect the type of technique used.

Example: Small Group Meetings

Of the listed techniques, small group meetings have proven to particularly effective and will likely be an integral component of most programs of community interaction activities. The following paragraphs illustrate the use of this one particular technique and indicate how fairly small and minor technique oriented decisions can be critically related to the success of the overall planning process.

Small meetings are an important way of communicating with interested groups or collections of unaffiliated citizens. The agency can provide detailed information on its activities, obtain opinions on the scope and timing of studies, and gather a variety of information which simply cannot be obtained through other techniques.

Meetings may be initiated by the agency or by an interested group specifically to discuss the transportation study. Alternatively, the agency may ask or be asked to attend a meeting whose primary focus is something other than the transportation study. The agency can start out by contacting known organizations to see if an initial meeting is desired, by publicizing its willingness to meet with interested groups, and by identifying communities or neighborhoods where a meeting might be appropriate. It is useful to ask known groups, local officials, and so on their opinions on who else should be contacted.

The timing of meetings is important. Initial meetings should be held early in the study so that people can comment meaningfully on the scope of the study, alternatives and impacts which should receive attention, and so on. The agency should be prepared to discuss its initial proposals for the study design so that there is something to focus on. On the other hand, proposals that are highly detailed may lead some people to conclude that decisions have already been made de facto and that their comments will have no effect. It is important to explain what decisions have been made and which options are still open.

The places where meetings are held can affect their success. The location should be well known and easily accessible to all interests, and it should have facilities suitable for holding meetings (comfortable seating, large enough rooms, perhaps blackboards). Local schools often will be a good choice.

The time of day the meetings are held is another important variable. Daytime meetings will not be attended by those who cannot get off from work. Conversely, evening meetings may be inconvenient for a businessman's association whose members go home to scattered suburbs.

The agency should consider other community events which might conflict with its meetings.

Finally, agency staff need to be sensitive to the effects that their dress and comportment will have on the people they are dealing with. Communicating well is a talent, but it also is a skill which can be learned and which improves with practice.

Effective communication at small meetings is aided if participants can sit at a table or in a circle rather than being divided into speakers and audience. This is true even when meetings approach a size of sixty persons, in which case two or three circular rows might be arranged around a large table.

Positive Personal Relationships

The definition of community interaction as a two-way communication process and the guidelines presented above for the conduct of small group meetings are each directed toward the establishment of positive personal relationships between agency staff and participants. It is of critical importance in achieving this objective that staff members be skillful listeners and good at bringing out issues without creating personal antagonisms. They must be careful to avoid "putting people down," intentionally or unintentionally, in the way they present information, respond to questions, or structure the interaction process.

It is often difficult not to answer anger with anger nor to become impatient in trying situations. But if people feel they are being belittled, ignored as "too ignorant" of transportation issues, or treated as adversaries, it will be impossible to develop the good working relationships necessary for effective community interaction. Staff members who undertake community interaction must have sufficient self-confidence to operate effectively in a sometimes threatening environment.

18 EVALUATION AND REPORTING

INTRODUCTION

Evaluation, narrowly defined, is the task of appraising alternative transportation plans to ascertain their acceptability and to develop a recommendation for a decision. In recent years, evaluation has moved from the field of "judgment" to a seemingly more rational, objective basis, with the use of "benefit/cost analysis" and similar economic criteria. Today, however, in the context of the wide range of social and environmental effects to be considered and the diverse views of different groups as to the values to be applied, it can be misleading if not dangerous to utilize solely economic methods for evaluation.

A broader, more subtle approach to evaluation of transport plans and projects is required. Such an approach is summarized in this chapter.

EVALUATION AS A BASIS FOR DECISION-MAKING

Evaluation is the process of periodically appraising the alternative transportation program packages to ascertain their acceptability, to identify issues that have arisen and to determine future tasks for the planning staff.

To be effective the evaluation and reporting process should be clearly and directly related to decision-making. The planning and design of transportation facilities and services can be characterized as a process that gathers certain data, makes decisions based on those data and communicates those decisions to an audience presumably affected by the decisions. Typically, key decisions are made throughout the course of studies. Decisions are made to collect certain data but not others, to develop particular alternatives but not others, to interact in certain ways with groups or individuals, to ask certain questions, and so on. Some of the most important decisions concern actions which are not to be undertaken.

The output of evaluation depends on what stage a study Is in. Early in project or system studies the output might be a clear definition of what kinds of alternatives are to be considered, what impacts are especially important and which interest groups desire to be involved. Later, as alternatives are

18.1 Basic activities of the evaluation method.

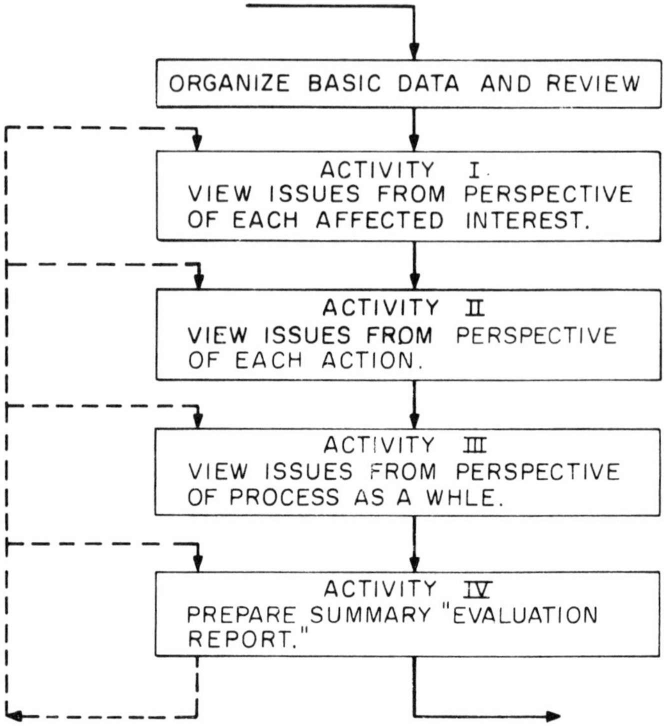

more clearly defined, emphasis shifts to documenting key impacts, and the preferences and potential acceptability of each alternative to various interest groups. As important decisions are made about which alternatives should be dropped, these decisions and the reasons behind them should be carefully documented and reported to the public and relevant agencies. It should be emphasized, however, that the output of evaluation should not be voluminous mounds of paper. Memos, brief reports, oral briefings, slides, maps and graphs are all legitimate mechanisms of communication. The goal of evaluation should be to convey the important information necessary to assist those responsible for making decisions, in as brief a space as possible.

The audience for an evaluation report will include planning staff, elected officials, other agencies, the general public, and other transportation agency staff not participating in the particular project being discussed. Some information will be sent to all of these people; some will be of interest to fewer persons. The periodic application of evaluation is emphasized as a means for coordinating various work activities and stimulating active citizen participation throughout the planning process.

In summary, evaluation and reporting should have the following attributes if decision-makers are to be provided with the necessary information in a comprehensive and timely manner:

- be performed periodically throughout the process;
- indicate the differential incidence of impacts resulting from different alternatives;
- highlight the trade-offs among alternatives;
- operate on qualitative, as well as quantitative, information;
- treat uncertain and incomplete information;
- recognize that different interests place different relative values on different objectives and impacts;
- guide a process by setting priorities for future staff work;
- be clearly related to the decision making process;
- document decisions made, and the issues considered;

APPROACH TO EVALUATION

The following evaluation method is proposed as satisfying the above requirements. It need not be followed in strict sequence, as it is more likely that the analyst will end up emphasizing different activities at different stages of the planning process. The method consists of four activities (figure 18.1):

Activity 1: View the Issues from the Perspectives of Each Identified Affected Interest

The thrust of this step is to view the consequences of proposed alternatives as they affect particular groups and individuals, and as those groups perceive them.

1. *Examine each identified affected interest, one by one*. Select each interest in turn and go through the following steps (2-4).
2. *Review the impact information*. How does each action affect that interest? If negatively affected, are reasonable compensatory measures included in the action?
3. *Review the expressed opinions, goals, and preferences for that interest, and the results of (2)*. What is the opinion expressed by the interest toward each action? Why? Is there uncertainty about the interest's attitude toward an action? What kinds of interaction with that interest would help to clarify their expressions of preferences? Can a ranking of alternatives by the interest be determined? Are there modifications of any action that would make it more acceptable to the interest? Are there any other actions that the individual or group has proposed for consideration or might be likely to propose?
4. *Review the impact data and their uncertainty*. Are there impacts for which the person wants more, or more accurate, in-

formation? Are there impacts not yet predicted that he or she considers important?

Activity II: View the Issues from the Perspective of Each Action

1. Examine each proposed action, in turn, and go through the following steps (2-5) for that action.
2. *Issues of community concern.* How does the proposed alternative respond to people's concerns? To transportation issues? To related issues? How can an alternative be changed to make it more responsive to such concerns? Do other agencies or governmental bodies have jurisdiction over some issues being raised?
3. *Issues of feasibility.* Is the action feasible technically, legally, financially? If not, what must be done to make it feasible?
4. *Issues of equity.* Who would receive benefits from this action relative to the do-nothing alternative? Who would receive adverse effects which cannot be adequately compensated?
5. *Issues of potential acceptability.*
• Which interests would be likely to support this action? To oppose it?
• For whom is there uncertainty as to their likely attitudes? Why?
• Would additional information on this action's effects be useful? Which effects? What are the relative priorities for impact prediction?
• Would additional information on anyone's preferences be useful? Whose? What are the relative priorities for community interaction?
• Would some modification(s) of the action increase its desirability? What modifications? What are the relative priorities for development of alternatives?

Summarize the major assets and liabilities of this action.

Activity III: View the Issues from the Perspective of the Process as a Whole

To gain further insights, all the actions are considered together, in a search for "patterns" or relationships extending across several, many, or all actions and/or interests.

1. Review the results of the preceding steps.
2. Is there any action(s) for which the disadvantages are sufficiently great that it should be removed from further active consideration?
3. *Issues of community acceptance.* Are there different groups of people who share similar views about each of a group of several actions? All actions?

• Who has opposing views about a single action? Some group(s) of actions?
• For interests in conflict over actions, are there other actions on which they would not conflict? Can compromises be reached by modifying alternatives or developing new ones?
• Is there any action(s) that could have or does have substantial community acceptance?

Activity IV: Summarize the Results of the Previous Steps

• What are the major arguments for or against particular alternatives?
• What are the major areas of conflict, both among alternative goals and/or among affected interests?
• Which actions have the greatest potential for public acceptance?
• What are the priorities for further work in the development of alternatives, the prediction of impacts, community interaction, and integration with system studies?

The results of using the evaluation method should be documented and reported to decision makers and all other interested persons. Such documentation and reporting could take a variety of forms, but would usually include some summary evaluation reports supplemented by briefings and meetings.

RELATION OF METHOD TO OTHER EVALUATION TECHNIQUES

The evaluation method described above made no mention of evaluation techniques such as benefit/cost ratios and linear scoring functions.[1] These techniques can serve useful purposes if used within the context of the broader evaluation approach being proposed. Too much emphasis on one construct, such as the benefit/cost ratio, however, works to the detriment of social and environmental concerns which cannot be readily valued in numerical or money terms. While these more quantitative evaluation techniques can provide useful information (even if only economic information) they must be used within a larger strategy aimed at illustrating all the issues which confront decision-makers; the severe limitations of these techniques must be continually kept in mind.[2]

The problems with using only benefit/cost analysis or linear scoring functions are, most fundamentally, that they attempt to "add together" things which cannot be legitimately added together by a technical or staff analyst. Many issues tend to be hidden due to the desire to reach a "total score" for each alternative. Within any "total score" there is a distribution of impacts on various groups, with some gaining and some losing. Thus, the "total score" purports to represent what is good for society, when not all groups would agree that society is served when they are adversely impacted.

Any benefit/cost analysis or linear scoring function uses "weights" for particular impacts, where values are assigned which represent the importance of those impacts. But whose "weights" should be utilized? Only a very naive group would agree to a compromise on a set of weights beforehand and then find that the resulting "highest score" alternative has disastrous results for them. Each group's values are important in the choice, and their importance should not be diluted by aggregating together the values of many groups.

The convenience of having one "total score" does not outweigh the loss of a wealth of information about the tradeoffs necessarily involved in the choices among alternatives. These techniques tend to hide the issues—who gains? who loses?—instead of bring them out.

To overcome these limitations, the evaluation approach recommended here stresses the use of discussions and analyses of alternatives as a way of stimulating people's understanding of what their actual goals and preferences are. In this way, the approach differs substantially from other approaches to evaluation which attempt to define goals or values first, then apply those abstract statements of preference to alternatives. The most basic difference between the recommended approach and other approaches is that the recommended approach does not assume that persons can consistently formulate their goals and values in an operational way without going through a *process* of looking at choices as they develop over time. Further, approaches to evaluation, such as benefit/cost analysis or linear scoring functions, are traditionally applied at one point in the process, and used to make a choice. In the recommended approach, such narrower techniques could be used but as only one of many ways to explore choices and values, not as a quick and easy basis of making decisions.

19 INTERRELATION OF SYSTEM AND PROJECT PLANNING

ENVIRONMENTAL CONSIDERATIONS IN SYSTEM PLANNING

Transportation system and project planning traditionally have been viewed as sequential activities, with system studies producing a transportation network "master plan" and project studies preparing detailed designs for particular links in the network—e.g., detailed highway project designs. While the two activities haven't been entirely divorced from each other, system plans usually have specified the total list of projects which could be considered without providing strong guidance for the scheduling and implementation of specific projects (i.e., priority setting and programming).

In most countries, the first steps taken to respond to the growing concern for environmental issues were at the project level—first to examine a wider range of impacts for an individual project, and second to involve a broader segment of the public in the project design process. While these efforts represent positive steps, there are a number of reasons why improvements only in project studies are insufficient for a truly systematic and effective treatment of environmental concerns.

First, there are environmental effects implied by a system plan that cannot be analyzed on a project by project basis. Land-use impacts, regional economic development, housing, air quality and changes in traffic patterns are examples of issues which must be handled on a system-wide basis.

In addition, decisions which are made during system planning studies may significantly establish some of the social, economic and environmental effects of a proposed project. Thus, the staffs doing location and design studies may be constrained in their ability to take actions to alleviate or avoid adverse economic, social or environmental effects because of system planning decisions. For example, the choice of technology, service characteristics, and general location of a particular link in the network may have been determined in system studies in adopting a network plan for the region. Thus, these choices are seen as fixed and unchangeable in project design studies, no matter how serious may be the issues raised by particular interests. For example, the project design team is usually not free to question the "need" for a facility, as that supposedly has been clearly established by system planning studies—even though project studies may develop information on possible adverse effects or sources of increased costs which call that "need" into question.

Finally, since careful environmental impact analysis is not undertaken until project studies, significant resources are expended for the design of projects which later may be delayed, extensively revised, or even dropped from further study, due to discovery of adverse environmental effects not anticipated in system-level planning. Such changes in project concepts or schedules create the need to revise implementation programs and system plans. While such delays or changes are never totally avoidable, anticipating project environmental impacts in system studies might result in fewer delays and disruptions to implementation programs.

There are, however, several reasons why all significant social, environmental and economic effects are difficult to anticipate during system planning studies:

1. Many impacts of system decisions are long term, and are diffused over wide areas of a region;
2. The impacts are difficult to predict because of poor knowledge of cause and effect relationships;
3. Some impacts are unknown because they are not really fixed until specific project planning, location, and perhaps even design decisions have been made;
4. Public involvement is useful in obtaining the views of different interests as to what impacts are important. However, effective public involvement has been difficult to achieve in system planning studies because of the remoteness of most system planning decisions (in that they concern actions to be implemented 10 or more years in the future).

Yet even with these difficulties in anticipating many social, economic and environmental effects during system planning, it is still necessary to make key system planning decisions.

Overcoming these difficulties requires careful and continual coordination among system and project planning activities.

A VIEW OF TRADITIONAL SYSTEM PLANNING

While many transportation agencies are beginning to recognize and address most of these issues, further steps can and should be taken. Any strategy for integrating system and project planning must recognize the need to address both area-wide and local issues throughout the process and specifically to deal

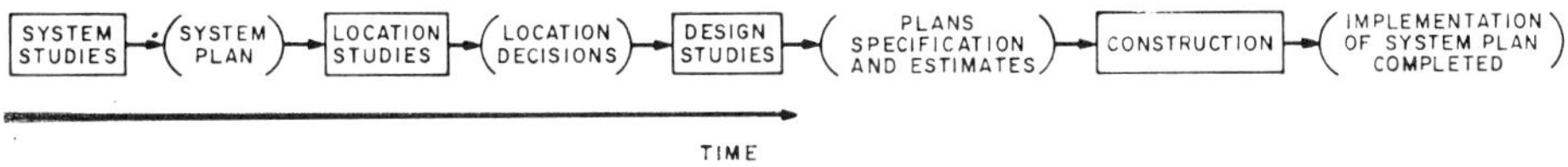

19.1 Traditional view of the planning process.

with the problems identified above at both the system and the project planning levels.

Two problems which must be overcome to achieve this goal of improved integration of system and project planning—and thus of increased environmental sensitivity in system planning—concern (1) the lack of an interdisciplinary approach, even among transportation professionals; and (2) the view that system planning is limited to the preparation (and possible updating) of network master plans of fixed facilities.

Personnel involved in system and project planning typically have very different perspectives on transportation problems. At the system level, area-wide issues are addressed and the planning profession is predominant. At the project level more detailed design work and local issues are emphasized and traditional engineering skills play a much larger role. Even more important than the difference in disciplines is the fact that system and project studies are generally carried on in different units within an agency, or even different agencies, compounding communication problems. These "cultural" and "institutional" gaps must be overcome through the use of a truly systematic interdisciplinary approach to planning in which all disciplines work with each other continuously on a day-to-day basis, developing an understanding and respect for the viewpoints of each of the other participants. Whether housed in one organization or in several organizations, it is essential that practical management procedures be utilized to insure that the total planning process is conducted as a fully integrated systematic interdisciplinary approach, with all relevant professional specialities influencing the course of studies and the final products.[1]

The traditional approach to planning has been that system planning preceded project planning, and ended with the adoption of a system "master plan." Then project planning activities developed the detailed designs of projects necessary to implement that "master plan" (figure 19.1). While the master plan gives a precise picture of what the future transportation system *might* be in some target year—e.g., 1995, in many cases the master plan has not been tied explicitly to the programming activities which determine how projects will be scheduled toward implementation. That is, rarely does the master plan include details of the staged implementation program that will be followed in bringing about the master plan. Thus, a master plan often represents an unrealistic goal because it does not take into account the problems and uncertainties in implementing the system on a step-by-step basis. Because completion of the plan (or completion on time) is assumed, year-by-year decisions on specific projects are distorted; if a more realistic perspective on completion of the system were taken, different project decisions would be taken in many cases.

More importantly, by focusing on only one future system, the master planning approach loses flexibility to revise plans. The implementation program is geared toward the construction of one plan for one "target year." Since it is assumed that all projects will be built eventually, projects may be selected which will operate effectively only when other interdependent projects are completed. Then, if community and environmental impacts discovered during project studies bring into question the need for one project or several links in the system, it is both technically and psychologically difficult to respond positively by considering new projects or new system concepts. Because no provision has been made in system studies for a range of project designs, or the potential deletion of a particular link, a large amount of resources are required to revise the plan later.

Uncertainties in funding, community preferences, and the impacts of a particular action place severe limitations on a master plan approach. Transportation options must be developed with the knowledge that today's decisions are based on an imperfect understanding of the future of a region. Unforeseen changes may require new responses and adaptations that are impossible to fully anticipate.

The important decisions are the near-term programming choices—the annual decisions on how to spend money next year. These are the only decisions which irrevocably commit resources to projects and studies. In system planning it is neither desirable nor necessary to make a choice of one target system for some future year. The choice of a system master plan is alterable and will in fact likely be altered many times before the target date arrives. By leaving future decisions open until more information is obtained, system planning can take into account possible future options and events and help to evaluate the most flexible direction for present programming decisions.

An examination of many transportation controversies leads to the conclusion that many of the problems are directly related to the inability of the present system planning process to explicitly deal with uncertainty and to effectively relate near-term programming decisions to longer-range system plans. *System planning must focus not only on desirable master plans but on implementation strategies as well.*

A STRATEGY FOR INTEGRATING SYSTEM AND PROJECT PLANNING

System planning should be viewed, not as a phase of planning preceding project studies, but rather as a framework within which project decisions can be made, serving to coordinate ongoing project studies (figure 19.2). System planning thus would *periodically* assign resources and priorities among the ongoing sub-area studies and project planning activities. Since project studies influence decisions about the overall system, they must be coordinated with system planning in an ongoing way. It is fair to say that traditional practice has not integrated project and system planning within the sort of time sequencing of activities proposed here, although some areas are certainly taking steps in this direction.[2]

Such an approach explicitly recognizes that transportation plans are not implemented in "one shot," but rather in a series of stages. Thus, system planning ought to examine a range of different implementation strategies. For example, a twenty-year time horizon might be divided into five-year stages. Each stage of a particular implementation strategy might include construction of a number of transit options, operating and pricing changes, as well as highway improvements, demonstration projects, and planning studies. No particular "end state" need be identified initially as a target system. Each action taken during a stage should provide immediate benefits (rather

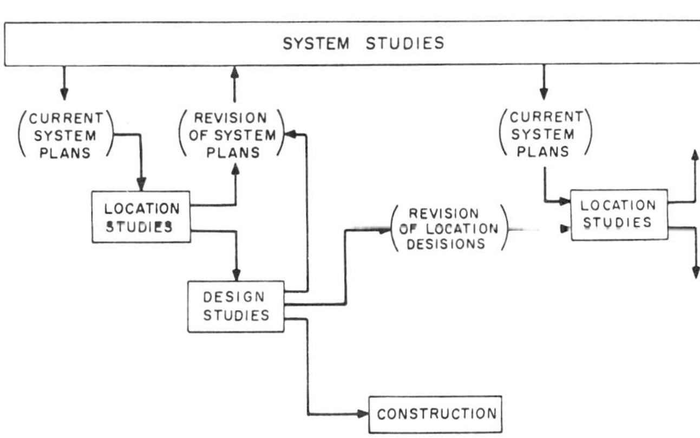

19.2 Suggested view of the planning process.

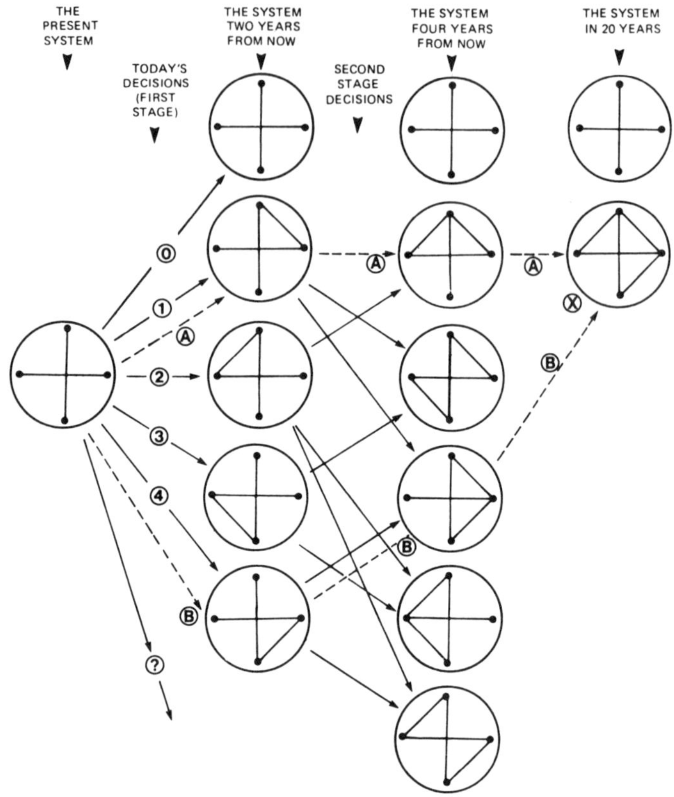

19.3 An implementation strategy approach to system planning.

than being dependent on as yet unimplemented steps), may fit into a variety of possible "end state" future systems, and is evaluated with a perspective of keeping future options open (figure 19.3). During implementation of the first stage, the subsequent stages in a strategy could be revised or updated in light of new information or changes that have occurred. The same future system may be reached by any of several paths—for example, in figure 19.3, both paths A and B reach system X.

This approach recognizes that many significant decisions affecting a system plan may be postponed until project environmental impact, corridor, and initial route studies are underway or completed. The service level, specific alignment, and indeed even the existence of a particular failicity may not be finally determined until later phases of planning. System plans should account for the possibility of a number of outcomes from these later studies.

By developing different sequences of actions on facility improvements, emphasis is placed on what choices are available over the planning time horizon, and how present decisions affect the range of choices available in the future. The different sequences can explicitly recognize uncertainty by evaluating the impacts of a number of potential outcomes from project negotiations or impact studies. Thus, implementation strategies provide a convenient framework for relating system and project planning by focusing on both short term decisions and longer-range plans.

Although the resources available for system planning will restrict the number of sequences and uncertainties that can be considered, attention need not be limited to one sequence over time. Implementation strategies cannot be developed for every possible event that may occur in the future, but they can represent what appear today to be the major choices facing the decision-making process.

The role of system planning in the context of alternative implementation programs is to carefully anticipate the choice issues that must be solved as planning continues, and devise tentative sequences of improvements based on potential outcomes of these choices. As new information is gathered, new options will be added, while others will be dropped from consideration.

In summary, system and project planning must be integrated so that a go/no go decision to implement any project or a particular design will not disrupt the ability to allocate funds smoothly to other high priority projects. Focusing on implementation strategies (rather than master plans) will allow and encourage a transportation agency to anticipate modifications so that when they occur they do not result in lost time. Only with ongoing integration can this problem of a great deal of lost time be avoided when something doesn't go according to schedule in the development of particular projects.

Naturally, both the master plan and a plan based on implementation strategies can be altered in future periods in response to changes. Neither irrevocably commits a region to one sequence of implementations. The two essential differences between the approaches are how initial decisions are made and the flexibility provided to revise the plan over time. Initial decisions with the master plan aim at one target year system. While the master plan can be revised, many alternatives are foreclosed prematurely by focusing on one target network. The implementation strategy approach considers uncertainty and a number of improvement sequences when initial decisions are made. By anticipating the changes which may occur and a range of the choices available in the future, this approach explicitly requires periodic evaluations and revisions and ongoing coordination with project studies.

PROGRAMMING DECISIONS AS A TECHNIQUE FOR INTEGRATION

One of the best opportunities for more effectively integrating system and project planning lies in strengthening programming as a decision-making activity. Programming is key for a number of reasons. First, the programming process is the focus for important decisions and negotiations concerning the commitment of money and manpower to various projects and studies. Second, programming provides an appropriate forum for consideration of both long-run and short-run actions aimed toward implementation of a system plan. Finally, programming is a periodic activity (often with yearly budgeting cycles) and thus is a convenient checkpoint to review the status of ongoing project studies and to revise the system plan in light of current project development activities.

Traditionally, the link between system planning and programming has been weak. System planning, primarily through functional classification and needs studies, provides lists of projects which are then assigned priorities in some manner. Programming then chooses projects until the budget is exhausted, subject to geographic and other constraints. One way to strengthen this linkage is to require a system plan format which encourages discussion and documentation of both system and project activities. The basic planning document should be a multi-year program package that combines both short-term and longer-run improvements while explicitly recognizing the resource and other constraints facing transportation plans. Such a document would represent an extension of the current multi-year implementation programs developed by many agencies to cover the entire planning period.[3]

Typically a target year plan for 20 to 30 years in the future is developed along with a short-range (up to five years) implementation program. The recommendation here is to combine the requirements for the target year transportation plan and the implementation program into one multi-modal planning program package that reflects reasonable resource assumptions and illustrates all the actions on transportation anticipated for an area over the entire planning horizon. Thus, the proposed planning document would contain all capital improvement projects (large and small) including maintenance, operating and policy changes, and planning studies for all modes within a region. In addition, the planning program should identify all

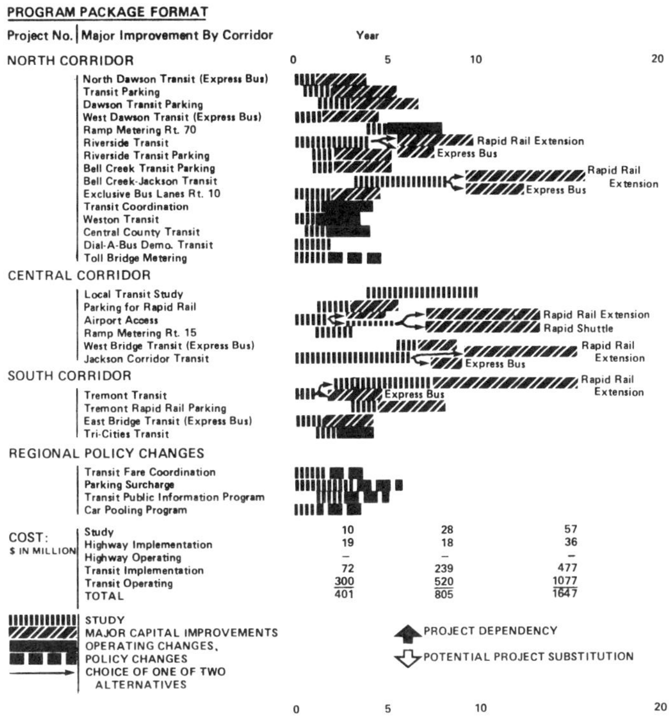

19.4 Example program package format.

sources of revenues and the implementing or operating agency responsible for each action contained in the program. An example of such a program is illustrated by figure 19.4. The key planning document thus would not be simply a series of maps, though maps might serve as supplementary information illustrating planned improvements for some categories of projects.

There are a number of advantages to defining the basic planning document as a multi-year planning program rather than a target year plan. Developing a planning program encourages the planning agency (or agencies) to focus on resources and other constraints early in the planning process. The result will be a plan that represents a proposal with a realistic potential for implementation and hence a better guide for transportation decision-making. In the past, system plans often have described only "desirable" or "wish list" target networks that have had little influence on and relationship to programming and implementation decisions. While the content of the plan can be expected to change, at any point in time the plan should reflect as realistically as possible existing or anticipated constraints on transportation decisions (i.e., resource, environmental, etc.) for at least the near-term future.

Combining short-run and long-range improvements in a program package increases the potential for the effective involvement of a wide range of different interest groups, particularly in system planning. Program packages define both immediately implementable steps (such as flow metering, priority bus lanes, dial-a-ride, and pricing policies) and the longer-range improvements (such as major new facilities and broad policy changes like funding, compensatory programs, peripheral parking schemes, or even new studies). Since the lead time for major transportation projects can be as long as ten years, program packages show interest groups how their concerns can be addressed in the near future as well as giving all participants a realistic sense of the time required for more major improvements.

The public should not be involved simply in deciding on the studies and projects to be included in the plan, but also on the relative priority of those studies and projects. The proposed plan format focuses explicitly on the scheduling of projects and planning activities.

Defining the plan as suggested here will make ongoing coordination of system and project planning easier. Since the desirability of some projects in a region, or their timing, may depend on other projects in the plan due to traffic or resource considerations, the program package can explicitly identify these interdependencies and indicate when or how changes ought to be made to the program package if the status of a particular project changes. In programming and project development, a project often is considered to be independent, for the most part, from other projects under study. Using a planning program format will facilitate the inclusion of system considerations in project development decisions. Also, combining plans and programs for all modes will encourage cooperation among agencies doing system planning and those responsible for implementation.

Contingency plans may provide additional flexibility within the basic planning program format, particularly in areas where a number of controversial issues are unresolved or for major projects with significant lead times. In these cases, in order to facilitate the orderly allocation of resources, it may be desirable to develop tentative implementation strategies for a range of potential future decisions. For example, if a major transportation terminal has a fifteen-year lead time but no assurance of its acceptability can be made that far in advance, it may be wise to examine and plan a contingency program for smaller scale satellite improvements in case at some point in project development the large single terminal concept is dropped. By anticipating such occurrences, the agency can provide for an orderly implementation program rather than reacting to crises only after they occur.

While developing contingency plans may require more resources for planning, it provides for a more realistic range of network and project choices when there is significant uncertainty in funding levels, community acceptability, the predicted impacts of proposed projects, and external events with implications for transportation (land-use control, air quality regulations, etc.). In the long-run, contingency plans may result in a more efficient use of resources by providing flexibility and keeping options open.

While alternative plans can be displayed particularly for the medium- to long-run, the first one to two years of the planning program should be decisive and represent what actually will be budgeted and implemented or studied during that time period. By making the first few years of the program decisive, but explicitly recognizing the range of choices available in subsequent time periods, the planning document can support and indeed encourage a periodic decision-making process. Every one or two years a new budget is prepared and the entire planning program updated and documented. The proposed format then encourages a planning process whose periodic output is a new budget reflecting the fiscal flows anticipated in the planning program.

The development of a multi-year program package should include explicit documentation of the social, economic, and environmental impacts through preparation of a system environmental report (SER). A system environmental report should neither approach the detail currently found in project environment impact statements as practiced in the United States nor simply summarize the current status of any ongoing project environmental analysis. Rather, the SER should present a summary of the plan's area-wide implications for, and impacts on, the environment as well as provide a framework within which later and more detailed project environmental analyses can occur.

Specifically, the SER might contain the following:

1. Identification (and possibly map overlays) of environmentally sensitive areas, land-use assumptions, prevailing air and noise pollution contours, and general topography. Also included would be basic demographic data and projections on population, income levels, and employment, and the range of uncertainty associated with these projections.

2. A summary of aggregate area-wide social and environmental effects implied by each of the system plan alternatives under consideration. Such a summary might estimate such things as the total open space and farm land likely to be taken or subsequently developed, the area-wide effect of displacements from all projects on the housing market, as well as the compensation program required to attempt to minimize adverse effects.

3. Identification of unresolved issues or further studies required in order to estimate the system-wide social and environmental effects.

4. Identification of the status of environmental studies for each project, including major unresolved issues, network implications (i.e., interdependencies with other projects) for projects currently being seriously questioned on environmental grounds, and the current status of the project environmental impact statement (underway, completed, approved, etc.).

20 IMPLEMENTATION

The preceding chapters have presented guidelines and techniques that transportation agencies can use, with the participation of other national, provincial, and county agencies and local citizens and officials, in developing and evaluating alternative transportation plans. Most of the methods and techniques described individually can be adopted without difficulty. However, when taken together, the proposals imply a basic strategy of transportation planning and decision-making that is significantly different from what has been traditional. The overall approach, then has far-reaching implications for the transportation professional.

APPLICATION EXPERIENCE

Elements of the proposed approach have been applied in recent years to system, location, and design-oriented studies, both in the United States and in other countries. As part of our own research, we have worked cooperatively with the states of California, Michigan, and Georgia on both urban and rural problems. Our primary conclusions from these efforts is that the approach is workable, is essential to the solution of transportation problems in today's society, and has been gradually accepted by the professionals with whom we have worked. Implementation, though, does take time; full and successful application is not immediately achieved.

The approach in its fullest extent involves more than just the introduction of one or two new techniques or procedures. Full implementation can have implications for personal attitudes, management capabilities, staffing, and the manner in which decisions are reached. Personnel from all levels and all departments of an organization need to be involved; implementation cannot be the sole responsibility of a single unit, such as the environmental staff.

For some individuals, the shift to a process approach is difficult; the traditional concern in most agencies has been with the development and implementation of specific, well-defined "products," and less concern has been devoted to the process by which these products are developed. It is our experience that implementation is best accomplished with a carefully co-ordinated and sequenced set of changes in training, policy, personnel, and practice executed over a period of time. In brief, the process of implementing change is as important as the recommendations themselves.

Determining what change should occur (what to do), however, is easier than determining how to bring about that change (how to do it). Implementation requires much more than just talking and reporting. While these clearly can be components of a larger change strategy, changes in behavior don't always follow, even when a person is saying the correct words. A number of organizational barriers exist which work against change in general and, in particular, the introduction of increased environmental sensitivity within transportation agencies. A social system tends to react to new changes in ways that keep old policies and practices stable. Major changes such as those being recommended tend to provoke complex and compensating reactions. Unless the interrelationships between different parts of an organization are accurately identified, the organization may react to subvert the intended change. A coherent diagnosis of the organization can help those who initiate change to anticipate these reactions and to plan for them.

UNITED STATES FEDERAL HIGHWAY ADMINISTRATION "PROCESS GUIDELINES"

The focus of many current environmental changes in United States highway agencies has been U.S. Department of Transportation, Federal Highway Administration Policy and Procedure Memorandum 90-4, the Process Guidelines, which requires each state highway agency to prepare an Action Plan describing the procedures and assignments of responsibility to be followed in considering social, economic, and environmental effects in all stages of developing a highway project, including network-wide system planning.

A fundamental intention of the Process Guidelines was that each state should systematically identify and review its current organizational structure, decision-making processes, and technical procedures utilizing a set of general guidelines that had been developed cooperatively by the states themselves, the federal government, and environmental organizations. The Process Guidelines are consistent with the issues and techniques that are discussed in this paper, and include specific sections on the consideration of alternatives; identification of social, economic, and environmental effects; involvement of the public and other governmental agencies, and interdisciplinary staffing.

NEXT STEPS

Major advances have been made in the understanding and development of methods for incorporating social and environmental considerations in transportation decision making. Even so, while research needs still exist, routine operations of most transportation agencies now lag behind current state-of-the-art understanding. Priority emphasis therefore should be given to implementation at all levels of government: national, provincial, county, and local.

Although many transportation organizations agree with this priority and recognize the desirability of adopting a planning process that more effectively incorporates means of considering social, economic and environmental impacts, there are several factors which retard implementation. The major need, then, is for development of means to aid the implementation of improved procedures.

Staffing

Many transportation agencies lack sufficient personnel with expertise in non-engineering disciplines and are hard pressed to obtain the authority to hire them. Salary scales and advancement opportunities for non-engineers often are relatively unattractive, making it difficult to find and keep qualified people. Arrangements with other agencies for the lending of specialists may be limited or nonexistent.

Because it is important to have an adequate in-house staff of social, economic, and environmental specialists, needs exist for

- ways of authorizing and funding non-engineering personnel;
- appropriate salary scales and career opportunities for non-engineers;
- means of obtaining the services of experts available in other governmental agencies.

Training Programs

Both engineers and professionals in other disciplines need training which updates their capabilities in their own discipline and which increases their understanding of, and capacity to work with, other disciplines. Development of methods and programs for on-the-job training and of opportunities for further formal education are essential.

Funding

Fiscal constraints that may inhibit the freedom to take a truly multi-modal and service oriented approach to transportation should be removed. Funding should be available on an equal basis for the construction, operation, maintenance, and planning of all modes and services. Compensatory programs should be broadened to achieve greater equity, such as the compensation of persons beyond the immediate transport right-of-way for adverse social, economic, or environmental effects which may be directly incident upon them.

Applicability to Public Policy Decisions

The approach described was developed originally for highway decisions at both the project and system level. Although the research was initially oriented to the construction of freeway type facilities in urban areas, subsequent phases have examined smaller scale projects such as secondary road upgradings, rural settings, and different geographic regions within the United States.

We feel that the approach is also directly applicable with few or no modifications to other transportation modes: transit route (or station) location and design; operation, parking, scheduling, and other regulatory policies; bus routing; railroad development; airport location and design; and, in fields other than transportation: power plants, refineries, urban development, flood control projects, and similar public works decisions.

In principle, the concepts underlying this approach, and the approach itslef, seem applicable to a wide spectrum of public policy problems. We believe the approach described is essential if transportation agencies are to have the confidence of the public. The issues of "the environment" cannot be handled by simply hiring a few architects or ecologists; nor can issues of equity be resolved simply by a program of citizen participation. To retain the confidence of the public that we are seriously responding to their concerns for environmental issues, we must reexamine and restructure our total processes of transportation planning and decision-making.

21 SELECTED REFERENCES

Alonso, William, "Beyond the Interdisciplinary Approach to Planning," *Journal of the American Institute of Planners*, Vol. 37, No. 3, 5/71.

Altschuler, Alan A., *Community Control: The Black Demand for Participation in Large American Cities*, New York: Pegasus, 1970.

Arnstein, Sherry, "A Ladder of Citizen Participation," *Journal of the American Institute of Planners*, Vol. 35, No. 3, 7/69.

Bennett, Elizabeth, and John H. Suhrbier (eds), "Proceedings of a Panel Discussion on the Consideration of Alternatives in Highway Planning and Design," in consultation with the U.S. Department of Transportation, Federal Highway Administration, Office of Environmental Policy, 12/4/73.

Bishop, Bruce A., *Public Participation in Water Resources Planning*, Alexandria, Virginia: Department of Army, Corps of Engineers, 12/70.

Bleiker, Hans, J. H. Suhrbier and M. L. Manheim, "Community Interaction as an Integral Part of the Highway Decision-Making Process," *Highway Research Record 356, Social and Environmental Factors of Transportation*, Washington, D.C.: Highway Research Board, 1971.

Colcord, Frank C., Jr., "Decision Making and Transportation Policy: A Comparative Analysis," *Southwestern Social Science Quarterly*, 48:3, 12/67.

——, "Transportation Systems Planning in California: Institutional Arrangements of State and Local Governments," Cambridge, Massachusetts: Massachusetts Institute of Technology USL Report 72-2, 6/72.

Federal Highway Administration, Office of Environmental Policy, "Progress Report on Implementation of Process Guidelines," 5/10/74.

Fielding, Gordon J., "Group Dynamics in the Urban Freeway Decision Process," Sponsored by California Division of High-

This list contains a number of reports and papers resulting from the M.I.T. Transportation and Community Values research effort, plus selected other references of a general nature.

ways and U.S. Department of Transportation, Bureau of Public Roads, Irvine, California: School of Social Sciences, University of California at Irvine, 9/71.

Geiser, Kenneth R., Jr., "Political Processes of Urban Freeway Controversies," Cambridge, Massachusetts: Urban Systems Laboratory, M.I.T., 1970.

Giel, Robert, "Transportation Decision-Making and Organizational Structure," Cambridge, Massachusetts: Massachusetts Institute of Technology, USL Report 72-9, 11/72.

Highway Research Board Conference Report, "Organization for Continuing Urban Transportation Planning," Washington, D.C.: Highway Research Board, 1971.

Highway Research Board, *Transportation and Community Values*, Special Report 105, Washington, D.C.: Highway Research Board, 1969.

Hill, Stuart L., "Century Freeway (Watts)," *Joint Development and Multiple Use of Transportation Rights-of-Way*, Special Report 104, Washington, D.C.: Highway Research Board, 1969.

Lockwood, Stephen, "The Boston Transportation Planning Review: A Case Study in Community/Technical Interaction," *Planners Notebook*, Vol. 2, No. 4, American Institute of Planners, August 1972.

Manheim, M. L., "Reaching Decisions About Technological Projects with Social Consequences: A Normative Model," presented to the Seventeenth North American Meeting, Regional Science Association, November 6-8, 1970. Professional paper P70-78. Also published by the Elsevier Publishing Company, Amsterdam, The Netherlands.

Manheim, Marvin L. and Elizabeth D. Bennett, "The Air Quality Standards: A Potential for Improving Local and State Decision-Making," paper prepared for a workshop on transportation & air pollution control by the U.S. Library of Congress for the Subcommittee on Air and Water Pollution, Committee on Public Works, U.S. Senate, 1/74.

Manheim, Marvin L., and J. H. Suhrbier, "Community Values in Transport Project Planning," *Proceedings—Twelfth Annual Meeting, Transportation Research Forum*, Oxford, Indiana: The Richard B. Cross Company, 1971.

——, "Incorporating Social and Environmental Factors in Highway Planning and Design," *Environmental Considerations in Planning, Design and Construction*, Special Report 138, Washington, D.C.: Highway Research Board, 1973.

Manheim, M. L., et al, "Community and Environmental Values in Transportation Planning: Summary of Findings and Recommendations," USL Report 72-2, Cambridge, Massachusetts: M.I.T., 6/72.

Ministry of Transport (United Kingdom), "New Roads in Towns, Report of the Urban Motorway Committee, 7/72.

Neumann, Lance A., and W. M. Pecknold, "Application of the Time-Staged Strategic Approach to System Planning," *Highway Research Record 435*, Washington, D.C.: Highway Research Board, 1973.

Orski, C. Kenneth, "Vehicle-Free Zones in City Centers," U.S. Department of Housing and Urban Development, Office of International Affairs Brief 16, Washington, D.C., 6/62.

Pecknold, W. M. et al, "Transportation System Planning and Community Environmental Values," U.S.L. Report 72-3, Cambridge, Massachusetts: M.I.T., 6/72.

Petersilia, Michael and J. H. Suhrbier, "Community Values in Highway Planning: A Case Study," USL Report 72-5, Cambridge, Massachusetts: M.I.T., 12/72.

Reno, Arlee T., and B. Richardson, "Integration of System and Project Planning: The Role of Transportation Corridor Studies," USL Report 72-4, Cambridge, Massachusetts: M.I.T., 9/72.

Reno, Arlee T., "Interaction Procedures in the Transportation Systems Planning Process," *Highway Research Record 394*, Washington, D.C.: Highway Research Board, 1972.

"Report to Congress on Section 109(h), Title 23, United States Code—Guidelines Relating to the Economic, Social and Environmental Effects of Highway Projects," Committee on Public Works Print 92-45, U.S. Government Printing Office, Washington, D.C., 8/72.

Suhrbier, John H., and Elizabeth Bennett (eds.), "Proceedings of a Panel Discussion on Community Involvement in Highway Planning and Design," in consultation with U.S. Department of Transportation, Federal Highway Administration, Office of Environmental Policy, Washington, D.C., Cambridge, Massachusetts: M.I.T., 1/26/73.

Suhrbier, John H., Elizabeth Bennett and Deborah Card (eds.), "Proceedings of a Panel Discussion on the Systematic Interdisciplinary Approach in Highway Planning and Design," in consultation with U.S. Department of Transportation, Federal Highway Administration, Office of Environmental Policy, Washington, D.C., 3/21/73.

Shurbier, J. H., and E. Bennett (eds.), "Proceedings of a Panel Discussion on the Interrelation of Transportation System and Project Decisions," in consultation with U.S. Department of

Transportation, Federal Highway Administration, Office of Environmental Policy, Washington, D.C., 11/1/73.

Thomas, E. N., and J. L. Schofer, *Strategies for the Evaluation of Alternative Transportation Plans*, Washington, D.C.: National Cooperative Highway Research Program Report 96, 1970.

Transportation Research Board, "Highways and Air Quality," Special Report 141, Washington, D.C., Transportation Research Board, 1974.

United Kingdom Department of the Environment, "Development and Compensation—Putting People First," London, 10/72.

NOTES TO PART IV

Notes to chapter 16

1. "Urban Public Transport: Service Innovations in Operations, Planning and Technology" by Robert Burco, Inquiry into the Impact of the Motor Vehicle on the Environment, Working Paper No. 2 (OECD, Paris, 1972).

2. The Urban Motorways Committee in the United Kingdom examined in depth the consideration of alternatives in the design process, examining the relation of urban road design to the planning of urban areas as a whole and developing methods of alleviating adverse effects through the use of geometric design practice, protective measures, and compensatory practices. A summary of the committee's work is provided by R. A. Brown and W. Deakin of the U.K. Department of the Environment, "Development: People and Compensation," *Proceedings of the International Conference on Transportation Research, Bruges, Belgium, June 1973*, Transportation Research Forum, Chicago, Illinois, 1974, pp. 72–77.

3. An excellent case study illustrating the preoccupation with the "hardware" of a transport system and the neglect of the necessary "software" is presented by Samuel P. Mauch (Baselhard, Hoffman; Zurich, Switzerland). "Origins and Consequences of the Overemphasis on Hardware in Transportation Planning," *Proceedings of the International Conference on Transportation Research; Bruges, Belgium, June, 1973*, Transportation Research Forum, Chicago, Illinois, 1974, pp. 62–66. Dr. Mauch examines expressway and subway planning in Zurich and proposed transit through the Swiss Alps, identifying institutional factors which bias the consideration of alternatives towards only physical planning.

4. Several examples of the specific kinds of contributions that can be made by the public are contained in the Boston Transportation Planning Review papers by Ralph Gakenheimer, John G. Wofford, Stephen C. Lockwood, Walter G. Hansen, Jason M. Cortell, and Allan K. Sloan in "Citizen Participation in Transportation Planning," Special Report 142, Transportation Research Board, Washington, D.C., 1973, pp. 105–131.

5. The technical aspects and procedures associated with the prediction of environmental impacts are addressed in numerous publications and are not treated as part of this report. Publications of the OECD Inquiry Into the Impact of the Motor Vehicle on the Environment which address these technical issues include: "The Motor Car and Natural Resources" by Gerald Leach, Inquiry into the Impact of the Motor Vehicle on the Environment, Working Paper No. 4 (OECD, Paris, 1973); and "Automobile Air Pollution and Noise: Implications for Public Policy," Report of an Ad Hoc Group to the Environment Committee (Published as Part II of Environmental Implications of Options in Urban Mobility, September 1973).

6. Donald Schon, *Beyond the Stable State*, Random House, New York, 1971.

7. Robert Burco, *op. cit.*, p. x.

Notes to chapter 18

1. A linear scoring function assigns weights to various attributes of alternatives (such as environmental amenity, user benefits, etc.) and adds up a "total score" for each alternative. It is one step broader than traditional benefit/cost analysis but suffers from virtually similar deficiencies.

2. A more in-depth discussion of other evaluation techniques is contained in Martin Wachs, Barclay M. Hudson, and Joseph L. Schofer, "Integrating Localized and Systemwide Objectives in Transportation Planning," School of Architecture and Urban Planning, University of California at Los Angeles, January 1973.

Notes to chapter 19

1. In the U.S., the National Environmental Policy Act of 1969 *requires* the use of a systematic interdisciplinary approach. One possible organizational approach to interdisciplinary planning is through the use of "area teams" where the team is responsible for one sub-area of a larger metropolitan area or region and is composed of individuals representing a variety of disciplinary backgrounds and functional responsibilities. The use of this approach by the Greater London Council is described by John Parker and Warren Panther, "Traffic and Environment: The Multi-Disciplinary Team and its Working Methods," *Proceedings of the International Conference on Transportation Research, Bruges, Belgium, June 1973*, Transportation Research Forum, Chicago, Illinois, 1974, pp. 78–85.

2. Pennsylvania and California in the United States are two examples where major changes in this direction are being implemented.

3. Such a document will shortly be required for state and metropolitan transportation plans in the U.S. See 39 Federal Register 39665 (1974) Transportation Improvement Program.

V

AUTOMOBILES AND CITIES: Strategies for Developing Countries

Wilfred Owen

ACKNOWLEDGMENTS

This report on the automobile in developing countries is an exploratory effort made possible through the cooperation of the Brookings Institution, the World Bank, and the Organization for Economic Cooperation and Development. Statistics on transportation in the developing world were prepared by Inci Ak Ozusta of the World Bank. Additional data on transportation in specific cities were supplied by Donald Strombom, R. Venkateswaran and Inai Bradfield. OECD reports providing inputs on technology, resources and traffic include those of Kenneth Orski, J. Michael Thomson, Robert A. Burco, Gerald Leach, Jörg Kühnemann, and Robert Witherspoon. A draft of the report was discussed at the first course on Guiding Urban Growth, conducted by the World Bank's Economic Development Institute, providing an opportunity to benefit from the views of policy-makers from major cities in Africa, Asia and Latin America. The conceptual framework of the report is based on the Brookings book, *The Accessible City*.

The attempt to present a broad-brush view of the automobile around the world reveals many areas of further inquiry that need to be explored. Among them are questions of the possible form of future urban settlements, the technology of the automobile itself, the potentials of transport innovations that might someday replace the private car, and the costs of different transport mixes for cities of varying physical characteristics and stages of growth. Extensive research and development are needed for long-term resolution of the problems of urban mobility and urban living.

Special thanks for support of the study are extended to Kenneth Orski of OECD; to Robert Sadove, Kenneth Bohr and Harold Dunkerley of the World Bank; to Andrew Kamarck and Raymond Frost of the Economic Development Institute; and to Kermit Gordon and Joseph A. Pechman of the Brookings Institution.

SUMMARY

Automobile ownership, which has been closely associated with the economic growth of the United States, Europe and Japan, is now a global phenomenon that is having major impact on developing countries. Cars are generally incompatible with big cities, and in the developing world half to three-quarters of all cars are concentrated in major metropolitan areas. There are more automobiles in Mexico City and Sao Paulo than in Philadelphia or Dallas. Congestion is already acute, yet the outlook in developing countries is for six times as many cars by the year 2000—well over 100 million altogether.

The most effective immediate answer to automobile congestion would be a concerted effort to improve public transportation. This service on which most people depend is often grossly inadequate. Yet if the number of buses in cities of over half a million population could be doubled, and public transit given priority on city streets, it would be possible to provide 135 million more rides per day for an investment of $4,000 million. The outlay would be far less than the cost of new roads or of continuing congestion, and would be less than 10 percent of the cost of an equivalent capacity in subways.

A public transport solution might also include a better organization of group-riding taxis or public automobiles. One taxi can substitute for six private cars and reduce automobile traffic by 80 percent. Both taxis and buses would need to be protected from the congestion caused by private cars in central areas, on high-density routes, and in peak traffic hours. Physical constraints on the automobile might include auto-free zones, all-bus streets, and bans on curb parking. Price constraints would include charges on automobiles to pay for urban highways and parking, and to cover the social costs of accidents, pollution and congestion.

But cities suffer from traffic congestion with or without the automobile, and the reasons lie beyond the technology of transport, in planless cities that generate unnecessary movement. The haphazard location and arrangement of urban activities, the poor environment, and the lack of physical relationship between housing, jobs, and services mean that conventional cities must rely on travel to compensate for disorder.

Transportation problems, therefore, will have to be cured in part by non-transportation solutions. In existing cities a better balance between transportation supply and demand calls for redevelopment that organizes and reduces travel requirements at the same time that improvements in public transport are increasing capacity. For new urban growth areas, the best way to deal with transportation problems is to deal effectively with other urban problems through community design and city-building. The most effective attack on urban congestion may be an attack on accidental cities.

The developing countries need help in planning and financing land acquisition and public infrastructure for the large-scale urban systems needed to accommodate urban growth.

Transportation should not be used exclusively to support congestion, but to move away from congestion and to help build communities that work. A positive approach to congestion will de-emphasize the problems of traffic movement and focus on using transport technology to help create new forms of urban settlement.

22 TRENDS AND OUTLOOK

THE AUTOMOBILE EXPLOSION

The desire to live in cities and ride in automobiles, generally viewed as a special attribute of affluent societies, has now spread to most of the developing world. Cities already overcrowded and suffering critical shortages of housing and public services will double their population by the end of this decade and triple their ownership of automobiles. While the results of the automotive invasion have included important economic and social change, the unwanted side effects are causing critical urban problems. Traffic congestion has reached staggering proportions, the environment has become increasingly polluted, public transit is grossly inadequate, and pressures are mounting for costly expressways and subways which may divert attention from other more basic urban problems and fail to relieve congestion.

The problems created by the automobile are not the most urgent problem facing cities, but mobility and accessibility have important impacts on urban form, on the availability of economic and social opportunities, and on the quality of urban life. Transportation has such pervasive effects that it dictates the success or failure of many aspects of community development.

The efficiency of the city and therefore the contribution of urbanization to development will depend to an important degree on how the problems of mobility are resolved. A key factor will be the degree to which the automobile can be accommodated or contained in the interest of a satisfactory total transportation system. The growth of car ownership and use on the western pattern may be inevitable, but it is also to envision successful efforts to make public transit work, which may require economic and physical restrictions on downtown auto use. It is also increasingly possible that urban growth policies and community designs may be able to reduce the causes of congestion.

The importance of a combined strategy for transportation and urban development stems not only from the critical problems of the moment but from population projections that assume nearly a thousand million more people in the cities of developing countries between now and the year 2000.[1] The increase in urban population will be accompanied by increases in income and presumably by a continuing expansion of car ownership. In the face of these trends, the question is whether prompt public action might help to avert the congestion and disorder that now threaten the cities of the developing world, or whether these cities must follow the trends and suffer the consequences.

Motorization and the recent disillusionment with cars in cities took place over a period of half a century in the United States and a quarter century in Europe. In the developing countries, however, the periods of motorization and of serious misgivings about it are happening almost simultaneously. The situation suggests the adoption of policies now that might avert some of the undesirable consequences of the automotive age. The hope of narrowing the income gap between the developed and the less developed countries will depend to an increasing degree on how the growing problems of the urban sector are met, and reducing the conflict between cities and their transportation seems to be one of the keys to an economically successful urbanization process.

Between 1960 and 1970, while the number of people in the world increased less than 20 percent, the number of motor vehicles increased more than 100 percent. In the single decade of the sixties the number of motor vehicles increased by 120 million. Almost half the gain was in Europe, and nearly one-fourth in the United States, but close to another fourth was registered in Asia, Africa and South America.

Table 22.1 World Motor Vehicle Registrations, 1973

	Millions of motor vehicles	Persons per vehicle
North and Central America	139.0	2.3
South America	11.0	18
Africa	5.0	69
Asia	32.5	64
Europe	102.0	7
Oceania	7.0	2.9
Total	296.6	12

Source: Motor Vehicle Manufacturers Association, *Automobile Facts and Figures*, 1975, pp. 26–27.

Table 22.2 Urban Concentration of People and Cars

	Percent of national total	
	Population	Automobiles
Abidjan	12.5	70.4
Bangkok	7.9	72.8
Bogota	11.6	35.6
Bombay	1.1	14.5
Caracas	19.3	36.1
Djakarta	3.8	36.6
Nairobi	4.7	59.4
Sao Paulo	6.1	25.8
Teheran	10.8	49.1

Source: IBRD Urbanization Sector Working Paper, March 1972, Annex I, p. 9.

Compared to Europe's one vehicle for every 7 people, there is only one for 18 people in Latin America, 69 in Africa, and 64 in Asia. Individual countries reveal greater extremes. There is one vehicle for 9 people in Singapore, 4 in Japan, 8 in Argentina, 445 in India, and 1,178 in China.[2]

But the imbalance between car ownership for the rich and the poor has begun to change. For whereas the number of motor vehicles in the United States increased 35 percent during the 1960's, the increase was 100 percent in South America and 300 percent in Asia. The number of vehicles increased nearly five-fold in Thailand and tripled in Pakistan and the Philippines.

The upward trend of registrations is accounted for largely by the private car, which becomes an increasing proportion of the motor vehicle total as national incomes rise. Passenger cars are still only 56 percent of the total in developing countries, but moving rapidly toward the 80 percent mark already achieved in Europe and North America. Of the 50 countries in the world with 100,000 or more passenger cars, about half of them are generally considered to be developing countries. Brazil has over 4 million cars and Argentina 2 million. Several others have between 300,000 and 600,000, including Venezuela, India, Puerto Rico, Yugoslavia, Philippines, Iran, Indonesia and Malaysia.

The impact of these trends on the cities has been magnified by the fact that automobiles in developing countries are heavily concentrated in a small number of large metropolitan centers. Bangkok has 74 percent of all the automobiles in

Table 22.3 Marginal Growth in Motorization and Population, Selected Urban Areas

	Growth Period	Net additional growth	
		Population (000)	Vehicles
Lima	1970–1980	2,100	388,745
Caracas	1970–1990	1,900	430,341
Sao Paulo	1967–1990	13,400	1,988,937
Buenos Aires	1969–2000	4,200	1,550,000
Istanbul	1970–1990	2,600	240,030
Kuala Lumpur	1970–1990	600	110,642
Dar Es Salaam	1967–1989	728	50,900

Source: Transport Plan Studies and Urban Reports, IBRD.

Thailand, Teheran has 49 percent of the total in Iran, and Nairobi accounts for 59 percent of the cars in Kenya. Brazil as a whole has only one car for every 61 people, but its largest city, Sao Paulo, has one for every 6 persons, not much different from the 5 persons per car in New York City.

What this means is that while the big cities of developing countries have fewer vehicles in relation to people than the typical western city, they often have more people and therefore more vehicles than some of the largest cities of the more developed countries. For example, Sao Paulo and Mexico City have more motor vehicles than Philadelphia, San Diego, or Dallas. Large cities in the developing world are coping with just as much automobile traffic as some of the most motorized cities in the United States, and doing so with far less roadway capacity, much higher levels of bus and taxi riding, and generally no subways.

The outlook is for still more congestion. Motorization, urbanization and affluence go together, and the world is becoming larger and more urban, and its cities more affluent. The 1970 world passenger car total of 187 million units was nearly double the 1960 figure, and car ownership is now expected to double again by 1985. By the end of the century the world is expected to have 524 million cars in operation.[3] Developing countries, which accounted for fewer than 17 million cars in 1970, may own over 100 million by the year 2000.

The impact of these growth trends will continue to be felt most acutely in the primate cities. Caracas is expected to double its population and triple the number of its automobiles in less than 20 years. In Lima the number of cars is likely to quadruple between 1970 and 1985, and there will also be 2 million more people. Istanbul may have six times as many cars in 1990 as in 1970. Buenos Aires is expected to have 4 million more people by 2000, and 1.5 million more cars, while Sao Paulo will have 13 million more people and 2 million more automobiles. If these projections turn out to be right, today's traffic conditions are only a foretaste of the congestion to come.

URBAN TRAVEL PATTERNS: RICH VERSUS POOR

Since most people in the majority of the developing countries are extremely poor, the growth of car ownership will still leave a large proportion of the people carless, and the automobile will continue to play a limited role in urban transportation.

Most people will depend on walking, cycling, and public transport. In Seoul, Bombay, Calcutta, Lima, Bogota, and Mexico City, 80 to 95 percent of all trips are by public transit, mostly bus. Autos account for less than 10 percent of trips in Seoul and Calcutta. At the other extreme, in Kuala Lumpur 47 percent of all motorized trips are by automobile, and in Caracas 46 percent. The figure is 37 percent for Teheran and 29 percent for Bangkok.

The bicycle is also playing a significant role and more recently the motor cycle, motor bike and motor scooter. In a selected group of 22 countries outside the United States, there were 40 million motorized two wheelers on the registration lists in 1970. In most cities, according to estimates, there are generally two to four times as many non-motorized two-wheelers.

Cities that estimate the number of pedestrian trips confirm the substantial dependence on walking. The percentage of total trips made on foot was as high as 67 percent in Dar Es Salaam, 63 percent in Kinshasa, 48 percent in Taiperi, 27 percent in Seoul, and 22 percent in Madras. The very poor operate outside the mechanized transport system or do not travel at all.

In Bogota 55 percent of all trips taken by the poorest families in 1969 were on foot, another 40 percent were by bus, and only 5 percent were by taxi. No one drove. In the middle income group most people were able to pay the bus fare, and 70 percent used this method of travel. But for high-income families the pattern was quite different: 80 percent rode in automobiles and only 10 percent in buses. Bombay travel patterns are even more closely income-related. For the poorest families it was 91 percent public transport (mostly train)

Table 22.4 Travel Methods in Selected Urban Areas (%)

	Motorized trips				
	Public Transport	Private Car	All Private Transport[a]	Walking	Total
Seoul[b]	95.2	4.8	4.8	–	100.0
[c]	67.2		5.9	26.9	100.0
Hong Kong	70.6	21.7	29.4		100.0
Taipei[b]	71.3	2.5	28.7		100.0
[c]	38.0		13.5	48.5	100.0
Bangkok	66.0	29.0	34.0		100.0
Singapore	67.3	24.2	32.7		100.0
Kuala Lumpur	39.8	46.9	60.2		100.0
Madras	56.0		22.0[d]	22.0	100.0
Bombay	89.0	10.7	10.7		100.0
Calcutta	90.4	8.2	9.6[e]		100.0
Teheran	42.0	37.0	58.0		100.0
Istanbul	43.3		56.7		100.0
Caracas	54.4	45.6	45.6		100.0
Bogota	83.3	16.7	16.7		100.0
Mexico City	81.0	19.0	19.0		100.0
Buenos Aires	72.5	15.4	18.4	9.1	100.0
Lima-Callao	89.0	11.0	11.0		100.0
Sao Paulo	73.7	26.3	16.3		100.0

[a]Includes private motorcar and two-wheeled motorized vehicle trips.
[b]Walking excluded.
[c]Walking included.
[d]Includes trips by public motorcar.
[e]Includes commercial vehicle trips (1.4%).

Table 22.5 Private Motor Cars, Motorized 2-Wheelers, and Non-Motorized 2-Wheelers (per 1,000 people)

	Private Cars	Motorized 2-wheelers	Non-motorized 2-wheelers
Casablanca	72.9	84.9	223.4
Singapore	73.0	51.6	182.0
Djakarta	18.0	22.6	46.4
Teheran	44.4	11.1	41.5
Lima	36.5	5.9	16.3
Seoul	4.2	0.4	8.0
Istanbul	15.6	2.4	3.1

Table 22.6 Urban Travel Patterns by Income Groups

	Percentage of home based trips (Bogota, 1969)				
Income Group	Walk	Bus	Taxi	Car	Total
Low	55	40	5	0	100.0
Medium	5	70	15	10	100.0
High	0	10	10	80	100.0

	Percentage of all trips (Hong Kong, 1966)		
Monthly income (1966 HK $)	Public Transport	Taxi	Private Car
0-300	88.4	5.6	6.0
1,001-1,500	67.6	9.7	22.7
2,001-3,000	37.3	9.9	52.8
Over 5,000	14.9	12.8	72.3

Monthly income (1969 NT $)	Mode utilized (Taipei, 1969)
Below 1,000	Bicycle
2,000-3,000	Bicycle or bus
3,000-5,000	Bus, motorcycle, or taxi
5,000-20,000	Taxis, or cars
Above 20,000	Cars

while in the highest income category 95 percent traveled by car.

In Hong Kong it is estimated that for an indefinite period three out of four people will be unable to afford a car. The 4 million residents of Hong Kong owned 90,000 automobiles in 1970, and registrations were increasing by 17 percent per year. Yet only 10 percent of the people were car owners, and 90 percent of travelers depended on bus and taxi. As a result, the number of trips taken per Hong Kong household by non-car owners in the low income brackets is only 0.3 per day, while middle and upper income households with cars make an average of 6 to 8 trips a day.

Altogether, Hong Kong streets accommodate 5.5 million passenger trips per day, 5 million by public transit and half a million by car. In peak hours, buses account for only 3 percent of the vehicles, but they carry 36 percent of the passengers. Cars account for 35 percent of the vehicles and carry only 15 percent of the people.[4]

Bangkok, with 290,000 motor vehicles (150,000 of them cars), is heavily congested, and Bangkok's 21 private bus companies are notorious for their poorly maintained equipment and bad service. On an average day about 2 million passengers are transported in the city, which means there are about two-thirds as many motorized trips per day as there are people. Of this total, 22 percent go by car, 15 percent by motor cycle and motor tricycle, 13 percent by taxi, and 46 percent by bus. Again, it is the bus and taxi that account for the major movement by motorized transport.

Istanbul, with 2 million people and a per capita income of US$400 (as of 1972) is somewhat smaller and less affluent than Bangkok. From 10,000 vehicles in 1952, the total has increased tenfold to 105,000 in 1970, 60,000 of them automobiles (of which 16,000 are taxis). Istanbul public transport is highly important in the total picture, but the conventional bus and trolley bus systems have been steadily losing ground to the taxi, the jitney taxi (dolmus) and the minibus. These various forms of "public automobiles" now carry more than the buses. The regular transit system accounted for 42 percent of total revenue passengers in 1952, but this proportion had fallen to 28 percent 15 years later. The decline is due partly to an antiquated route system and poor organization of bus operations.

The situation is quite different in Cairo, with its 6 million people and low average family income. There the private automobile is an insignificant factor, with about one motor vehicle per hundred people, most of them taxis and trucks. The public transit system carries over 3 million passengers a day, 2.5 million of them by bus. (The best ride the train or trolley bus.)

Cairo's 1,450 buses each carry 1,700 passenger per day—well above the 1,200 passenger level of most heavily traveled bus systems. With normal levels of bus use it would take another 1,200 buses to carry the present daily load, and in addition the 10 percent annual growth of traffic requires an additional 1,200 units over the next five years, even at present loadings. The dilemma is that surface streets might not be able to move more buses than they do now.[5] With per capita income of only $275, Cairo faces the difficult question whether a subway system is an economically feasible alternative.

A city that has already completed a subway is Mexico City, a metropolitan area of 8.5 million people, with 650,000 private cars, or one for every 13 people. Despite the large number of cars, the majority of trips taken in 1971 were by public transport. The city's 10,000 buses accounted for 6.8 million daily passengers, or 55 percent of the total, while autos carried 2.9 million people, or 24 percent. Altogether 12.3 million persons rode by various methods on an average day, including 1.2 million on the new Metro. The Metro has made commuter journeys faster and more pleasant, and even with a fare double that of the bus, part of the system is now operating close to capacity. Plans are being made to increase the network from 25 miles to 50 miles, but the extension also threatens to become overloaded as the rapidly growing population continues to spread beyond the city but commutes to the center where employment is concentrated.

From these urban travel patterns several observations can be made about the conditions of transportation, the public policies that are called for, and the basic questions that remain to be answered.

1. In most cities of the developing countries, transport demand is about one trip per person per day. Public transit accounts for most of this travel: generally about three quarters of one trip per person per day.

2. The bus is the principal method of mass transportation at the present time, but it is seriously deficient in many respects. If transportation is to be improved for the poorest people, priority needs to be given to acquiring more buses and operating them more effectively. Most cities need to consolidate bus operations, rationalize routes and fares, and introduce traffic engineering improvements geared to the needs of transit.

3. The public automobile or jitney taxi has become a significant factor in public transport. In some cities such as Manila, taxi operations are extremely effective, while in Istanbul the public automobile is adding to traffic congestion and disrupting bus service. Better organization of public automobiles and their integration into the public transit system could make automobile service more widely available to the carless at reasonable cost.

4. Better traffic management and control could greatly increase the capacity of existing streets. Among the possibilities are computerized traffic signals, street parking bans, and the segregation of different types of traffic. All-bus streets, cycle paths, walkways, and car-free zones could all assist in organizing traffic.

5. Where population densities are high and surface congestion has reached a critical level, rail rapid transit on separate rights of way appears to be a necessary but costly means of relief. But even this will fail to overcome congestion on the sur-

face without appropriate land-use controls to avoid further overcrowding.

6. In every city the severity of the transportation problem and the possibility of its solution appears to depend on the future role of the automobile. Are present car ownership trends inevitable, or could improvements in public transport reduce dependence on the private car? Is the automobile sufficiently important economically and politically to make car ownership and use desirable and acceptable, and if it is, what must be done to reduce the conflict between cities and cars? Finally, will shortages of materials and energy impose a barrier to automobile transportation and increase its cost to the point where public transport is the only alternative?

IMPLICATIONS FOR ECONOMIC AND SOCIAL DEVELOPMENT

The greater availability of private cars is clearly a boon to those who can afford this method of transportation. In many ways the automobile is the means of getting things done in cities. But the congestion of automobile traffic also means lower standards of public transit, and the emphasis on accommodating cars is often at the expense of better transit, which may mean no transport at all for large segments of the population. Are the benefits of automobile ownership worth the costs to society? For while the economic and social impacts of the automobile can be substantial, so can the costs of congestion, accidents and pollution, which are often paid by those who realize little or nothing of the gains from motorization.

The trend toward more automobiles in the developing countries has also been attacked on the grounds that the private car is socially undesirable, since it is dividing society into two separate classes of the mobile and the immobile.[6] The latter, who have no car, are denied many of the economic and social advantages of the city. Moreover, housing needs continue to be unmet while autos proliferate, not because of any attributes of the car, but because of an institutional defect that makes it easy to finance a car and difficult to finance a house.[7]

A more desirable course, it is claimed, would be to focus on the consumption requirements of the typical citizen—food, clothing, minimum shelter, education, and various basic services such as water and sewage disposal. If developing countries put into practice such a "popular consumption criterion" for development, the automobile would not be very high on the list of what is needed.[8]

But if the automobile is in fact undesirable, efforts to convince consumers have not been notably successful. In Europe, automobiles were long viewed unsympathetically by governments. Cars were heavily taxed, along with fuel, little was done to build and improve highways and parking facilities, and both rail and transit facilities were heavily subsidized to help them compete with the automobile. These negative policies were the opposite of the approach to the automobile in the United States, where taxes were kept low, highway programs heavily subsidized, and transit and rail facilities neglected. The results have not been very different. Europeans adjusted by making smaller and more economical cars, by suffering higher accident rates and greater traffic jams, and by parking on the sidewalk. But they did not give up the idea of driving.

In the middle fifties, the Japanese were viewing the possibilities of motorization in that country with alarm, fearing that lack of metals and petroleum would invite serious balance of payments difficulties and that the resources required to create a road net for an economy based on rail and water transport would stifle postwar recovery. Yet events over the past twenty years have made Japan the world's second largest producer of motor vehicles, and being a motorized nation has not prevented the Japanese from enjoying the world's highest sustained economic growth rate.

It can be said of the United States, Western Europe and Japan, that the economic threat of the automobile was exaggerated. In fact all of the high-income urbanized countries of the world are also countries with a large number of cars in relation to population. Similarly, there is a close relation in developing countries between per capita income, percentage of the population urbanized, and levels of car ownership. Car ownership is high in high-income highly urbanized countries such as Mexico, Venezuela and Argentina. But there are relatively few motor vehicles in Nigeria, India and Indonesia, where per capita income is $100 or less, and fewer than 20 percent of the people are living in cities. In the first group of countries motor vehicles number between 39 and 111 per 1,000 of the population, while in the latter group there are only 2 to 6 vehicles per 1,000 people.

Differences in ownership and use of the automobile in cities are not determined solely on the basis of income, however. Also important are car prices and availabilities, size of cities, density of population, space constraints, and tax policies. But in general the wealthier the city the more cars in relation to population.

Table 22.7 Relationship of Urbanization, Motorization, and Income (1969–1970)

	U.S. dollars of GNP per capita	Percent of population in cities	Vehicles per 1,000 persons
Nigeria	100 >	20.9	2.5
Tanzania	100 >	6.1	7.3
Zaire	100 >	15.6	2.9
India	110	18.8	2.5
Indonesia	110	17.9	5.7
Kenya	130	9.7	12.4
Thailand	160	14.4	19.8
Senegal	200	27.3	12.0
Korea	210	48.8	4.0
Brazil	270	56.0	32.5
Colombia	290	57.2	13.3
Taiwan	300	64.4	55.3
Peru	330	52.5	30.6
Malaysia	340	45.0	61.1
Iran	350	40.6	17.0
Turkey	350	35.1	10.7
Mexico	580	59.5	38.9
Venezuela	1,000	76.4	76.0
Argentina	1,060	76.8	110.6

Sources: *World Bank Atlas, IBRD, 1971*; Davis, Kingsley, "World Urbanization", 1950–1970, Vol. I, University of California, Berkeley, 1969; and Highway Statistics for 1970, International Road Federation.

In densely populated cities that also have relatively high per capita incomes, car ownership may be extensive but use of the car for work trips is limited. This is the situation in New York, Tokyo, and London, where there are millions of cars but 90 percent of travel into the city is by public transit. Mexico City, Buenos Aires and Sao Paulo have also arrived at levels of income and population that encourage car ownership but discourage car use for commutation, especially where extensive suburbanization has not yet generated work trips away from the city center.

Yet the evidence is also clear that there is no necessary connection between automobiles and economic growth. In Western European countries with a high per capita GNP of $1,000 or more, for example, motor vehicle ownership is almost universal, with one car for every 3 to 6 persons. But in Commu-

Table 22.8 Relation of Car Ownership to Per Capita Income and Population Density, 1969 or 1970

	No. of persons per private car	Per capita Income (US $)	Density (persons per sq. km.)
Caracas	11.0	1,600	16,053
Mexico	13.2	1,274	5,733
Buenos Aires	13.5	1,800	8,272
Casablanca	13.7	817	12,357
Singapore	13.7	800	12,346
Sao Paulo	16.1	784	9,804
Nairobi	19.0	494	10,147
Kuala Lumpur	19.6	660	9,653
Bangkok-Thonburi	20.1	524	10,669
Teheran	22.5	950	10,286
Lima Callao	27.4	865	22,001
Dakar	33.3	580	18,750
Hong Kong	38.2	850	46,336
Bogota	45.5	450	19,099
Djakarta	55.5	325	12,191
Istanbul	64.2	810	12,000
Lagos	77.9	555	10,007
Taipei	97.6	555	26,870
Seoul	236.9	440	14,993

nist countries that have limited the production or importation of cars on economic or political grounds, there are many examples of high levels of economic activity without high levels of car ownership. Hungary, with per capita GNP of $1,000 in 1970, had only one motor vehicle for every 61 people. Per capita GNP figures for Portugal and Albania were in roughly the same range, ($510 and $430), yet the number of persons per vehicle was very different (19 for Portugal, where cars were plentiful, contrasted with 200 for Communist Albania where cars were officially limited). Nations can be relatively prosperous without motor vehicles, and relatively poor in spite of them. But those without them are also typically countries with poor roads and lack of communications between the rural and urban sectors. The cities, however, cannot be said to suffer for lack of cars, congestion, or pollution.

The fact remains that increasing car ownership in the developing countries is going to continue, and that most emerging nations will be viewing the automobile as the symbol of a

Table 22.9 Relation of Motor Vehicle Ownership to Economic Progress

	GNP per capita (1971 US $)	Persons per vehicle	Growth in GNP per capita (%)
Sweden	2,920	3.4	3.4
Switzerland	2,700	4.5	2.6
France	2,460	3.7	4.8
Luxembourg	2,420	3.1	3.3
Denmark	2,310	3.8	3.7
Germany (W)	2,190	4.1	3.7
Norway	2,160	4.5	4.0
Belgium	2,010	4.4	3.5
Finland	1,980	6.2	3.9
United Kingdom	1,890	4.0	1.8
Iceland	1,850	4.7	2.6
Netherlands	1,760	5.1	3.1
Germany (E)	1,570	14	4.1
Austria	1,470	6.0	3.9
Italy	1,400	5.4	4.7
Czechoslovakia	1,370	18	3.9
Ireland	1,110	7.3	3.5
Hungary	1,100	61	5.5
Poland	940	48	5.1
Bulgaria	860	176	6.7
Romania	860	182	7.5
Greece	840	34	6.2
Spain	820	12	6.5
Malta	710	6.7	4.1
Yugoslavia	580	29	4.6
Portugal	510	19	4.9
Albania	430	200	4.9

Source: World Bank Atlas, 1971, and Motor Vehicle Manufacturers Association, *Automobile Facts and Figures*, 1971.

better life. Under the circumstances it is a dubious position for nations already motorized to take the view that poor countries should forego the car for economic reasons that have already proved groundless elsewhere. Just as urbanization is global, we can also expect the ownership of the motor car to become fairly universal.[9]

Given the present state of transport technology, it must be concluded that the automotive revolution is irreversible and that the question posed for developing countries is how to achieve the most advantage from the car while reducing its unwanted side effects. It is necessary to reduce automobile congestion in big cities, but this does not mean that the automobile cannot be part of the urban transport system, that it can make no contribution to the modernization of industry and agriculture, and that it has no role in the creation of new patterns of urban settlement or in the integration of urban and rural life.

The role of the automobile in linking the urban and rural sectors of developing countries has already had considerable economic and social importance. The movement of technicians, teachers, doctors, agricultural extension workers and other specialists has been greatly facilitated by the automobile. The private car has also made it possible to expose the urban entrepreneur to opportunities in the agricultural sector and to overcome the isolation of the farm.[10] It is not the truck and bus alone that establish the economic justification for hard-surfaced roads and intercity highways, but the urban based business cars and other automobiles that often comprise over half the nation's mechanized road traffic. Automobiles help to justify and pay for the highway projects that are linking countries and regions, and the costs of truck and bus transportation are being reduced by joint use of the roads with private cars.

Developing countries may also reap considerable benefits from economic activities created by the automobile. World motor vehicle output, about 33 million units annually, has stimulated activity in the steel and other metals industries, in tire production, petroleum, refining, and in a wide assortment of accessory manufacture. Most of this impact of date has been in the developed countries. In the United States one out of seven jobs can be attributed to the motor vehicle, and Europe is enjoying comparable economic impacts as 1970 output of vehicles rose above that of the United States. A great complex of steel, petroleum, and other industrial plants has become a major element in European economic expansion,

Table 22.10 World Motor Vehicle Production, 1971

	Passenger cars	Trucks and buses	Total
Argentina	193,385	60,255	253,640
Australia	391,242	55,514	446,756
Austria	650	6,228	6,878
Belgium	278,687	17,077	295,764
Brazil	342,214	173,824	516,038
Canada	1,094,631	275,970	1,370,601
Czechoslovakia	149,016	27,846	176,862
Finland	–	1,500	1,500
France	2,693,989	316,305	3,010,294
East Germany	132,000	26,000	158,000
West Germany	3,696,779	285,943	3,982,722
Hungary	–	9,500	9,500
India	41,746	47,364	38,110
Italy	1,701,064	115,955	1,817,019
Japan	3,717,858	2,092,916	5,810,774
Mexico	153,598	47,242	210,840
Netherlands	78,087	13,225	91,312
Poland	86,200	59,500	145,700
Portugal	–	200	200
Rumania	8,000	27,000	35,000
Spain	452,921	79,437	532,358
Sweden	287,398	29,897	317,295
Switzerland	–	800	800
United Kingdom	1,741,940	456,206	2,198,146
United States	8,583,653	2,088,001	10,671,654
U.S.S.R.	518,000	612,000	1,130,000
Yugoslavia	114,477	17,891	132,368
Total	26,457,535	6,953,596	33,411,131

Source: Motor Vehicle Manufacturies Association of the U. S. Inc., *1971 World Motor Vehicle Data.*

while increasing volumes of automotive production in Japan have played an important part in domestic employment and in the foreign exchange earnings of the Japanese economy.

Most developing countries still lack markets large enough to benefit from the scale economies necessary to produce automobiles at reasonable cost. An internationally competitive motor vehicle assembly requires production of 40,000 to 60,000 units of a given model per year. The manufacture of major components such as engines and transmissions, and the stamping of sheet metal body components call for much higher annual outputs.[11] But the situation is changing. Several developing countries have moved into the mass production phase, including Mexico, Argentina, and Brazil, and automotive plants are beginning to shift out of countries with high labor costs into Asia and Latin America. The probability of much larger vehicle production programs in developing countries is increased by current efforts aimed at designing an automobile specifically for countries with limited resources and an undeveloped road network. For example, a new all-purpose vehicle has been designed for production and use in Southeast Asia, built for primitive roads, capable of being produced within the region with a high local labor content, and able to serve as both freight and passenger mover. It can serve as a farm runabout, a source of power for rice husking, water pumping and wood sawing, and a power unit for agriculture and light industrial purposes.[12]

The conventional car that is still relied upon involves imports that create a heavy drain on foreign exchange. In many countries automotive imports are over 10 percent of the value of all imports, and petroleum requirements often add substantially to the import bill. But the number of automotive plants being erected in Asia and Latin America is rapidly increasing, and the economic effects of the industry have already become of very sizeable proportions. These impacts cannot be overlooked in considering the restrictive measures that might be needed to deal with the special problems of automobile traffic in big cities.

The economic effects of the automobile are not limited to its production, for in most countries the operation and upkeep of vehicles has stimulated numerous small businesses and created substantial numbers of jobs. But economic impacts go far beyond car production and operation, involving the location of industry and housing, the organization and management of business and agriculture, changes in personal horizons and living styles, and the expansion of economic opportunities, job options, and recreational activities. On balance it appears that the positive aspects of the private car are of sufficient magnitude to warrant serious attention to overcoming its deficiencies and dealing effectively with side effects that ought to be preventable.

THE QUESTION OF RESOURCES

The advantages of the automobile as a method of transportation and as a creator of industry and jobs have to be weighed against the insatiable appetite of today's cars for nonrenewable resources. If the 1972 global fleet of some 200 million passenger cars should increase to more than 500 million by the end of the century, will there be enough fuel and metals to go around? Already there is disturbing evidence that petroleum supplies at acceptable costs may be exhausted before the end of the century, and that a shortage of metals used for motor vehicles will cause sharp increases in production costs.

Underlying the question of resource availability for the automobile is the question whether the world's resources will be sufficient to support the total needs of the 7 billion people expected in the year 2000, to provide them with reasonably adequate standards of living. The answer is a conditional one, and depends on how resource policy decisions are handled. The outlook differs according to whether present consumption trends continue or whether there is more recycling and relcaiming of materials, new designs to decrease the size and increase the durability of the car, new power sources and operating economies, and changes in community design, technology and life styles that alter the demand for movement.[13] The limits to growth are set by many tangible items, but also by certain social factors such as peace, social stability, education, and technology. It may be of little resource consequence, for example, for the world to have one car for every ten of its 7 billion people in 2000 if there has been a substantial reduction in armaments by that time.

At present there is an obvious state of disequilibrium in the use of resources that call for a global assessment of the automobile and its future. Countries already highly motorized, with some 700 million people, or one-fifth of the world's population, are absorbing large amounts of materials and energy for their mobility. The rest of the world is still dependent to a large extent on human and animal power. Unless the nations are to be divided into the mobile and the immobile, some other method of movement may have to be introduced or some materials-saving approaches to automobile design and operation may have to be taken to reduce per capita consumption for travel. Lacking such innovations, materials may not be available for extending today's automotive technology to the developing four-fifths of the world, and if this so, how long will the developed countries be permitted to draw on resources that developing countries might otherwise use to meet their own priority needs?

In the early period of automobiling in the United States, the vehicle was simple in design, easy to operate and repair, and surprisingly inexpensive. Today's large, expensive and complicated cars are the only ones available, and they are ill-adapted both to urban use and to low-income families. The filter-down process provided through the used car market has been the means of providing low initial payments for car purchase, but operating costs and upkeep increase with age. Great importance should be attached to efforts that might lead to production of a simple and economical vehicle specifically designed for developing countries, along the lines of the Southeast Asia experiment.

These issues that seem at the moment to pose unsolvable problems might respond to a number of remedies in the long run. Changes in technology outside the transport field may substantially alter the nature of the problem. Communications may prove to be an important factor in reducing urban concentration, and changes in the social system may so alter life styles that society would be less subject to home-to-work peak-hour transportation. Mass travel by high-speed ground transport, plus predominantly local use of cars, might reduce per capita automobile travel or per capita material consumption for travel. So might a system of jitney taxis, more opportunity to walk, and more chance to use bicycles. New concepts of urban design may move society in directions that reduce the proportion of local movement by vehicle and especially by individual cars. The effect on energy requirements could be far-reaching if the micro-bus or public automobile were to be used extensively, providing three times as much travel per unit of energy as a subway.

Based on projections of vehicle registrations, and assuming present fuel consumption patterns, the fuel requirements of the world automobile population could cause a serious shortage before the end of the century unless high estimates of world oil reserves prove to be correct.[14] But a wide variety of responses are possible, including improved car performance, the

Table 22.11 Efficiency of Various Transport Modes

	Passenger miles per U.S. gallon of fuel (or 130,000 BTU)
Helicopter	7
Supersonic plane	10
Boeing 707	20
Boeing 747	30
Most cars (2 people riding)	33
Small car (2 people riding)	60
City bus (rush hour)	95
Small car with 4 passengers	100
City metro, rush hour	120
Suburban bus	250
Microbus (7 passengers)	375
Train (broad gauge)	390
Electric bicycle	400
Bicycle	1,000

Source: Gerald Leach, "The Motor Car and Natural Resources," Working Paper No. 4 (Paris: OECD, 1972), p. 6, based on report by Richard Rice, "Historical Perspective in Transport System Development," Carnegie Mellon University, Transportation Research Institute, 1970.

use of alternative fuels, the development of an electric vehicle, or greater use of mass transportation. In any event, energy costs can be expected to be a major factor in the design of transport systems, even taking an optimistic view of world oil reserves.[15]

The problem of energy supplies can be alleviated to a considerable degree by conservation measures. The United States, with 6 percent of the world's population, accounts for 36 percent of the world's energy consumption. It is not feasible, in the face of long term trends in petroleum supply and demand, that one country should continue to absorb such a high proportion of this source of the world energy. There is urgent need for acceptable public transportation substitutes for the automobile, for greater fuel economy in the car, and for changes in the motive power of the automobile. The debate over petroleum supplies has revolved around various restrictive measures that might be taken to cut fuel consumption, including increased fuel taxes and prices, or alternative rationing. But very little has been said about the positive measures that need to be taken to assure the continuation of essential private transportation services on which a large number of the world's cities continue to depend.

According to studies of the U.S. Department of Transportation and the Environmental Protection Agency, it would be possible, between now and 1980, to make changes in the fuel efficiency and in the proportion of small and large cars in the United States that would increase the overall performance of passenger cars from an average of 14 miles per gallon in 1974 to 22 miles per gallon in 1980—an increase of 60 percent.[16] The efficiency of individual cars can be further increased by weight reduction, reduced aerodynamic drag, improved engines and better transmissions. Engine improvements offer the largest single potential for improvement.

It may be necessary to accelerate global efforts for greater fuel efficiency by imposing taxes on new cars that do not meet specified standards. It was suggested by *Project Independence* that a minimum auto efficiency standard for new cars of 20 miles per gallon should be established and all new vehicles failing to meet such a standard be taxed on the basis of each mile per gallon below the standard. Taxes would range from $350 per car getting 18 miles per gallon to $2,100 for a car getting only 8 miles per gallon.[17]

But along with an energy-efficient car it will be necessary to create a more energy efficient transportation system, and it is in this area that public sector programs to aid mass transportation could prove highly effective.

It is clear from urban tavel data that in all cities a reduction in automobile use could be accomplished without sacrificing essential mobility, and that even trips to work are subject to considerable reduction through carpools, shifting to bus transport, and in some cases moving closer to work. With rising prices and taxes, car owners would have no difficulty deciding how best to cut non-essential driving, and many short trips could be made by walking. And a higher tax on fuel could also generate revenues to support a transportation fund to help finance the modernization of transit. In Brazil in late 1975, the price of gasoline was increased 25 percent and the proceeds dedicated to mass transit and oil exploration.

But higher fuel prices and fuel rationing are essentially negative approaches to energy efficiency that could have adverse effects on the economy, including inflation and unemployment. To avoid these impacts a basic aim should be to make the automobile more efficient, and to make public transportation more inviting. Public policy for highways and transit

Table 22.12 U.S. Automobile Use

Trip purpose	Total travel (%)
Work	
To and from work	34.1
Business related to work	8.0
Personal	
Medical	1.6
Shopping	7.6
Other	10.3
Education, civic, and religious	5.0
Social and recreational	33.4
Total	100.0

Source: Motor Vehicle Manufacturers Association, *Automobile Facts and Figures*, 1973–74, p. 37.

needs to be closely linked to the encouragement of energy-efficient automobile designs and smaller models. There is an obvious need for integrated transportation and energy policies.

With respect to other resource requirements for the automobile, estimates by the OECD indicate that there are no difficulties ahead with respect to models for the half billion vehicles anticipated by the year 2000.[18] The reason for this is that much of the material used in car manufacture is recycled by scrap dealers. In the United States, for example, the number of cars scrapped each year has increased to some 8 million units, which is more than 80 percent of the number of new cars produced. This percentage of scrapped cars to new cars has increased as the car market has become primarily replacements. Thus only a small fraction of each year's new car output requires fresh metals.

On a global basis, car production is expected to double from 23 million units in 1970 to 52 million in 2000. At the same time, however, the number of vehicles scrapped, which is now about 11 million a year, can be expected to quadruple to 43 million. Thus the number of vehicles requiring new metal will remain about the same as today, at 10 to 12 million cars per year. To the extent that some metals run short—and these will not be iron and steel, which are plentiful—substitute materials can be expected to take their place.

Estimates of future vehicle registrations in developing countries, however, may be too conservative. If more rapid economic growth supports a much larger number of vehicles, there would be a period when metal requirements for new cars would be very high without compensating scrappage. For example, the West European average was one car per 5 persons in 1970. If this figure were to be achieved in the developing countries by the year 2000, total cars on the road in these areas would number over a billion, and metal requirements in a period of rapid expansion and low scrappage rates would be very high. Even on the most optimistic assumptions the fuel costs to support a fleet of this size would require smaller cars and other methods of energy conservation.[19]

What these rough calculations indicate is that from the viewpoint of resources, the world could afford to be far more motorized than today. But a more critical problem for the year 2000, with some 7 billion people and a majority of them living in cities, is the conflict between the city's space limitations and the automobile's space requirements. In a rapidly urbanizing world, is it possible to count on cars in the long run? Already, and without any further increase in automobile ownership, it is physically impossible to use the car for substantial percentages of peak-hour travel in the central areas of most big cities. Are there public transportation alternatives and potentials that ought to be exploited to keep the cities mobile? Or do we face a new era of urbanization that will reject the extreme crowding of earlier times in favor of the more extensive regional city which modern transport and communications have now made possible?

23 IMPROVING URBAN TRANSPORT

IMPROVING BUS TRANSPORT: A FIRST PRIORITY

The immediate task is to overcome transport deficiencies in existing cities as quickly as possible and with the least expenditure. For this objective it is the bus, despite its inadequacies, that offers the principal hope that mobility can be improved for the majority of urban residents. In most cities of the world bus operations suffer from traffic congestion created by other uses of the street. What is needed is an integrated approach to the highway, traffic control and vehicle components of the bus transport system so that a complete transport service is provided. Focusing on the equipment alone, without concern for the roadway and traffic engineering, has failed to supply an effective system.

The potential capacity of bus transport is indicated by traffic volumes provided by buses on urban expressways. For example, on the George Washington Bridge in New York it takes 136 buses to carry 6,900 passengers in the peak hour, but it takes 27 times as many cars to carry the 6,200 passengers that drive over the Bridge at the same time. On the San Francisco-Oakland Bay Bridge, 216 buses are used in the peak hour to transport 7,800 people. There are 30 times as many cars, but they carry only 20 percent more people.[1]

Experience with a 2.5-mile exclusive bus lane into the Lincoln Tunnel in New Jersey shows a peak-hour traffic volume of 400 buses per hour, with 25,000 passengers moving over a single highway lanes. But in some peak periods movement is at the rate of 700 to 800 buses per hour, which would mean perhaps 45,000 or more passengers being transported per hour. Time savings for bus riders average 10 minutes per trip as a result of the traffic speed-up.[2]

The possibility of capitalizing on this traffic carrying capacity of buses has begun to be exploited in Europe and the United States by designating the curb lane for buses only, and by providing exclusive rights of way for bus operations. The creation of exclusive busways may be an important new rapid transit potential and might be furthered by the construction of short sections of bus subways in the central city. While the bus would have considerably less capacity than rail subways, it would have the advantage of being able to serve a dense network of routes and would be able to distribute passengers off the main route at both ends of the trip.

Cities in both developed and developing countries might supply better public transport service by designating city-wide networks of streets for the exclusive use of buses during peak hours. A combination of new buses, bus roadways, traffic control, and traffic regulation that exclude automobiles from specific routes or from downtown areas might achieve an effective system of transit which could be implemented quickly and supplied at low fares.

If an effort to improve public transport by these methods were undertaken for the larger cities of the developing world, the cost of the program would be well within practical limits. Under existing practice, cities are operating about 500 buses per million urban residents. To avoid overcrowding and to operate enough routes to provide adequate geographic coverage suggests that perhaps double this number of buses might be required to furnish acceptable standards of service. This assumes that a speed-up of service would be achieved at the same time through preferential treatment of buses and traffic engineering improvements.

It is estimated that developing countries in 1970 had about 270 million people in cities of over 500,000 population.[3] The size of their bus fleets is estimated at 135,000 units, assuming 500 vehicles per million people. Doubling this fleet to 270,000 units at a cost of $30,000 per bus would require an outlay of $4 billion. This is about equal to the investment required for eight subway systems comparable to the Mexico City Metro. The additional buses (assuming 1,000 passengers per bus per day) could carry 135 million passengers per day compared to a combined total capacity of 10 million passengers for the eight subway systems.

The bus alone, however, is unlikely to meet all of the transportation requirements of the city, and in particular it cannot provide the flexibility and door-to-door service of the automobile. It may be possible to supplement the bus, however, by organizing the taxi as a more integral part of the public transportation system.

THE IMPORTANCE OF THE PUBLIC AUTOMOBILE

Urban traffic patterns show that the group-riding taxi, or public automobile, is rapidly expanding in popularity in such cities as Seoul, Teheran, Singapore, Caracus and Bogota, and that in many cases the bus has already lost considerable ground to the more convenient jitney taxi. The nature of these trends suggests that the taxi offers a relatively unexploited means of supplying an improved public transport service. If extensive and better organized public automobile fleets were used as a supplement to the bus, or as a substitute, they would supply service comparable to that of the automobile, would eliminate parking problems, and would reduce the social gap that separates car owners from the carless by providing all urban residents with physically comparable equipment.

The importance of the taxi in urban transportation is not fully appreciated. In U.S. cities, even with the extensive ownership of private cars, a fleet of 170,000 taxis operating in 1970 carried 2.4 billion passengers during the year, 28 percent of all public transport riders. Taxis carried more people than rail rapid transit, and more than half as many as the bus. Taxi revenues in the United States exceeded the total revenues of buses and rapid transit combined. These figures were attained despite the fact that there was generally less than one taxi per 1,000 persons and taxis were rarely used for group-riding. But the average taxi achieved considerable efficiency, covering 40,000 miles a year and carrying an average of 14,000 people.[4]

The potential of the jitney taxi is far greater than that of the conventional metered cab. In Teheran, for example, 11,000 taxis carry 290 million people a year, or about 26,000 people per taxi. But the jitney taxis, running on fixed routes in that city, carry an average of 50,000 passengers per year. Their patronage increased 250 percent during a five-year period in the 1960s, while ordinary taxis were experiencing only a 50 percent rise in patronage and bus use was being cut in half. The so-called collective taxi numbered only 2,000 units but was carrying 100 million passengers per year. If a fleet of jitney taxis were to be substituted for the bus, which in 1972 carried 700 million passengers a year, it would take some 14,000 jitney taxis to assume the entire burden of the Teheran bus system.[5] But the public automobile would hardly be a solution with private automobiles congesting the streets.

A shift from 1,665 buses to 14,000 jitney taxis (plus the 11,000 taxis already operating) would mean a total of 25,000 taxis in Teheran. This fleet of vehicles would have to compete for street space with 200,000 other motor vehicles, most of

Table 23.1 Public Automobiles in Selected Urban Areas

Cities outside the U.S.	Public automobiles per 1,000 persons (1970)[a]	U.S. Cities	Public automobiles per 1,000 persons (1969)
Caracas	6.8	Washington, D.C.	13.3
Istanbul	5.7	Honolulu	3.2
Taipei	5.6	Atlanta	2.7
Sao Paulo	4.0	Boston	2.0
Bogota	3.6	New York	1.5
Singapore	3.4	Miami	1.5
Teheran	3.3	Chicago	1.3
Lima-Callao	3.1	Kansas City	1.1
Bangkok & Thonburi	3.0	Pittsburgh	1.0
Buenos Aires	2.8	Tulsa	0.9
Mexico City	2.7	Detroit	0.8
Bombay	2.5	Philadelphia	0.7
Casablanca	2.4	Birmingham	0.6
Nairobi	2.2	Minneapolis	0.5
Dakar	2.2	Los Angeles	0.4
Seoul	2.1	Phoenix	0.2
Hong Kong	1.1	Tucson	0.2
Kuala Lumpur	1.0		
Lagos	0.7		
Djakarta	0.1		

[a]Includes "jitney" cabs, taxis, and hired cars.

them private cars. A public automobile network, therefore, could not be fully effective without a limit on the use of private vehicles, at least during rush hours in the central city. But if there were to be shift of passengers from private car to public taxi, the daily work performed by Teheran's 160,000 private automobiles could probably be accommodated by as few as 10,000 collective taxis, assuming 6-person trip per car per day and 100 passengers per day per taxi.

In summary, therefore, Teheran could be supplied with a complete system of public transportation with about 35,000 public automobiles, which would be about 20 percent of the fleet of vehicles now performing the urban passenger transportation task. In highway capacity requirements, parking space, and cost, the public automobile system would have much in its favor.

In the Philippines the 6- to 10-passenger jeepney, or group-riding taxi, shares responsibility for public transportation with the bus at the same fares. For the country as a whole there were 32,500 jeepneys in 1970, of which 14,900 were in the Manila area. Manila buses still numbered 3,800, but the superior services of the jeepney has made this the preferable low-cost method of getting around.[6]

A special advantage of the jeepney is its ability to maintain high average speeds that are not much lower than that of the automobile. Jeepneys fill up quickly and their trips involve fewer stops, although the service suffers from lack of terminal parking and waiting space. From the standpoint of employment, jeepneys that work around the clock provide jobs for a large number of drivers and the many shops for assembling, maintaining and overhauling these vehicles have become an important industrial complex.

Jeepneys cost an average of about $2,500 and their ownership varies from a few who operate their own vehicles to some fleets of 100 or more, owned by corporations. Drivers rent the vehicles for a fixed amount and time period and pay for gasoline and oil, but maintenance is the responsibility of the owner. Operations are confined to routes specified by the Public Service Commission, and in 1972 rates were set at about 2.5 centavos per passenger kilometer with a minimum charge of 15 centavos (about 3 cents U.S.). Application for permission to operate must indicate the proposed route, number of vehicles to be used, and the schedule, and a public hearing provides an opportunity for competing operators to be heard.[7]

In many cities there have been various efforts to eliminate group-riding taxis, partly because of their competition with the bus and partly because private automobile operators charge that the large number of such taxis creates congestion. Actually the group-riding taxi needs to be freed from the congestion created by private cars, and the sharing of exclusive transit streets with the bus may be the way to promote a more universal adoption of the jitney taxi as a kind of community re-cycling of the automobile.

Based on experience in Teheran, Manila and other cities, a moderate-sized fleet of group-riding taxis could provide high quality transport at low cost if operated in zones freed of private automobiles. If a bus costing $32,000 can carry an average of 1,200 passengers per day, 8 public automobiles priced at $4,000 apiece and carrying 150 persons per day could perform the same transportation task for the same cost. The jitney taxi would provide a seat for all, frequent scheduling, and nearly door to door service. And compared to a private car, assuming as many as 10 people are carried per day (5 trips with an average of 2 people) the public automobile could do 15 times as much work. The services of 100,000 private cars could be accomplished with 6,600 public cars.

REDUCING CONGESTION: TRAFFIC MANAGEMENT

The use of the streets and the prices charged for such use indicate a kind of global perversity that rewards those who waste space and discourages those who conserve it. The facts have been well documented, but the political implication of some of the recommended solutions has made public officials reluctant to act.

Methods of limiting traffic include three types: restraints, such as pricing urban street use and charging motorists more for parking; restrictions, such as the designation of vehicle-free zones; and avoidance, such as designing urban activity systems that reduce the need for vehicle movement. The last of these measures to limit or to organize traffic will be considered in a later section.

Pricing urban road services, banning street parking, and reflecting social costs in the fees paid by motorists could help reduce congestion quickly, with relatively minor investments in buses or other public transport. Pricing policy could act as a rationing device that would reduce the peaks and shift some travel from private cars to public vehicles. One approach would be to use the proceeds of the social costs paid by the motorist to subsidize transit in recognition of its social benefits. By keeping transit fares low, or eliminating them altogether it might be possible to minimize the total transportation bill.

The high standards of public transportation in European cities reflect generous subsidies that recognize the wider benefits to the community that cannot be charged to the user alone. Public transport undertakings in Europe charge their deficits to the public in a variety of ways, including general taxation and the merging of transit operations with electricity, gas, water supply, or other public utilities.

At the present time automobiles in developing countries are generally able to park on the streets without charge. They are not paying the social costs of air pollution, accidents, and environmental destruction. In addition, automobile user charges are uniform regardless of where and when the travel is performed, so that peak-hour commuters are being subsidized by off-peak drivers. The simplest remedy appears to be the banning of free parking, which is a burden on the city, and the introduction of parking charges for off-street facilities that reflect both the cost of parking and the costs incurred for peak-hour driving.[8]

Another low-cost remedy for traffic congestion is the banning of automobile traffic in limited areas of the city, and dedication of selected streets to pedestrian traffic. About 100 vehicle-free zones and pedestrian shopping streets have been created in European cities with moderate financial commitment and physical change, often by simply paving over streets and installing lights and landscaping. Another means of regulating vehicle use has been introduced in Göteborg, Sweden, where the division of the city into separate sectors and the prohibition of direct movement of cars from one sector to another has had favorable effects on traffic flow and on the environment.

A 1971 survey of opinion in Sweden, the most highly motorized country in Europe, shows that vehicle-free zones are gaining acceptance among car owners. Two-thirds of all persons interviewed considered driving essential to get to work, to the store, or to vacation, but a surprising 77 percent expressed the view that cars are unnecessary in central city areas. High value is placed on the automobile by farmers and residents of smaller cities and towns, but in the big city there is an outspokenly negative attitude forwards the automobile."[9]

It was also found that 74 percent of those questioned were satisfied with Stockholm's public transport, and primarily with the modern subway system. If public transit were improved and private cars barred from the center, 70 percent of respondents said they would travel there to make purchases as often as they do now, and 14 percent would travel more often. Only 9 percent said they would visit the city center less often. The deterring effects of charging more for entering the central city appear to be substantial, for even a low toll of 15 cents led 44 percent to say they would seldom if ever go by car to the center if a congestion charge were levied. A higher charge of 75 cents would deter 63 percent.[10]

The Swedish survey reflects a growing realization that the big city center cannot accommodate large numbers of automobiles, that cars are not necessary in such a setting, that supplying good public transport is essential, and that this may require barring the automobile from some parts of the city.

Unfortunately there is not much appeal to traffic engineering solutions, and the greater visibility of expressway and subway undertakings gives these high-cost remedies greater acceptance. Yet an integrated traffic management strategy could have impressive results. It would include pricing policies that charged the motorist his full costs, preferential treatment for buses, computerized signal systems and other uses of electronics, street parking prohibitions, pedestrian zones, strict enforcement of traffic rules, routing of through traffic away from the center, staggered work hours and staggered days of work. Such a strategy could be reinforced by investments in bus and taxi equipment, grade-separation of selected highway intersections, relocation of bus and truck terminals, and off-street truck loading and unloading facilities. Thus far no city has set about to design and carry out such a strategy.

SOLUTIONS THROUGH RAIL RAPID TRANSIT

The case for a good public transport system is not simply that a large part of the urban population is too poor to own cars. European cities indicate that even with relatively high incomes the limited street capacity of high-density urban centers makes it impossible for everyone to drive. Thus in London in 1966, approximately 84 percent of all persons commuting to the central business district went by public transport. Other British cities with much lower populations but with strict limitation of street capacity also depended principally on public transit. The figures were 74 percent for Liverpool, 73 percent for Manchester, and 68 percent for Birmingham.[11] These cities except London relied for public transport primarily on bus, but typically the largest cities of the world depend on rail.

The fact that rail systems predominate in the largest cities in borne out by traffic patterns in Paris and Tokyo as well as London. In Tokyo in 1970, 76 percent of all passengers moving within a ten-mile radius of the center were using railways or subway—21 million people altogether.[12]

In Paris, of the 1.1 million daily commuters into the city in 1965, about 80 percent moved by public transport, and three-fourths of them used rail facilities.[13]

In Europe, approximately 700 miles of urban rapid transit lines are either in the planning stage, under construction, or newly completed. In South America the first line of the Sao Paulo Metro is operating, and Rio de Janeiro's subway is being built.

Toronto's 21-mile automated subway system was carrying 147 million passengers a year in 1972, about one-fourth of all

Table 23.2 Characteristics of North American Cities with Rail Rapid Transit Systems and Number of Route Miles in 1975

	Population, 1970 (millions)[a]		Population in central city per square mile, 1970 (thousands)[a]	Employment in central city, 1970 (thousands)[b]	Rail rapid transit route miles, 1975
	Urbanized area	Central city			
New York	11.6	7.9	24.4	2,885	237
Chicago	7.0	3.3	12.3	1,349	90
Philadelphia	4.8	1.9	15.2	772	29
Boston	2.8	0.6	13.9	374	64
Cleveland	2.1	0.7	9.9	392	32
Toronto	2.6	0.7	17.8	n.a.	21
Montreal	2.7	1.2	20.9	n.a.	16
San Francisco	3.1	1.1	10.0	556	71
Washington[c]	2.9	0.8	12.3	492	98
Atlanta[c]	1.4	0.5	3.8	288	50[d]

Sources: American Institute of Planners and Motor Vehicle Manufacturers Association of the U.S., *Urban Transportation Factbook* (Detroit: MVMA, 1974), pp. I-9, I-10, I-42, I-82; John Paxton, ed., *The Statesman's Year-Book, 1973-1974* (St. Martins Press, 1973), p. 240; *Encyclopaedia Britannica* (1973), vols. 15 and 22, pp. 798, 90, respectively; and information provided by the individual transit systems.
n.a.: Not available.
[a]1971 for Toronto and Montreal.
[b]Number of residents in the urbanized area that are working in the central city.
[c]Under construction.
[d]The rail rapid transit system will have an additional 14 miles of exclusive busways.

public transit riders. In Montreal 16 miles of subway have been constructed at a cost of $13 million per mile. Trains with pneumatic tires running on a concrete track carry 126 million passengers a year in 1972. In both cases operating revenues have been sufficient to cover operating expenses, exclusive of depreciation.

San Francisco's Bay Area Rapid Transit System, comprising 75 miles of subway and surface rapid transit, has recently opened and Washington's 99-mile Metro (25 miles underground) is in process of construction. The cost of the BART system will be $24 million per mile and the Washington system will cost about $45 million per mile.

In the developing world Mexico City has a 25 mile subway that was carrying 1.2 million passengers per work day in 1972, or 8 percent of total daily rides. For the year as a whole, the number of metro riders was 389 million. Cost of Mexico's subway was $458 or about $17 million per mile, and as in the case of Toronto, revenues were covering operating expenses but are not enough to amortize the investment.

Surveys are being conducted in a number of countries in the developing world for various types of rapid transit systems, including Bogota, Caracas, Teheran, Singapore and Hong Kong, while Sao Paulo, Santiago and Rio de Janeiro are in various stages of construction.

Should developing countries moving into the automobile age start now to build rapid transit, rather than waiting until they have invested heavily (and in vain) in automobiles and highways? What are the conditions that would suggest turning to this solution?

The conditions of population, land use and traffic that have been specified as prerequisites for economically justified rapid transit solutions include a population of 2 million, a density of 14,000 persons per square mile, employment of 100,000 people in the central business district, and 100,000 people moving across the boundaries of the central business district in the peak hour. On these criteria many cities in developing countries readily qualify for subways. The number of cities with 2 million population and 14,000 persons per square mile in 1970 was at least 20.[14] Based on a cost of $20 million per mile the total cost of supplying an average of 50 miles of subway rapid transit per city would be US$1,000 million. For 20 major cities the cost would be $20,000 million.

The high cost of travel by rapid transit and automobile compared to bus is reflected in data for urban travel in the United States. Driving a car is the most expensive method of making a 10-mile trip in the city, based on the prorated annual cost of the vehicle and urban highway, plus parking cost. A rail rapid transit trip is only slightly lower in cost, while a conventional bus trip costs only one-third as much as either car or

Table 23.3 Conditions Suggesting Rapid Transit

	Magnitudes	
Primary Determinants	Desired	Minimum
Urban area population	2,000,000	1,000,000
Center city population	700,000	500,000
Center city density (people/sq. mile)	14,000	10,000
CBD employment	100,000	70,000
CBD daily destinations/sq. mile	300,000	150,000
Peak-hour persons leaving CBD	100,000	70,000

Source: Wilbur Smith and Associates, "Urban Transportation Concepts," Center City Project, for the U.S. Department of Transportation, 1970, p. 95.

Table 23.4 The Economic Cost per Person of Traveling Ten Miles in a City

	Public and private costs ($)	Cents per passenger mile
Car and driver[a]	2.64	26.4
Rail rapid transit	2.46	24.6
Conventional bus	0.86	8.6
Car with two occupants[a]	1.31	13.2
Car with three occupants[a]	0.88	8.8
Car with four occupants[a]	0.66	6.6

Source: Marshall F. Reed, "Comparison of Urban Travel Economic Costs," Highway Users Federation, Washington, D.C., February 1973, p. A-20.
[a]Includes estimated economic cost of $1.16 per car for off-street parking and 5 cents per vehicle mile for urban street cost.

rapid transit. But clearly the automobile has an advantage when more than one person in a family is traveling. When there are four car occupants, the cost per passenger is about the same as for the bus.

The foregoing comparisons indicate a total economic cost of about 25 cents per passenger mile for rail rapid transit, but in actual practice most rail transit systems show higher costs for operation alone, exclusive of depreciation. For example, New York subway operating expenses in 1972 were 36 cents per passenger mile. The Montreal and Toronto subways have the lowest operating costs (23 and 19 cents in 1972) but these figures do not include depreciation.

Politically a subway has considerable attraction despite its cost. It is a monumental achievement and an impressive means of transport for those in a position to use it. But they are likely to be a small percentage of total trip makers. Surface transport will remain congested for the majority who still depend on it unless major programs for improved bus transport are also undertaken, which may involve measures to curtail the use of the automobile. The subway itself may become heavily congested unless action is taken to restrict building heights and densities, increase housing in downtown areas, and guide new growth beyond the built-up areas in communities where a combination of jobs and housing can reduce the volume of commuter trips.

The nature of the problem is illustrated by Caracas, an automobile city with such severe congestion that only a subway seems capable of relieving it. Caracas is well suited to mass rapid transit, having a population density of 38,000 persons per square mile, which is expected to increase to 61,000 by 1990. The metropolitan region is contained by the ocean and bordering escarpment on one side, and by mountain barriers inland, so there is little chance of spreading out. The city itself offers a hospitable climate and has a strong attraction relative to the hot low-lands. This is why it is expected that by 1990 seven out of ten people in the region will reside in the capital city, and that the present population of about 2.1 million will rise to 4 million. The proposal for alleviating the massive traffic jams in Caracas is to build a 30-mile rapid transit system, most of it underground.

The rapid transit proposal and the estimated effects on travel in Caracas suggest that the hoped-for relief of congestion may fail to emerge from the subway program. In 1966 the daily number of trips by surface transit totaled 1.3 million, while daily trips by automobile were 1.2 million. Thus a total of 2.5 million trips were being accommodated on the surface, with heavy congestion. But with the doubling of the city's population by 1990, total trips are expected to increase to 6.4 million a day, with half of them still provided by automobile. Only 500,000 of the rapid transit trips will be all the way without transfer, while 1.7 million metro passengers will transfer to and from surface transit. As a result, the number of transit passengers using surface streets will still be 2 million greater than 1972, despite the rapid transit, and they will have to contend with automobile traffic that is heavier than ever. For automobile trips are expected to increase from 1.2 million daily in 1966 to 3.2 million in 1990.

Table 23.5 Rail Rapid Transit Revenue and Operating Expenses per Passenger, 1970 ($)[a]

	Revenue	Expenses
New York	0.31	0.36
Chicago	0.47	0.52
Boston (MBTA)	0.31	0.41
Toronto	0.25	0.19
Philadelphia (SEPTA)	0.32	0.33
Montral	0.28	0.23
Cleveland	0.24	0.32

Source: Institute of Defense Analysis, "Economic Characteristics of the Urban Public Transportation Industry," U.S. Department of Transportation, 1972, pp. 6-83, 6-85.
[a]Excluding depreciation.

The accomplishment of the metro will be to make the doubling of population possible, but it is not clear that it will make Caracas any easier to move around in. For along with those who enjoy the ability to move with reasonable speed in the subway (but under highly crowded conditions) the majority of people will suffer increasing delays as they make surface transfers or as they attempt to drive the 650,000 cars expected by 1990.

The success of the subway, therefore, will depend on improved bus transport and on measures to restrict automobiles to the extent necessary to enable the feeder bus system to operate. But if it is politically feasible to introduce such restrictions with the subway, an alternative might be to restrict car use without the subway. The money saved could be used to provide a system of buses on exclusive bus streets, fares could be considerably lower than for the subway, and automobiles could be confined to areas outside the center and to off-peak hours. The cost of the subway could be avoided, and the resources used to remedy the critical housing conditions that compel one-fourth of the population to live in squatter settlements.

The Caracas dilemma reflects the subway problem now confronting public officials in Mexico. The Mexico Metro is heavily congested on lines close to the city center. Plans are being made for extending the subway another 25 miles, meanwhile increasing the capacity of the existing 25 miles of line by purchasing more equipment and reducing headways.

It is probable that the subway extension will further overload the existing system, and may itself become overloaded in a short time as the possibility of crowding more jobs into the center generates more traffic. One way to avoid this would be to raise fares, but this is considered politically unfeasible. A second possibility would be to build a by-pass transit route around the congested downtown area, and charge less for those who divert their trips away from the center. But the absence of a land use plan and land use controls means uncertainty about future transport demand and suggests that the intensity of land use will continue to increase. If so, the new subway capacity will be quickly absorbed and the city will be no less congested despite heavy transit investments.

The Mexico situation highlights the importance of arriving at some reasonable assurance that there will be an equilibrium between transport supply and demand that will permit a lasting transport solution. If such a balance cannot be assured, then the subway will confront the same problem so well known to the builders of expressways: without a workable land use plan, the demand for transport tends to increase until it nullifies the investment in new capacity.

The guiding principles that cities need to follow are illustrated by the World Trade Center in New York. This twin-towered building can be thought of as an urbanized area (daily population 130,000) in which the transport system had to be designed to carry the traffic that the building would generate, and the building had to be designed for no more activity than a practical transport solution could serve. At first it seemed impossible to design a viable building with 110 stories because a conventional elevator system would absorb too much rentable space.

The building was then "laid on its side" and a rapid transit system built into it based on the express and local train system of the New York subway. The final system was 190 elevators with transfer points between express and local carriers that reduced the space devoted to transport from 37 percent to 16 percent of the area of the building. At this point transport supply and demand were in balance, and the total system was economically vaible.

If apartment living had been provided in the Trade Center, along with work places, a better balance between urban activities and transportation might have been accomplished. The point, in any case, is that a building is like a great city, and cities need to balance transportation and activities like a great building.[15]

REDUCING CONGESTION BY INFLUENCING DEMAND

The alleviation of traffic congestion will require not only a transportation strategy designed to move the traffic, but an urban design strategy aimed at altering the conditions under which the traffic moves. When interrelated city functions are located without reference to each other, travel volumes will be greater than when easy access is provided to the daily needs of the family. Transportation solutions are typically introduced into a setting where there is no concern over how the location, density or arrangement of housing and economic activity are likely to affect the demand for movement.

It follows that transit and highway programs will have to be part of a community development strategy if they are to stand any chance of furnishing a lasting transportation solution. In existing cities, where land use arrangements are already established, changes will have to be brought about gradually through redevelopment, and meanwhile emphasis will have to be on introducing whatever transportation improvements are possible to accommodate the existing system. In newly urbanizing areas, however, where transport demand is not given, it is possible to achieve a workable transport system by creating an urban environment in which housing, job opportunities, community services and transportation are physically integrated and designed to function together.

In developing countries the fact that millions more people will be moving into cities means that extensive new areas will be devoted to urban living—in effect new cities will be built, whether planned or unplanned. If unplanned, the difficulty of moving around will be magnified, the satisfactions of the urban community will be forfeited, and excessive resources will have to be devoted to traffic relief. The challenge is to assure that the location of housing, industry, and community services produce convenient arrangements that solve transportation problems instead of creating them.

Planned Cities in Developed Countries

Planned cities in the United Kingdom provide some basic lessons for other countries. The conditions that promoted them are familiar to developing areas of the world at the present time. The old cities of Britain were in urgent need of reconstruction, and millions of low-income families needed better housing. These changes meant moving people out of the slums to reduce densities, and supply new urban facilities to accommodate the people displaced. In addition, economically depressed areas were encouraging migration to the big cities where there were jobs, and the new arrivals were making rehabilitation of the slums increasingly difficult. To discourage the flow of migrants an effort was made to support other growth points by subsidizing industrial plants in areas of high unemployment. At the same time new satellite communities were built near big cities to provide jobs and housing in communities designed to reduce congestion and pollution, to minimize transportation, and to take the pressure off the central city. British new cities are now being designed to provide good accessibility to all parts of the city by both public and private transportation.

Cities that focus on a single high-density urban center will inevitably face high transport infrastructure costs because the convergence of many people and goods means high-capacity and high-cost solutions—expressways or subways. The multicentered community that attempts to avoid such over-concentration is illustrated by the new city of Runcorn, which is designed to disperse concentrations and cluster related activities. Local neighborhoods have their own schools, playfields, shops, and social center, and each neighborhood is connected to the others by motor way in the form of a figure 8, with four separate industrial areas located outside the 8.[16]

A feature of British new towns especially applicable to developing countries is the separation of motor vehicles from other types of traffic. Stevenage has 25 miles of cycle ways and parallel walkways, the town center is confined to pedestrians, and shoppers can reach the center by pathways from the housing areas. The effort to segregate traffic and to get parked cars off the streets has paid off in efficiency and safety: in a recent five-year period motor traffic in Stevenage increased 40 percent, but accidents declined 25 percent despite a rising national trend.

In the United States, events that could have an important effect on urban transportation are the initial moves that have been made to establish national urban growth policies, federal aid for new communities, and community development corcorporations for large-scale city building. One idea proposed in New York, for example, is a plan to halt present development trends in South Richmond by substituting a planned community for normal growth trends. Under present conditions a projected 250,000 more people in South Richmond before the end of the century promises heavy congestion, environmental pollution, and a community that will be undesirable for both living and working. The alternative of planning South Richmond's future has been rejected by the state legislature, but the proposal illustrates what might be achieved through improved growth policies.

The alternative offered was a planned city for 450,000 inhabitants which would combine economic viability with extraordinary liveability. The new city would include a pedestrian-oriented town center, a comprehensive park system, employment in planned industrial park and commercial centers, and 12 communities of 35,000 people, with the highway system forming the boundaries of each cluster. These communities would have their own schools, convenience shopping, and recreation facilities within three to four minutes walk for everyone. The city would have a major center for entertainment, high-rise apartments, offices, a college, a government center and a transportation terminal, with internal transit from city center to town and neighborhood centers, and with connections to New York City by rapid transit. What is important for developing countries is that public revenues from the planned city between 1970 and 2005 are estimated at nearly four times what would be realized if present planless growth is allowed to govern. Large-scale building operations would permit major cost savings, and instead of 30 percent of the area being devoted to streets, the street system would be reduced to 16 percent. Land-use controls would prevent overcrowding and unreasonable traffic volumes. The plan demonstrates that devoting more land to streets in conventional cities does not help decongest traffic and that good circulation can better be accomplished by the clustering of related activities, separating people from vehicles, and using streets to delineate urban functions and organize economic activity.[17]

Hartford, Connecticut is another metropolis which has attempted to reverse current growth trends for a more cost-effective development pattern. It has created two public corporations, one responsible for physical, economic, and social planning, and another for the public and private financing of the actual development. A large percentage of Hartford residents are victims of poor housing, dismal neighborhoods, overcrowded schools, prohibitive medical costs, unemployment, inadequate training, crime, drugs, and racial tensions. Automobile congestion and poor transit are problems, but they are not the basic problems. The trouble with Hartford, like most other places, is that the present fragmented approach to building and operating the community will never lead to acceptable solutions. No one is responsible for attacking community problems on a scale large enough to make a difference. The alternatives are either to let growth run its uncharted course, or to organize and direct the forces of growth to improve economic and social systems.[18]

The first phase of the redevelopment program covers half the area of Hartford and 40 percent of the population. It aims at improving the financial techniques for home ownership and rehabilitation, creating cohesive neighborhoods of about 10,000 people, and supplying for each neighborhood vocational training, day care centers, shops, clinics, community meeting places, recreation, and open space. The physical and economic models of the Hartford program indicate that the public investments programmed to achieve these facilities and services could provide enough payback through land sales and taxes to repay the front-end public investments over a 20-year period and build a surplus, despite higher standards. But Hartford has been the victim of depressed economic conditions, unemployment, and a reduction in federal support for housing. The program is not moving ahead, and its future appears to depend as much as anything on changes in national policy in support of more effective city building efforts.

What these programs teach is that the solution of transportation problems is often a secondary effect of solving other problems. The success of such a strategy is already being demonstrated in Mexico, Malaysia, Brazil, The Philippines, Singapore and other developing countries.

Developing Countries: Focus on Community Development

Developing countries have already moved beyond Europe and the United States in exploiting the potentials of planned urbanization, and in doing so are demonstrating basic approaches to solving transportation problems. Brasilia, with 650,000 people in the metropolitan area, is a city in which living quarters are closely related to schools, stores, recreation and community services. The central city has a city-wide limited access road system that eliminates cross-streets from 80 percent of the network and provides for the flow of motorized traffic without signal lights. The city has totally neglected the pedestrian in the downtown area, however, and it has greatly increased travel volumes by the separation of housing from jobs and by segregating low-income families in satellite towns remote from the planned community.

In Mexico the new city of Cuautitlan Izcalli, now under construction 25 miles from Mexico City, will accommodate 1.6 million people by 1980. It is being built in paralled strips of housing, commercial establishments, and industrial developments, using the main road and the limited access highway as major structural components of the linear pattern. The purpose of Cuautitlan Izcalli is to provide a self-sufficient satellite beyond the congestion of Mexico City which will provide the space needed for new housing and industry and attract part of the inward migration that would otherwise move into the already overcrowded capital.

In Mexico City itself, the planned community of Independencia provides a microscopic example of how it is possible to achieve an integrated housing, industrial development and transportation solution for an urban cluster. This was designed for 16,000 people and was built by the social security organization for families working in nearby industries. Pedestrian pathways from housing to church, stores, recreation and schools, and inconspicuous parking areas demonstrate a synthesis of transportation and community that has made this lower middle-income settlement one of unusual character and beauty.

In Malaysia, which now has an Urban Development Authority, the earlier experience with the new satellite city of Petaling Jaya demonstrates how the building of a new community with limited capacity for self-containment can create rather than resolve traffic problems. The good housing in Petaling Jaya matched the high-paying jobs in Kuala Lampur, while the factory jobs in Petaling Jaya were filled by low-income workers from the center of Kuala Lumpur. The result has been an interchange of commuters that has created massive traffic jams on the main highway connecting the two cities. This could have been avoided by a satellite city with a good mix of housing types and a broader range of job opportunities. The lesson has been learned and applied to the new city of Shah Alam, being built as the new state capital for 200,000 people. Shah Alam will be one more cluster in Malaysia's Klang Valley, which is becoming a multi-centered urban region that could provide the setting for economic development in a planned environment, with traffic congestion reduced by design.

There are other examples of how it is possible to move out of the congestion of the pre-automobile city and to create an environment in which a variety of problems, including transportation, can be solved. The new commercial city of Makati, adjacent to Manila, was built in ten years for 200,000 people. A major weakness has been the lack of housing for low-income people, but the city has demonstrated that congestion can be minimized by a combination of building controls, open space, housing, shopping centers, parking restrictions, boulevards and good public taxi service.

Singapore, with a population of 2.4 million in 1975, is probably the best illustration that instant city-building on a giant scale is physically possible, and that moving out of the congestion is the way to get people adequately housed. In the late 1950's the city had 550,000 squatters and slum dwellers living in some of the most congested slums in Asia. Instead of focussing on what to do about the congestion, it was decided that housing was the top priority, and that by moving out of the congestion all low income families could be provided with decent shelter. Ten years later 600,000 people had been located in modern apartments with all the related public services, in housing estates and satellite communities surrounding the city. In the 1971–1975 five-year plan, Singapore stepped up its housing program to produce a second 600,000 homes in five years rather than than ten, and continued the work of development.

The importance of the Singapore program is that instead of attempting to provide the transport infrastructure needed to overcome the congestion of a single center, steps were taken to disperse activity in a series of concentrations. Transportation was used to connect the old built-up area with the new satellites, and daily transportation for shopping, school, recreation, and other family purposes was made easier by the convenient location of these services in the housing areas. Efforts were concentrated on overcoming the most urgent needs for shelter and sanitary facilities, and in the process many jobs were created in the construction and supply industries. At the same time, a high-quality urban environment was created by tree planting, landscaping, recreation programs, and the preservation of low-density land uses around the housing estates.

There are now eight separate clusters of housing, factories and community facilities surrounding the old city of Singapore, most of them located five to six miles from the center. Queenstown, for example, with a target population of 150,000, is divided into seven separate neighborhoods, and has an extensive pedestrian town center with shops, recreation, and convenient parking. Frequent bus service covers the area, and buses provide transportation to downtown Singapore and other satellites.

Farther out, the new industrial city of Jurong is being constructed for 450,000 people 12 miles from the center of Singapore, with an extensive network of roads and a mix of industry and living that is predicated on a fully motorized community. Woodlands is more distant from the city, and will also to a large extent be independent of it, but Telok Blangah is just beyond the port of Singapore, and will introduce a planned environment, with designed-in transport solutions, just two miles from the center.

Although it might be thought that attempts to accommodate the entire population of a city in quality housing and environment would entail a heavy sacrifice of other community goals, the Singapore experience suggests a different conclusion. Not only has the city been transformed and low-income families provided with immeasurably better living conditions, but thousands of new jobs have been created and Singapore's per

Table 23.6 Some Measures of Economic Growth, Singapore, 1960–70

	1960	1970
Population (millions)	1.6	2.1
Gross domestic product (US$ million)	700	1,855
Per capita GNP (US$)	414	894
Manufacturing output (US$ million)	556	1,650
Government revenues (US$ million)	103	347
Private cars	85,668	142,568
Consumer price index	100	111.5
Private consumption expenditure (US$ million)		
Goods	516	774
Services		
Housing	53	97
Transportation & communication	78	163
Other	91	130
Total consumption expenditures	738	1,164

Source: Government of Singapore, "Year Book of Statistics," Singapore, 1970.

capita gross domestic product has doubled in a decade. Consumers spent nearly twice as much on housing in 1970 as they did in 1960, and they were also spending more than twice as much on transportation, mostly private cars. More housing and more automobiles turned out to be complementary rather than conflicting goals.

Singapore is unique in geography and in other special circumstances, but it helps to demonstrate approaches that begin to cope with the causes of congestion. First is the integration of related land uses and activities within clusters to reduce the volume of daily travel, especially commuting. Second is the potential of large-scale city-building to reduce unit costs and rescue whole populations from the slums in a short time. Third is the guiding of growth into lower-cost land on the urban periphery, and the use of public revenues from increasing land values to support community services. Fourth is the ability of urban systems-building to cluster housing, transport, jobs, recreation, and sites for modern industry, all in a more satisfying total environment.

The focal point of all planned cities and downtown redevelopment projects is the multi-purpose center, or planned unit development. A center having such multiple uses as apartments, offices, shops and recreation creates a large volume of internal traffic which in turn reduces the traffic outside the center. This has two effects. One is that the internal trips are short, and are therefore made on foot, and the automobile is excluded. The other is that if the city is gradually developed into a series of multiple-use centers, external traffic can be organized along specific routes.

The multiple use center in effect makes walking competitive with riding, for studies show that in downtown areas when consideration is given to the time to walk to an automobile, unpark it and park again, a distance of 2,200 feet can be walked or driven in the same time. If 1,100 feet is considered the walking limit, this distance can be covered on foot in 4 minutes and by car in 7 minutes.[19]

For large urban clusters, new technology may provide local transport systems or people movers that are essentially horizontal elevators which can supply auto-free circulation over a large area. Some method of rapid transit can then provide the connection between clusters. A combination of planned unit developments and internal transport solutions may yield the combination needed to supply viable forms for community growth in a rapidly urbanizing world.

THE POSSIBILITIES OF NEW TRANSPORT TECHNOLOGY

The possibilities thus far considered for reducing the conflict between cities and cars include measures that would improve public transport performance and longer range measures for urban planning and development. But is there not the further possibility that innovations in transport technology may introduce substitutes for the automobile that will make it easier to bring transportation supply and demand into balance?

At the present time there are several technological potentials that have promising application, including high speed ground transport, new guidance systems, new sources of power, changes in automobile design, and new applications for electronic communications. An important coincidence of some of these innovations seems to be emerging. One is the concept of urban clusters with reliance on walking, aided by people-mover systems. A second is the concept of personal rapid transit operating on a guideway and serving individual trip needs over a wide urban network. Another is high-speed ground transportation combining air cushion suspension with new methods of propulsion to aid in the greater dispersal of urban concentrations. Still another may be the dual mode automobile system which provides electronic guidance for vehicles under heavy traffic conditions. The combination of these innovations may make possible regional networks of cities with internal pedestrian systems and fast inter-urban connections.

Personal rapid transit (PRT) in the process of development relies on networks of one-way guideways that would derive their economy by operating on relatively lightweight elevated structures, perhaps over the streets and in buildings. The system would include high-speed main routes and a larger mileage of collection and distribution routes. Small cars without wheels would provide non-stop service from loading stations to destinations. Cost of the structures might be one-third the cost of ordinary elevated railways, and the structure itself would be more in harmony with the urban environment. A comparison between conventional rapid transit and a PRT system for about the same price illustrates the economic advantages of the latter, and suggests that systems now being installed in some cities may be obsolete in the near future.

A conventional rapid transit system for Los Angeles would cost $1.42 billion, would provide 189 miles of one-way routes, and serve 67 stations. For about the same cost, a PRT system would provide 468 miles and serve 455 stations. According to this estimate, PRT would offer more than double the seating capacity of a conventional system of the same cost and cut the time of trips in half.[20] Another appraisal of personal rapid transit, for Boston, has shown the advantages of a 400-mile system of PRT compared to a conventional 62-mile rapid transit network.[21]

An application of high-speed ground transport for city use was a proposed train between Paris and the new city of Cergy-Pontoise, 18 miles from the center of Paris. This planned city for 330,000 people (with 140,000 jobs) was to be served by non-stop air cushion train to the new Paris subcenter of La Defense, a high-density cluster of offices and apartments surrounding a mile-long pedestrian mall. At La Defense an internal circulation system was to have been provided by a people-mover installed on the deck for pedestrian traffic. The people mover is a three-speed belt-conveyor that moves seated passengers at a speed of about 11 miles per hour, and will furnish transportation to and from office building and apartments. At Cergy-Pontoise, internal circulation for the low-density city will be by car and bus.

Personal rapid transit and people mover experiments reflect the fact that intensive international efforts are under way to find a successor to the automobile for the big city. For developing countries an innovative approach already available may be a more organized use of public automobiles. But there is also the possibility that high speed ground transport by fast public conveyance will permit intercity transport methods to reduce urban transport problems by linking together networks of separate urban concentrations.

Whatever the prospects for innovative technology that may some day complement or substitute for the car, the lessons of Japan suggest where the important innovations may lie. Japanese planners now take the view that despite the unparalled subway and rail network serving Tokyo, congestion will not be overcome either by building subways or by controlling automobiles, but only by curtailing over-concentration and planless urbanization. By 1985 about 70 percent of the population of Japan will be living in cities, with per capita incomes four to five times higher than in 1965. Twice as much land will have to be devoted to urban living, and this fact has focussed on the need for a broader distribution of urban growth throughout the country, making the entire nation function as a single metropolitan area. Such a development pattern is now made increasingly feasible by the introduction of nationwide information systems, high-speed rail connections, new air transport networks, the automobile and expressway, and the growing interest in less commutation and more satisfying living conditions.[22]

Japanese economic plans have not counted the automobile out. Rather, they suggest that 2.5 times as much area will need to be allocated to highways in 1985 compared to 1965, and that the proportion of consumer outlays used for transportation and communications will be quadrupled over the 20-year period—by far the most significant change in spending habits anticipated for the Japanese economy. There is expected to be an increase in car ownership from 20 million vehicles in 1970 to at least 40 million in 1985, with 15 percent of Japanese families having more than one car and only 15 percent being carless. These figures assume a decentralized pattern of urban settlements, with much more attention to housing and amenities.

The evolution in Japanese thinking re-emphasizes the need for national urbanization planning and for urban design solutions that begin to cope with the underlying causes of congestion. The lesson is of special importance for developing countries, where the key reality is that much of the growth in urban population is yet to come. Unless this growth can be accommodated in a satisfactory way, which includes acceptable standards of shelter, water, sanitation, and other services, no amount of transport investment, transport controls, or concern over the automobile will save the cities.

The fact that transportation solutions alone will not resolve urban transportation problems is better illustrated by Bombay than by Tokyo. Metropolitan Bombay, with an estimated five million people, has few cars and a very extensive public transport system. The surburban railways carry two million people into the center every day, and another 2 million move by bus. With population density at 115,000 persons per square mile in the central area, these commuters make an enormous crush of traffic inevitable, whatever the method of transport. If more transport outlays were to make it easier for more people to live in the city, the result would magnify many of Bombay's problems, including water, sanitation, food, supplies and shelter. Seventy-five percent of the population lives in grossly overcrowded conditions, with 5 to 10 people per room. Only 15 percent of the dwellings have a toilet, and only 28 percent have water. Building more transport capacity is not a solution. A more promising strategy may be the decision to build a new satellite community for a million people across the harbor, or to divert some of the additional growth into more distant communities that could be connected to Bombay by good transport and communications.

Innovations in transportation that hold out the greatest long-term promise, therefore, appear to be those that combine transport with planned urban design and with regional planning of urban growth.

- High-density clusters within existing cities for living, working, shopping and recreation, dependent for local transportation on pedestrian systems, aids to walking, and improved public transit.
- Planned new communities to accommodate growth, providing lower densities and mixed land uses that permit the operation of private cars along with buses and public cars but reduce the necessity for long commuting trips by the proximity of jobs and housing.
- High-speed ground transport connections and communications systems that link both the new and old communities and reduce trip time for many journeys to levels where the private automobile is no longer competitive.

24 CONCLUSIONS

Increases in automobile ownership in developing countries will continue despite higher energy costs, and total registrations of over 100 million cars can be expected by the end of the century, six times the present number. There are many economic and social advantages to be gained through the automotive revolution, but the conflict between cities and cars creates major problems for most countries.

First, car ownership is heavily concentrated in a small number of giant cities where the density of development is already so high that more cars mean fewer opportunities to use them. Second, the automobile is interfering with the effective operation of public transit, on which the majority of people depend. Third, the millions who are very poor will continue to walk, the automobile will continue beyond their reach, and the economic and social gap between car owners and the carless will continue to widen.

The first task confronting policy makers is how to assure satisfactory standards of public transportation and reasonable cost, at the same time recognizing the political necessity of being responsive to the growing numbers of people who own or aspire to own an automobile.

Whatever public policy is adopted, the physical limitations of most big cities necessitate major reliance on public transit for the trip to work. Despite its key role, however, public transportation is nearly everywhere grossly inadequate. Most cities need to double or triple their bus fleets to reduce crowding, increase trip frequency, and expand route mileage. A $4 billion program to double the number of buses in cities of the developing countries with over half a million population could provide over 135 million more rides a day. The same amount invested in subways would supply less than 10 percent as much transportation.

The fundamental weakness of the bus, however, is the conflect with private cars. To make surface transit more effective would require constraints on passenger car use in congested areas by pricing policies that reflect social costs and by physical regulation of traffic and parking in central business districts. Effective measures might include vehicle-free zones, pedestrian streets, all bus streets, parking bans, and a shift of parking responsibilities from the public to the private sector, with charges for off-street space at rates that cover costs.

Improvement of traffic conditions for both transit and automobiles will require a variety of street and traffic engineering solutions, including by-pass routes, one-way circulation, grade separation of important intersections, signal systems and other traffic controls, enforcement measures, elimination of curb parking, and use of the curb lane for buses.

The construction of high-capacity rapid transit systems, in subways or on elevated structures, will be necessary in some cities where incomes are sufficiently high to permit the large investment required. The fixed nature of rail transit, however, makes the subway decision irreversible, and in the face of uncertainty over technology and the urban future, the risk of miscalculation is high. Measures will need to be taken to control the demand for transportation and to prevent the subway from creating new congestion underground. Complementary efforts will be needed to improve public transport on the surface, which will still be required for the majority of travelers. Bus transport systems might be supplemented by well organized jitney taxi or public automobile service. A public automobile can do the work of at least six private automobiles.

Given the rapid growth of urban population, however, transport measures for the relief of congestion will not be enough. Expanding urbanization presents the necessity and the opportunity to renew old cities and guide new growth into urban settlements designed to make transportation problems manageable.

There will need to be a new focus on controlling the factors that generate the demand for transportation. They are no less important than the supply of transport capacity. These factors include the location of urban structures and activities, their densities, their physical arrangement, and the interrelations among housing, work places, shopping, recreation, and other services. Relying on more transport is often an attempt to compensate for the disorder and disarrangement of planless communities.

It is difficult to correct these conditions in existing cities. The land uses are there and the demands on the transport system in the short run are fixed. But the city is in a constant state of change, and a start needs to be made to guide the process of growth in ways that will reduce traffic problems rather than make them worse. This can be done by providing housing and a good environment in the central city close to places of employment. Clustering of related land uses will increase the proportion of the population having the option of a short ride or an easy walk to stores, schools, jobs and recreation. The opposite course of concentrating employment in a single city center multiplies commuter traffic, destroys the concept of community, and puts the emphasis on dealing with traffic rather than with the more basic problems of urban life.

Fortunately what needs to be done to make the city more liveable is generally what needs to be done to make the transportation problem more manageable. A program to reconstruct obsolete city centers and improve the urban environment also attacks the underlying causes of congestion.

It will not be enough, however, to focus on cities as they now exist, for the impact of growth can defeat the objectives of redevelopment. There is simultaneous need to guide new growth into planned additions to the metropolis, and to formulate national urbanization plans that aim at dispersing urban settlements. Planned satellite communities built to be partially self-sufficient can avoid adding to the problems of the old center, help absorb some of its population pressures, and furnish new sites for industry and commerce. Automobile and public transport will both have a role to play in the support of new growth centers and can help cities to move out of the congestion rather than support and reinforce a degree of crowding that is no longer necessary.

APPENDIX V A:
Basic Data on Cities and Motorization in Developing Countries

Table 1 Base Information on Selected Urban Areas

	Date	Population (in millions)	Annual Rate of increase (%)[a]	Income per capita (US$)
Seoul	1970	5.5	8.5	440[b]
Djakarta	1970	4.8	5.3	325[c]
Hong Kong	1971	4.1	2.9	850[b]
Bangkok & Thonburi	1970	3.1	6.2	524[b]
Singapore	1970	2.1	2.6	800[b]
Taipei	1971	1.8	3.1	555[d]
Kuala Lumpur	1971	0.7	6.5	660[b]
Calcutta	1971	8.6	3.5	270[e]
Bombay	1971	5.9	3.7	388[e]
Madras	1971	3.6	4.5	180[b]
Teheran	1971	3.6	7.0	950[b]
Istanbul	1971	3.0	6.0	810[f]
Caracas	1971	2.2	5.4	1,600[b]
Bogota	1970	2.5	7.3	450[b]
Lima-Callao	1970	3.6	9.0	865[g]
Sao Paulo	1970	8.4	6.4	784[c]
Mexico City	1970	8.6	5.8	1,274[h]
Buenos Aires	1970	8.8	2.4	1,800[b]
Cairo	1971	6.5	5.7	275[c]
Casablanca	1970	1.5	4.5	817[i]
Kinshasa	1970	1.3	12.1	660[j]
Lagos	1970	0.9	7.9	555[c]
Nairobi	1970	0.6	8.1	494[c]
Dakar	1970	0.6	5.3	580[b]
Dar es Salaam	1970	0.4	9.0	712[b]
Kampala	1970	0.4	9.2	600[c]

Source: Respective Transport Plan Studies and IBRD reports for urban areas.
[a]The annual growth rates one for the period 1960–1970.

Table 2 Gross Densities in Selected Urban Areas (persons/km^2)

	Date	Urbanized area (km^2)	Population (millions)	Gross density (persons/km^2)
Seoul	1970	366.84	5.5	14,993
Djakarta	1970	369.12	4.5	12,191
Hong Kong	1967	82.01	3.8	46,336
Bangkok & Thonburi	1970	290.56	3.1	10,669
Singapore	1971	170.10	2.1	12,346
Taipei	1971	66.99	1.8	26,870
Kuala Lumpur	1963	51.80	0.5	9,653
Calcutta	1964	126.91	4.2	33,094
Bombay	1968	437.00	5.5	12,586
Madras	1964	128.37	2.8	21,812
Teheran	1971	250.00	3.6	10,286
Istanbul	1971	250.00	3.0	12,000
Caracas	1966	112.13	1.8	16,053
Bogota	1970	130.90	2.5	19,099
Lima-Callao	1970	163.63	3.6	22,001
Sao Paulo	1970	856.75	8.4	9,804
Mexico City	1970	1,500.00	8.6	5,733
Buenos Aires	1970	1,063.79	8.8	8,272
Kinshasa	1970	210.60	1.3	6,448
Lagos	1965	69.95	0.7	10,007
Nairobi	1965	40.00	0.4	10,147
Dakar	1970	32.00	0.6	18,750
Dar es Salaam	1966	52.05	0.3	5,764
Kampala	1968	29.84	0.3	10,054
Cairo	1971	214.20	5.6	26,144
Casablanca	1970	113.30	1.4	12,357

(Notes to table 1 continued)

[b]Cities with reported incomes.

[c]Cities updated based on information in respective IBRD reports, other urban studies and upon conference with the responsible country operational and economic officers

[d]On basis of family income distribution survey of 1968.

[e]Estimated from the 1961 relationship of the u[illegible]an areas with the average national income per capita.

[f]Based on gross product estimates for the metropolitan area in 1972 and its relationship with national income levels in 1964 and 1968.

[g]On basis of regional breakdown of total GNP.

[h]On basis of family income survey of 1969.

[i]Estimated from Casablanca's contribution to national GOP which is 33.3%.

[j]Estimated from Kinshasa's contribution to national GOP which is 22%.

Table 3 Vehicle Registrations in Selected Urban Areas of Developing Countries

	Year	Total vehicles on register
Bangkok & Thonburi	1970	292,700
Singapore	1971	310,699
Hong Kong	1971	164,378
Seoul	1970	61,355
Kuala Lumpur[c]	1970	193,174
Djakarta	1969	198,754
Taipei	1969	132,866
Bombay	1969	135,467
Istanbul	1970	105,680
Teheran	1971	240,000
Caracas	1970	300,012
Bogota	1969	130,000
Lima Callao	1970	214,032
Sao Paulo	1967	516,079
Mexico City	1971	810,926
Buenos Aires	1969	914,464
Casablanca	1970	279,201
Lagos	1969	33,561
Nairobi	1965	41,616
Nairobi	1970	66,894
Dar Es Salaam	1967	15,520
Kampala	1970	n.a.
Kinshasa	1970	54,624
Dakar	1971	n.a.

[a]Includes taxis, "jitney" cabs and hired cars.
[b]Includes only motorized two-wheeled vehicles such as motorcycles, motortricycles, scooters since information on bicycle registration for all cities was not available.
[c]Registrations are for Selangor State. 85% of the vehicles are concentrated in the Kuala Lumpur area.
[d]There are no motorcycle registrations available for Colombia in I.R.F. publication for 1970 either.
[e]The data for two-wheeled vehicles are for 1967.
[f]Includes buses.

Motor cars			Motor buses	Two-wheeled vehicles[b]	Commercial vehicles
Private	Public[a]	Total			
151,600	9,100	160,700	4,100	78,200	49,700
154,089	7,126	161,215	2,687	108,988	37,809
105,874	4,290	110,164	8,455	17,507	28,252
23,365	11,505	34,870	4,805	2,355	19,325
82,958	1,659	84,617	1,562	86,506	20,489
77,670	200	77,870	4,595	97,639	18,450
17,320	9,440	26,760	640	98,716	6,750
74,459	13,999	88,458	1,542	20,293	25,174
43,591	15,930	59,521	2,933	6,607	36,619
160,000	11,950	171,950	3,600	40,000	24,450
200,000	15,000	215,000	1,300	23,712	60,000
55,000	9,000	64,000	3,500	–[d]	62,500
131,255	11,000	142,255	9,865	21,134	40,778
411,063	26,501	437,564	8,609	n.a.	69,906
650,089	23,000	673,089	11,515	48,840	77,482
650,000	25,000	675,000	14,200	n.a.	225,264
108,959	3,615	112,033	646	126,901[e]	39,080
19,191	600	19,791	800	6,009	6,691
21,000	877	21,877	600	2,939	16,200
n.a.	n.a.	n.a.	n.a.	n.a.	n.a.
n.a.	n.a.	9,100	n.a.	1,710	4,710[f]
13,846	n.a.	n.a.	n.a.	n.a.	n.a.
n.a.	n.a.	n.a.	n.a.	n.a.	n.a.
16,500	1,200	17,700	580	n.a.	n.a.

Table 4 Distribution of vehicles registered in Selected Urban Areas of Developing Countries by Type (%)

	Year	Motor cars		
		Private	Public[a]	Total
Bangkok & Thonburi	1970	51.79	3.11	54.90
Singapore	1971	49.59	2.29	58.88
Hong Kong	1971	64.41	2.61	67.02
Seoul	1970	38.08	18.75	56.83
Kuala Lumpur[c]	1970	42.94	0.86	43.80
Djakarta	1969	39.07	0.15	39.22
Taipei	1969	13.03	7.10	20.13
Bombay	1969	54.96	10.33	65.29
Istanbul	1970	41.25	15.07	56.32
Teheran	1971	66.67	4.98	71.65
Caracas	1970	66.67	5.00	71.67
Bogota	1969	42.31	6.92	49.23
Lima-Callao	1970	61.32	5.14	66.46
Sao Paulo	1967	79.65	5.14	84.79
Mexico City	1971	80.17	2.84	83.01
Buenos Aires	1969	71.08	2.73	73.81
Casablanca	1970	39.02	1.29	40.32
Lagos	1969	57.18	1.79	58.97
Nairobi	1965	50.46	2.11	52.57
Dar Es Salaam	1971	n.a.	n.a.	58.63

[a]Includes taxis, "jitney" cabs and hired cars.
[b]Includes only motorized two-wheeled vehicles such as motorcycles, motortricycles, scooters since information on bicycle registration for all cities was not available.

Motor buses	Two-wheeled vehicles[b]	Commercial vehicles	Total vehicles on register
1.40	26.72	16.98	100.00
0.86	35.08	12.18	100.00
5.14	10.65	17.19	100.00
7.83	1.49	33.85	100.00
0.81	44.78	10.61	100.00
2.31	49.18	9.29	100.00
0.50	74.29	5.08	100.00
1.15	14.98	18.58	100.00
2.78	6.25	34.65	100.00
1.50	16.67	10.18	100.00
0.43	7.90	20.00	100.00
2.69	–	48.08	100.00
4.62	9.87	19.05	100.00
1.67	n.a.	13.54	100.00
1.42	6.02	9.55	100.00
1.55	n.a.	24.64	100.00
0.23	45.45	14.00	100.00
2.38	17.90	20.75	100.00
1.44	7.06	38.93	100.00
–	11.02	30.35[d]	100.00

[c]Registration are for Selangor State. 85% of the vehicles are concentrated in the Kuala Lumpur area.
[d]Includes buses.

Table 5 Motorization Rates in Selected Urban Areas of Developing Countries (in number of persons per vehicle)

	Year	Total vehicles on register
Bangkok & Thonburi	1970	10.4
Singapore	1971	6.8
Hongkong	1971	24.6
Seoul	1971	90.2
Kuala Lumpur	1970	8.4
Djakarta	1969	21.7
Taipei	1969	12.7
Bombay	1969	41.0
Istanbul	1970	26.5
Teheran	1971	15.0
Caracas	1970	7.3
Bogota	1969	19.2
Lima-Callao	1970	16.8
Sao Paulo	1967	12.8
Mexico City	1971	10.6
Buenos Aires	1969	9.6
Casablanca	1970	5.4
Lagos	1969	25.1
Nairobi	1965	9.6
Nairobi	1970	7.7
Dar Es Salaam	1969	17.6
Kampala	1970	n.a.
Kinshasa	1970	23.8
Dakar	1971	n.a.

Motor cars			Motor buses	Two-wheeled vehicles	Commercial vehicles
Private	Public	Total			
20.1	335.3	19.0	744.1	39.0	61.4
13.7	296.2	13.1	785.4	19.4	55.8
38.2	943.0	36.7	478.5	231.1	143.2
236.9	481.2	158.8	1,152.1	2.340.7	286.5
19.6	981.9	19.3	1,042.9	18.8	79.5
55.5	21,559.5	55.4	938.4	44.2	233.7
97.6	179.0	63.2	2,640.6	17.1	250.4
74.1	200.0	31.7	3,600.8	273.6	111.2
64.2	175.8	47.0	954.7	423.8	76.5
22.5	301.3	20.9	1,000.0	90.0	147.2
11.0	146.7	10.2	1,692.3	92.8	36.7
45.5	277.8	39.1	714.3	–	40.0
27.4	327.3	25.3	364.9	170.3	88.3
16.1	249.0	15.1	766.6	n.a.	94.4
13.2	373.9	12.8	746.9	176.1	111.0
13.5	352.0	13.0	619.7	n.a.	39.1
13.7	413.6	13.3	2,314.2	11.8	38.3
77.9	2,491.7	42.5	1,868.8	140.1	214.8
19.0	454.5	18.2	664.3	135.6	24.6
n.a.	n.a.	n.a.	n.a.	n.a.	n.a.
n.a.	n.a.	29.9	n.a.	159.4	57.9
n.a.	18.1	n.a.	n.a.	n.a.	n.a.
n.a.	n.a.	n.a.	n.a.	n.a.	n.a.
33.3	458.3	31.1	948.3	n.a.	n.a.

Table 6 Motorization Rates in Selected Urban Areas by Order of Magnitude

No.	City	Total vehicles registered per 1,000 persons	Persons per total number of vehicles
1	Casablanca	186.8	5.4
2	Singapore	147.2	6.8
3	Caracas	136.4	7.3
4	Nairobi	130.4	7.7
5	Kuala Lumpur	118.6	8.4
6	Buenos Aires	103.9	9.6
7	Bangkok	95.9	10.4
8	Mexico City	94.3	10.6
9	Taipei	78.6	12.7
10	Sao Paulo	78.2	12.8
11	Teheran	66.7	15.0
12	Lima-Callao	59.5	16.8
13	Dar Es Salaam	57.0	17.6
14	Bogota	52.0	19.2
15	Djakarta	46.0	21.7
16	Kinshasa	42.0	23.8
17	Hong Kong	40.6	24.6
18	Lagos	39.9	25.1
19	Istanbul	37.7	26.5
20	Bombay	24.4	41.0
21	Seoul	11.1	90.2

Table 7 Number of Non-Motorized Vehicles and Ownership Rates in Some Urban Areas

	Year	Bicycles	Other	Total	Population	Non-motorized vehicles per 1,000 persons
Casablanca	1967	297,744	–	297,744	1,332.5	223.4
Singapore	1971	380,733	3,306[a]	384,039	2,110.4	182.0
Djakarta	1969	–	200,000[b]	200,000	4,311.9	46.4
Teheran	1964	100,000	–	100,000	2,410.5	41.5
Lima	1963	32,077	–	32,077	1,967.1	16.3
Seoul	1966	30,399	–	30,399	3,805.0	8.0
Dar Es Salaam	1967	30,000	–	30,000	272.5	110.1
Istanbul	1969	8,400	–	8,400	2,674.1	3.1

[a]Trishaws.
[b]Betjaks (three-wheeled pedicabs).

Table 8 Number of Persons per Car, 1970

Country	Passenger car	Country	Passenger car
United States	2	Mexico	41
Canada	3	Brazil	43
Australia	3	Bulgaria	52
New Zealand	3	Hungary	54
France	4	Chile 56	56
West Germany	4	Peru	60
Sweden	4	Poland	68
Italy	5	Kenya	95
United Kingdom	5	Iran	105
Kuwait	5	Saudi Arabia	117
Puerto Rico	6	Angola	118
Japan	12	Colombia	118
South Africa	12	Philippines	138
Spain	14	U. S. S. R.	147
Singapore	14	Thailand	175
Argentina	16	Turkey	255
Venezuela	17	South Vietnam	255
libya	17	U. A. R.	267
Israel	19	Indonesia	508
Lebanon	21	South Korea	524
Portugal	21	Pakistan	663
Yugoslavia	28	Nigeria	817
Hong Kong	36	India	902
Malaysia	39	China	5,711

Source: Motor Vehicle Manufacturers Association, *Automobile Facts and Figures*, 1972, pp. 28-29.

Table 9 Use of Public Transport in Selected Urban Areas by Order (%)

No.	City	Public transport[a]	No.	City	Public transport[b]
1	Seoul	95.2	1	Calcutta	85.5
2	Calcutta	90.4	2	Bombay	80.0
3	Bombay	89.3	3	Seoul	74.2
4	Lima-Callao	89.0	4	Mexico City	72.0
5	Bogota	83.3	5	Bogota	70.7
6	Mexico	81.0	6	Hong Kong	70.6
7	Hong Kong	78.3	7	Dakar-Pikine	70.2
8	Dakar-Pikine	76.5	8	Taipei	67.3
9	Sao Paulo	73.7	9	Buenos Aires	65.7
10	Buenos Aires	72.5	10	Sao Paulo	65.3
11	Taipei	71.3	11	Lima-Callao	61.0
12	Singapore	67.3	12	Bangkok &	
13	Bangkok	66.0		Thonburi	59.0
14	Teheran	63.0	13	Madras	51.0
15	Madras	56.0	14	Kampala	47.0
16	Caracas	54.4	15	Istanbul	43.3
17	Kampala	47.0	16	Singapore	42.4
18	Kuala Lumpur	39.8	17	Teheran	42.0
			18	Caracas	36.1
			19	Kuala Lumpur	34.7
			20	Kinshasa	23.6
			21	Dar es Salaam	17.7

[a]Including trips by public motorcar.
[b]Excluding trips by public motorcar.

Table 10 Major Automobile Owning Countries, 1970 (with over 100,000 cars)

	Passenger cars (thousand)	Percentage of total motor vehicles
United States	89,280	82.4
Germany (West)	14,376	92.1
France	12,290	85.3
United Kingdom	11,792	86.1
Italy	10,209	91.7
Japan	8,779	49.9
Canada	6,602	81.7
Australia	3,780	79.7
Netherlands	2,500	88.1
Spain	2,382	76.1
Sweden	2,288	93.6
Brazil	2,234	63.1
Belgium	2,060	87.6
South Africa	1,653	78.2
U. S. S. R.	1,650	26.4
Argentina	1,563	66.3
Switzerland	1,388	90.7
Mexico	1,234	68.9
Austria	1,197	90.4
Germany (East)	1,160	82.6
Denmark	1,077	81.0
New Zealand	899	83.2
Czechoslovakia	814	82.1
Norway	747	82.7
Yugoslavia	720	84.4
Finland	712	86.5
Venezuela	615	72.0
Inchia	610	56.4
Poland	479	62.0
Portugal	452	80.7
Ireland	389	88.4
Northern Ireland	287	86.7
Malaysia	279	77.1
Philippines	279	60.8
Iran	272	74.5
Indonesia	239	65.7
Greece	226	65.7

	Passenger cars (thousand)	Percentage of total motor vehicles
Peru	226	63.0
Morocco	225	72.8
Thailand	204	54.8
Hungary	192	73.3
Colombia	178	60.3
Chile	176	54.0
Pakistan	172	77.1
Bulgaria	162	80.5
Israel	151	68.0
Singapore	148	80.0
Turkey	147	46.5
Algeria	142	63.4
Kuwait	141	77.5
Lebanon	136	88.9
China	133	21.7
Romania	130	39.4
United Arab Republic	125	80.6
Uruguay	121	57.9
Rhodesia	120	71.9
Kenya	115	83.9
Hong Kong	114	79.2
Libya	110	67.9

Table 11a Present Modal Split in Selected Urban Areas (in percentages of number of person trips)

Transport mode	Kuala Lumpur, 1970	Bangkok, 1970: Memorial Bridge	Bangkok, 1970: Main approaches to core city
Private car	46.9	22.1	29.0
Two-wheeled vehicle[c]	13.3	5.0	5.0
Commercial vehicles	—	—	—
Private transport	60.2	27.1	34.0
Bus	34.7	68.9	59.0
Public car[f]	5.1	4.0	7.0
Others	—	—	—
Public transport	39.8	72.9	66.0
Walking	—	—	—
Total	100.0	100.0	100.0

Source: Respective Transport Plan Studies and IBRD reports for urban areas.
[a]Home-based trips.
[b]A: Excluding "walking" trips. B: Including "walking" trips.
[c]Includes motorcycle and bicycle trips.
[d]Includes 16.2% bicycle trips.
[e]Includes 8% bicycle trips.

Table 11b Present Modal Split in Selected Urban Areas (in percentages of number of person trips)

Transport mode	Madras, 1971	Bombay, 1968
Private car	n.a.[a]	10.7
Two-wheeled vehicle[b]	22.0	—
Commercial vehicles	—	—
Private transport	22.0	10.7
Bus	42.0	41.0
Public car[c]	5.0[a]	9.3
Other	9.0[d]	39.0[d]
Public transport	56.0	89.3
Walking	22.0	—
Total	100.0	100.0

Source: Respective Transport Plan Studies and IBRD reports for urban areas.
[a]Includes trips by private car.
[b]Includes motorcycle and bicycle trips.
[c]Includes taxis, "jitney cabs" and hired cars.

Singapore, 1968	Hong Kong,[a] 1965	Seoul, 1970[b] A	Seoul, 1970[b] B	Taipei, 1968[b] A	Taipei, 1968[b] B
24.2	21.7	4.8	5.9	2.5	0.5
8.5	–	–	–	26.2[d]	13.0[e]
–	–	–	–	–	–
32.7	21.7	4.8	5.9	28.7	13.5
42.4	55.2	74.2	64.8	64.3	34.0
21.7	7.7	21.0	2.4	4.0	2.0
3.2	15.4[g]	–	–	3.0[h]	2.0[h]
67.3	78.3	95.2	67.2	71.3	38.0
–	–	–	26.9	–	48.5
100.0	100.0	100.0	100.0	100.0	100.0

[f]Includes taxis, "jitney cabs" and hired cars.
[g]Includes 13% rail and 2.4% ferry trips.
[h]Trips by train.

Calcutta, 1966	Teheran, 1970	Istanbul, 1967	Caracas, 1966	Bogota, 1969
8.2	37.0	56.7[a]	45.6	16.7
–	–	–	–	–
1.4	–	–	–	–
9.6	37.0	56.7[a]	45.6	16.7
34.3	42.0	27.7	35.5	70.7
4.9	21.0	56.7[a]	18.3	12.6
51.2[e]	–	15.6[f]	0.6[a]	–
90.4	63.0	43.3[g]	54.4	83.3
–	–	–	–	–
100.0	100.0	100.0	100.0	100.0

[d]Trips by train.
[e]Includes trips 31.2% by tram and 20% by rail.
[f]Includes 3.0% rail and 12.6% ferry trips.
[g]Excluding trips by public motorcar.

Table 11c Present Modal Split in Selected Urban Areas (in percentages of number of person trips)

Transport mode	Mexico City, 1966	Buenos Aires, 1969
Private car	19.0	15.4
Two-wheeled vehicle[d]	–	3.0[e]
Commercial vehicles	–	–
Private transport	19.0	18.4
Bus	65.0	54.2
Public car[h]	9.0	6.8
Other	7.0[i]	11.5[j]
Public transport	81.0	72.5
Walking	–	9.1
Total	100.0	100.0

Source: Respective Transport Plan Studies and IBRD reports for urban areas.
[a]Survey confined to Kimban seke, Masina and N.djili communes.
[b]Reported to represent internal Dakar trips as well.
[c]Includes trips by private car.
[d]Includes motorcycle and bicycle trips.
[e]Mostly trips by bicycle.
[f]Includes 11.9% trips by bicycle.
[g]Includes 30.7% trips by public buses and 39.5% by privately run mini-buses.
[h]Includes taxis, "jitney cabs" and hired cars.
[i]Trips by tram.
[j]Includes 7.1% rail and 5.4% subway trips.
[k]Transport provided by the employer.

Lima-Callao, 1966	Sao Paulo, 1967	Kampala, 1969	Dar Es Salaam, 1967	Kinshasa,[a] 1970	Dakar-Pikine,[b] 1971
11.0	26.3	53.0[c]	2.2[c]	12.2[c]	19.4
–	–	–	13.2[f]	1.2	–
–	–	–	–	–	4.1
11.0	26.3	n.a.	n.a.	n.a.	23.5
54.0	60.2	47.0	13.2	21.5	70.2[g]
–	8.4	–	2.2[c]	12.2[c]	6.3
7.0	5.1	–	4.5[k]	2.1[k]	–
89.0	73.7	n.a.	n.a.	n.a.	76.5
–	–	–	66.9	63.0	–
100.0	100.0	100.0	100.0	100.0	100.0

NOTES TO PART V

Notes to chapter 22

1. See International Bank for Reconstruction and Development, *Urban Sector Paper*, 1972, p. 12.

2. Motor Vehicle Manufacturers Association, *Automobile Facts and Figures*, 1975, pp. 26–27.

3. Projections to 1985 are based on OECD data. Figures to the year 2000 have been provided by Gerald Leach, "The Motor Car and Natural Resources," Working Paper No. 4, OECD, Paris, 1972.

4. Data on Hong Kong from R. Butler and N. K. Chan, ECAFE Workshop on Urban Traffic and Transportation, December 1972, p. 142.

5. Report of Cairo Rapid Transit Study Review Mission, October 21–28, 1971, D. A. Stombom, Feb. 9, 1972.

6. National Planning Department, Government of Colombia, "An Urban Policy for Developing Countries," August 13, 1971.

7. The same.

8. J. K. Galbraith, *Economic Peace and Laughter*, 1971, p. 208.

9. Barbara Ward, "The Processes of World Urbanization," Background Paper No. 1, United Nations Symposium on the Planning and Development of New Towns, Moscow, August 24–Sept. 7, 1964, p. 12. In United Nations, "Planning of Metropolitan Areas and New Towns."

10. Leon V. Hirsch, The Littoral Highway in El Salvador, in *The Impact of Highway Investment on Development*, Brookings Institution, 1966, p. 100.

11. Baranson, Jack, "The Automotive Industry in the Republic of China," IBRD, July 31, 1970, p. 2.

12. The Fiera, by the Ford Motor Company. News Department, Dearborn, Michigan, May 24, 1972.

13. See D. H. Meadows, D. L. Meadows, J. Randers and W. H. Behrens III, *The Limits to Growth*, A Potomac Associates Book, New York, 1972.

14. Gerald Leach, "The Motor Car and Natural Resources," Inquiry into the Impact of the Motor Vehicle on the Environment, Working Paper No. 4, OECD, Paris, 1972.

15. See Morris Adelman, "Is the Oil Shortage Real?," *Foreign Policy*, Winter 1972–73, pp. 69–107.

16. *Potential for Motor Vehicle Fuel Economy Improvement*. Report to the Congress. U.S. Department of Transportation and U.S. Environmental Protection Agency, Oct. 24, 1974, p. 66.

17. U.S. Federal Energy Administration, *Project Independence*, November 1974, p. 163.

18. Gerald Leach, *Ibid.*

19. See Edmund Faltermayer "The Energy 'Joyride' is over," Fortune, Sept. 1972, p. 99, and C. William Savery, "Future Energy Sources for Transportation," Traffic Quarterly, Oct. 1972, p. 485.

Notes to chapter 23

1. Center City Transportation Project, "Urban Transportation Concepts," Wilbur Smith and Associates, p. 84.

2. A study of Ways to Use Existing Transportation Facilities More Effectively, Vol II Busway Applications. Prepared by R. H. Pratt Associates, Inc., for the U.S. Department of Transportation. July 1972, p. 37.

3. Based on UN population estimates contained in "Urbanization: Development Policies and Planning," International Social Development Review No. 1, 1968, p. 19.

4. U.S. Department of Transportation, "Economic Characteristics of the Urban Public Transportation Industry." Feb. 1972, pp. 8-1-8-5.

5. Data on Teheran transport from IBRD memo dated June 5, 1972, "Teheran Transport Study Seminar," by R. Venkateswaran.

6. Sigurd Grava, "The Jeepneys of Manila," Traffic Quarterly, October 1972, pp. 470–483.

7. Ibid.

8. For a fuller discussion of automobile and transit subsidies and user charges, see Wilfred Owen, *The Metropolitan Transportation Problem*, (Brookings, 1966), pp. 142-164, and *The Accessible City*, (Brookings, 1972), pp. 46–48.

9. "The Automobile in Swedish Society." A social research project for the Swedish Road Federation, Stockholm, 1971.

10. Ibid.

11. Peter Hall, "Transportation," in *Developing Patterns of Urbanization*, edited by Peter Cowan, 1970, p. 133.

12. Ryuji Yukawa, "Recent Status and Developments of Subways in Tokyo," presented at the 5th International Conference on Urban Transportation, Pittsburgh, Sept. 10, 1971.

13. Institut d'Aménagement de d'Urbanisme de la Région Parisienne, "Le Future Reseau de Transports en Région de Paris," 1966, p. 34.

14. Istanbul, Caracas, Bangkok, Manila, Bogotá, Lima, Santiago, Singapore, Teheran, Karachi, Delhi, Seoul, Cario, Djakarta, Rio de Janeiro, Calcutta, Bombay, Sao Paulo, Buenos Aires, Mexico City and Saigon.

15. It should be noted that the meticulous care in balancing transport and economic activity within the building was in contrast to the partial measures taken to avoid congestion in the interface between the building and the community.

16. Wyndham Thomas, "Britain's New Towns and Cities," prepared for the OECD, Paris, April 1972, Mimeo. 43 pp.

17. From a report to the City of New York, *Analysis and Development Trends in South Richmond*, The James W. Rouse Company, May 1970.

18. See The American City Corporation, "The Greater Hartford Process," April 1972, p. 9.

19. Peter W. Kwantes, "Transportation Aspects of Multi-Use Centers." Traffic Quarterly, Vol. XXVI, No. 4, Oct. 1972, p. 522.

20. Leon Bush, The Aerospace Corporation, El Segundo, Calif., "The Economics of High-Capacity PRT System," presented at the National Conference on Personal Rapid Transit, Minneapolis, Minn., Nov. 1-3, 1971.

21. William F. Hamilton and Dana K. Nance, "Systems Analysis of Urban Transportation," Scientific America, Vol. 221 (July 1969), pp. 19-27.

22. Economic Planning Agency, Government of Japan, "New Comprehensive National Development Plan," May 1969.

VI

AUTOMOTIVE AIR POLLUTION AND NOISE: Implications for Public Policy

Ad Hoc Group to the Environment Committee of OECD

25 OVERVIEW

This section considers strategies open to countries to abate air pollution and noise from motor vehicles. The countries involved differ in many ways from an environmental control point of view. They vary in climate and topography. They vary too in the extent to which they are industrialized. Some contain sprawling urban agglomerations, while others have only a few cities of moderate size set in expanses of relatively thinly settled countryside. Levels of car ownership differ widely too. They range from about one car for every forty people to one for every two.

These national variations influence the levels of air pollution and noise and the numbers of people that are exposed to them. But if they are considered in the light of forecasts of urban growth, car ownership and traffic density, one salient point emerges: if present trends continue, few countries are likely to escape the impact of mounting levels of air pollution and noise.

Environmental conditions are likely to deteriorate most seriously in metropolitan areas where traffic growth will be more pronounced than elsewhere. Within such areas, air pollution and noise will be spreading in space and time as well as in intensity under the influence of continued suburban growth. Rising numbers of people will be exposed to high levels of air pollution and noise more frequently. Variations in climate and in levels of car use between countries should not be allowed to obscure the universality of these trends and the inexorability of their growth.

The confidence with which these general trends can be forecast contrasts strongly with the uncertainty about their dimensions and their implications for public policy. Urban pollutant concentrations vary considerably, depending on time and location. In the absence of detailed data on air quality, reliable assessment of the scope and intensity of vehicle air pollution in individual cities cannot be made. As the Group's inquiry has clearly indicated, such data is seldom available except in isolated cases. More accurate and more complete monitoring of air quality is therefore needed both at the metropolitan scale and locally before a clear picture of the dimensions of the auto air pollution problem can be obtained.

Moreover, the health effects of automotive pollutants—particularly of low-level exposure over long periods of time—are not well understood. While it is generally acknowledged that these pollutnats are harmful in very high concentrations, there is no agreement on thresholds of harm. Even less is known about the long-term effects of noise on man. At levels normally encountered in cities traffic noise is likely to be more a nuisance than a threat to health. Nevertheless, a suspicion persists that chronic exposure to even moderate levels of noise may be harmful, although there is no agreement on the nature and extent of the damaging effects.

These uncertainties preclude a sweeping indictment, but they have not prevented governments from taking forceful steps to control the further growth of vehicle emissions. Action is being increasingly justified on the grounds that in matters affecting the public health and welfare even the suspicion of harmful effects is sufficient cause for taking preventive action. More and more importance is being placed on reducing the risk of damage even though its occurrence is not always an established fact. Ultimately, the case for limiting vehicle emissions and noise rests not on incontrovertible evidence of physical harm but rather on the growing feeling that it is in the public interest to minimize all emissions suspected of having adverse effects on man's health, well-being and environment.

VEHICLE EMISSION AND NOISE STANDARDS

Limiting vehicle emissions and noise at source by setting and enforcing standards is the prevailing form of action adopted by governments to combat the problem of automobile air pollution and traffic noise. The United States, where air pollution is already acute in many metropolitan areas, has chosen to proceed at a rapid pace by adopting standards for automotive emissions. Germany has adopted a target in its environmental protection programme of a 90 percent reduction in the vehicle emissions of 1969 models, such reduction to be effective by 1980. Other European countries intend to follow the standards they have agreed upon within the Economic Commission for Europe (ECE), and since promulgated as a directive by the Commission of the European Communities. They call for reductions in carbon monoxide and hydrocarbon emissions of about 35 per cent from uncontrolled levels.

As of 1972, noise emission standards were in effect in the six nations of the European Community as well as in several other European countries (Switzerland, Spain, United Kingdom). Legislation was passed in the United States in 1972 which required vehicle noise emission standards to be established within 18 months of the enactment of the law.

SHOULD STANDARDS BE INTERNATIONALLY UNIFORM?

Differences in the severity of air pollution due to variations in the level of car use and in climatic conditions between countries and differing national views as to the adverse effect of ambient levels of carbon monoxide, hydrocarbons and nitrogen oxides, explain the existence of different emission control requirements. Similarly, different perceptions of the nuisance of noise have led to the setting of different national noise standards.

Nevertheless, the presence in all countries of the same trends of increasing numbers of people living in the midst of rapidly growing numbers of vehicles, argues strongly for a progressive tightening of pollution and noise standards everywhere. While the pace at which this should be done may vary considerably from country to country because of different economic conditions and different perceptions of how urgent the problem has become, there is a good likelihood that, ultimately, vehicle emission and noise standards will become internationally uniform.

A strong case for such uniformity can be made on the grounds of convenience. Manufacturers exporting to many countries are already obliged to devote considerable resources to complying with numerous national requirements governing the construction of motor vehicles. To allow this mass of regulations to diversify still further would be to impose an additional and costly burden on the industry, a burden which in the end must be borne by the public.

Moreover the view has been expressed that differences in vehicle standards could distort international trade. It is said that the introduction by one country of more stringent standards than those found elsewhere could put producers of inexpensive cars in other countries at a disadvantage because the necessary modifications would add proportionately more to the cost of their vehicles than to those of rival makers of more expensive models. Some firms might also find that they had insufficient resources for effective research and development into emission control systems.

The occurrence of such inequalities is a possibility but it cannot be considered a barrier to trade. Few burdens on the motor industry are internationally uniform, and there seems little logic in making environmental standards a special case. Transport and wage costs, taxes, the price of materials, interest rates, labour productivity and many other influences on manufacturing costs vary widly from nation to nation. Setting emission and noise standards to a common denominator would not alone equalize competition. It would, however, be a barrier to the legitimate desires of nations to pursue environmental goals of their own choosing and at their own pace.

But above all, to lay too great a stress on the need for a single international set of standards might be to place undue importance on the requirements of the motor industry and too little on the fact that not all countries aspire to the same environmental quality objectives or, when their objectives are similar, require an equal degree of emission control. Agreement on common standards is therefore needed in the first place among countries whose environmental conditions and levels of car ownership are similar. Such regional harmonization already exists and is of great value in reducing costs and promoting the free flow of trade. The achievement of worldwide uniformity in standards can only be the result of a more gradual process in which evolving technology, industrial strategies and public attitudes will all play a major role.

THE NEED FOR UNIFORM TEST PROCEDURES

Acceptance of the necessity for the time being for two or more sets of emission standards in no way reduces the need for seeking international agreement on uniform procedures for testing emissions from new vehicles. A car equipped to meet an air pollution standard set in terms of one test procedure may fail to meet a similar standard expressed in terms of another procedure. The only way for manufacturers today to avoid this prospect is to test their vehicles according to all prevailing procedures, a costly process that must ultimately be borne by the public. Hitherto it has been assumed that differences in driving conditions in different countries, and the need to have such variations reflected in the driving cycle, precluded the achievement of uniformity in this field. Recent evidence, however, suggests that driving patterns in metropolitan areas may be sufficiently similar everywhere to facilitate the design of a driving cycle that would be acceptable to all countries. Even fewer obstacles stand in the way of reaching agreement on common measuring methods and instrumentation.

A detailed assessment of the feasibility of uniform test procedures is desirable. Such an assessment should include an evaluation of the technical requirements that would be necessary to determine the driving patterns in a representative number of metropolitan areas, an estimate of the cost of carrying out the necessary tests and proposals for funding.

There already is a widely accepted method of testing noise emissions, developed within the International Standards Organisation. Nevertheless, it now seems necessary to revise the existing procedure in order to take into account presently available technology and the wishes of certain countries. Different opinions about the form which these modifications should take have so far prevented the adoption of a revised procedure. The main issue is the proper weights to be given to vehicle noise both at maximum engine power and at engine speeds typical of normal driving. The ISO working group on vehicle noise measurement should be encouraged to seek an early resolution of this issue.

THE CHOICE OF STANDARDS AND TIMETABLES

Evidence obtained from manufacturers and other sources suggests that the costs of reducing emissions and noise will rise steeply once a certain level of emission control has been attained. Just how far beyond such point it is appropriate to go may, in part, depend on assessments of the incremental benefits that would flow from the additional expenditures. Present and contemplated standards should be regarded as modifiable in the light of any new insights that further analysis of these marginal costs and benefits may reveal.

Examination of past and expected costs of emission control in Europe, in any event, suggests that the point where the cost curve steepens has not yet been reached. Small increases in the cost of emission control in European cars could therefore be expected to yield substantial additional emission reductions. In the United States, on the other hand, small relaxations in the contemplated emission standards, particularly for nitrogen oxides which is the most difficult vehicle pollutant to suppress, could significantly lower the cost of compliance.

Similar observations can be made in relation to vehicle noise. Evidence from several countries indicates that, while some reduction in the engine noise output could be achieved at a modest cost, these costs are likely to increase sharply beyond the first few decibels, especially in the case of cars. What is more, there remains the very important challenge of finding effective ways of reducing noise from tires which, at high speeds, becomes the dominant source of traffic noise.

In the abatement of air pollution, the pace at which controls are introduced is also of importance. In the United States and Canada, but to a smaller extent also in Europe and Japan, the rate of decrease in the levels of emissions until about 1980 depends more on the retirement of older, uncontrolled vehicles from use than on further reductions in the emission rates of new cars. Thus, the effect on total emissions of a modest stretchout in the implementation of any proposed new standard might be comparatively small. The foregone emission reductions may be well worth contemplating if the delay in applying the standards can provide additional time to improve the reliability of emission control systems and spread out the increases in car prices, thus lessening dislocations in the economy.

THE EFFECTS OF MAJOR INCREASES IN THE PRICE OF CARS

Concern has been voiced that severe emission and noise control requirements could provoke sharp increases in automobile prices and produce serious reverberations throughout the economy. There is, however, little evidence to substantiate such fears.

Despite a diligent search, no satisfactory studies of the long-term effects of rising prices on demand have been uncovered. One reason could be that with so many factors—disposable income, availability of credit, consumer confidence and appeal of new styling—affecting the demand for cars, it is very difficult to single out the influence of higher prices.

Common sense however suggests that a rapid surge in prices could in the short term prompt people to keep their existing cars longer before trading them in, a decision reinforced by dissatisfaction with the poorer performance of controlled cars. At the same time increases in the cost of the cheapest cars could be expected to oblige some buyers to turn to second-hand cars. The combined effect of these two phenomena would be to reduce the supply and to increase the demand for used cars. This would force their prices upwards and tend, in turn, to encourage the owners of older vehicles to go on repairing and holding on to them.

The effect of these developments could be a temporary slowing down in the abatement of air pollution. More older, high-polluting cars would be kept on the roads. Sales of new, low-emission models would be depressed. But this could only be a short-term effect. Old cars cannot be kept on the road indefinitely. The replacement of them by cleaner models could only be temporarily delayed.

Although rapidly rising prices may cause some people to "trade down" to less expensive models or to second-hand cars, they seem unlikely to deprive sizeable numbers of people of automobility. Those who already own cars will not give them up simply because their prices are rising, although they may decide to hold on to them longer. Among those who come on the market for the first time some low-income buyers may conceivably be deterred by higher car prices, though the existence of installment plan credit would tend to minimize the impact of price increases. Some slackening in demand could also be expected for second cars. The effect of these forgone purchases on the trend in the level of *car ownership*, however, would be quite modest.

The effect of rapidly rising prices on trends in *new car sales* may be more pronounced. The cost of complying with current and contemplated European Communities emission and noise directives are estimated by industry sources to lead to average price increases in real terms of about 2 per cent a year up to 1974, with possibly greater increases after that. Although such price rises seem unlikely to depress the absolute volume of car sales, they may slow down the rate at which these sales have recently been growing. In the United States, the cost of foreseeable emission controls may be large enough to cause other market changes as well. Price increases on the order of $350 may not only slow down the rate of replacement and purchases of second cars, but may also alter significantly the present mix of cars bought—though even there car sales are expected to continue to grow.

Increased costs due to emission and noise control requirements would fall unevenly on cars of different kinds and to some extent on different manufacturers. Small cars of 1000 cc engine capacity and 800 kg in weight or under would be most affected. The cost of incorporating emission and noise control devices in them would be proportionately higher. One effect could be to narrow the difference in price between low- and medium-priced cars and to make the former seem less attractive in value, but the likelihood of the trend towards trading down already referred to makes it hard to assess the net result on sales.

EFFECTS OF EMISSION CONTROLS ON DRIVING PERFORMANCE AND FUEL ECONOMY

No adverse changes in driving performance are expected from present European emission controls. In the United States, however, retarded ignition timing and leaner fuel/air ratios have already affected the drivability of some models, causing engines to stall more often and vehicles to hesitate during rapid acceleration. The use of exhaust gas recirculation on future models may further affect acceleration capability and overall driving performance. In addition, a penalty in fuel economy of 5–8 per cent might be incurred as a result of lowering compression ratios to accommodate the lower-octane lead-free fuel.

EMISSION CONTROLS IN THE CONTEXT OF OTHER VEHICLE MODIFICATIONS

Measures to reduce air pollution and noise should not be looked at in isolation. Other changes are in prospect that could affect the design and the cost of vehicles. Modifications aimed at reducing injury to car occupants in accidents are foremost among these. Some countries require design changes that will reduce damage to vehicles from low-speed collisions. The rising cost of car repairs and changes in the lead content of fuel could be the cause of further modifications. All such developments would affect the price and performance of vehicles, and it is therefore important to bear them in mind when considering emission and price control strategies.

The need for a comprehensive view of the effects of prospective controls is particularly great in countries where regulating the motor vehicle is the responsibility of more than one agency. Where this is the case, performance standards and design specifications may be set by individual agencies without regard to what other requirements are in prospect. The resulting cumulative impact on vehicle cost and performance may cause severe repercussions on car users, industry and international trade. Effects of this kind could be avoided if agencies coordinated their moves to produce a staged plan of action.

CONTROLLING EMISSIONS FROM VEHICLES IN USE

Air Pollution

The rate at which vehicle air pollution can be reduced is significantly governed by the rate at which new, low-emission cars replace old, less-controlled ones. On the average only about one vehicle in ten is replaced every year.

In densely developed metropolitan regions where air pollution is likely to be a hazard to health, control over vehicles that are already in service may therefore be needed to slow down the rate of growth of emissions until normal attrition eliminates the older, more polluting cars from use. Such control can also be used to reduce smoke and smell from special classes of vehicles, such as diesel engined trucks and buses, a nuisance of which the public is becoming increasingly aware.

Another reason for controlling in-use vehicles is to counteract the tendency for emissions to increase with age and mileage. At present such deterioration seems to occur in vehicles both with and without exhaust control systems (see Chapter 27). This may change in the future as emission control devices become more durable and reliable, and particularly if standards for new vehicles were to be coupled with requirements on manufacturers to guarantee the performance of control devices for a specified time or mileage.

Noise

The limited evidence available suggests that only slight increases in noise occur with increasing vehicle age and mileage. Greatly increased noise can, however, result from deliberate tampering with silencers, from the rusting of exhaust systems and from lack of concern on the part of drivers to anchor tailgates and other parts of the bodywork of commercial vehicles.

Methods of Control

Efforts to reduce air pollution and noise from in-use vehicles can take various forms. One possibility, which is more suited to controlling air pollution than noise, is to oblige owners of older models to install devices designed to bring emissions down to or towards the level of those from new ones. Such retrofitting could be made a condition of renewing an annual licence or required when a vehicle changed hands.

Mandatory maintenance of emission and noise control systems is another possibility. Periodic adjustments could be required to ignition timing and fuel mixture; devices for controlling crankcase emissions could be compulsorily serviced or replaced. Vehicle owners could be obliged to have such maintenance performed at regular intervals and to show proof of it on applying to renew their annual licences.

Another and complementary approach is periodic inspection and testing of emission and noise control systems to assure that emissions and noise levels of vehicles in use remain within allowable limits. The practice of testing carbon monoxide emissions from engines at idling speed is spreading. More sophisticated exhaust emission tests, involving dynamometers and the measurement of hydrocarbons and carbon monoxide emissions during simulated drives, are being investigated in a number of countries. Noise checks are already obligatory in some countries for heavy commercial vehicles. Research into a "noise cycle" that would reflect different typical patterns of urban driving has also been undertaken.

Practicality and Effectiveness of Mandatory Emissions Inspections

The practicality of any programme for monitoring emissions from in-use vehicles rests on the availability of a quick, accurate and inexpensive static test giving results akin to those that would be obtained from a full driving cycle test. Test procedures which reveal what is wrong with the emission control systems that fail inspection and which indicate what needs repairing, are equally essential.

The potential emission reductions from inspection and maintenance cannot be determined precisely until more information is available about the rate at which emission control devices deteriorate in the intervals between maintenance. Pending the availability of such data, as well as of a quick emission test, inspection programmes other than tests for carbon monoxide at idle are being treated with caution. Some countries believe that full-scale inspection of emission control systems in vehicles already in service would be practical only if combined with inspection for safety, and then only as part of a supplementary control strategy in regions with serious pollution problems.

DUAL STANDARDS FOR THE CONTROL OF VEHICLE EMISSIONS

Air pollution is invariably most severe and the numbers of people exposed to it greatest in densely settled areas. Tighter control over vehicle emissions is accordingly more desirable in urban regions than in rural areas.

The distribution of excessive noise follows different principles. Noisy or badly silenced vehicles are offensive in places where traffic is concentrated but can also be irritating individually. Variations in noise control are therefore inappropriate from place to place.

Justification for applying more stringent levels of emission control to vehicles operating primarily in urban regions can be made on economic as well as on equity grounds. In France, for instance, fewer than 30 per cent of the total vehicle population is registered in the metropolitan areas of Paris, Marseilles and Lyons. In Sweden only 40 per cent of all vehicles are registered in the three metropolitan areas of Stockholm, Göteborg and Malmö. Even in the United States, it has been estimated that some 30 per cent of the total vehicle population operate in areas which have no significant air quality problems. If cars used primarily outside such regions were permitted to comply with less stringent emission requirements, the result would be not only a financial saving for large numbers of car owners but a major conservation of national resources.

However, these considerations do not outweigh the drawbacks of a dual standard strategy. The temptation for residents of urban regions to buy and register cars in less stringently controlled areas in order to avoid the cost and poorer performance of highly controlled vehicles would be great. The scope for evading enforcement based on place of registration or garaging would be large. And the proposal for dual standards does not take account of the mobility of cars which would be bound to lead to uncontrolled vehicles circulating in high pollution regions. Even if such vehicles belonged only to visitors or were only in transit, residents who were obliged to pay for stringent emission control would inevitably resent them. An alternative would be to deny access to vehicles from less controlled areas, but it is unlikely that this would be practical. An example, though probably an extreme one, of what might be involved can be found in Switzerland where 40 million foreign cars enter the country every year.

In these circumstances the Group felt that it would be impractical as far as air pollution is concerned and undesirable in the case of noise to set different control standards for vehicles based on *place of use*, though they felt that there was a potential for doing so on the basis of *vehicle type*. Controls of the latter kind could be limited to special types of vehicles such as buses, taxis and delivery vans. They could be applied at the manufacturing stage and thus would pose no difficulties of enforcement.

This principle may be applied both to the control of noise and air pollution. Buses and taxis (at least in cities where they are diesel powered) are inherently noisy; they are also high emitters. There is therefore a strong case for imposing more severe controls on them than on other vehicles in the city.

LIMITING THE CONTENT OF LEAD IN MOTOR FUEL

A growing number of countries have decided to limit the content of lead in motor fuel. Their plans range from modest reductions to near-total elimination of lead additives. In some cases there are companion proposals to introduce a grade of lead-free low-octane fuel.

Two factors have prompted countries to embark on this course of action: first, the effect of airborne lead as a pollutant and its impact on public health; and second, the adverse effect of lead on the performance of catalytic converters. There is disagreement as to whether present levels of airborne lead constitute a danger to public health. Nevertheless, many governments believe it is desirable that today's levels of airborne lead should not be exceeded and, if possible, that they should be reduced. At the same time, there is evidence that lead, even in minute quantities, can poison some catalysts, thereby rendering them ineffective for purposes of emission control. It is not yet clear, however, whether catalytic devices will be required in Europe and many other areas of the world; this will depend on future vehicle emission control policies.

The purpose of lead additives is to raise the octane rating of motor fuel. Fuel with a sufficiently high octane number is needed to prevent a phenomenon known as detonation or "knocking", which can result in considerable power loss, severe engine damage and increased nitrogen dioxide emissions. Knock can occur at either high speed or low speed, depending on the octane number of the different components used in the gasoline. The total fuel blend must be designed to avoid both of these conditions. Current premium grade fuels contain from 0.1 to 0.9 grams of lead per litre, while regular grade fuels contain from zero to 0.8 grams per litre.

As a modest reduction in the octane rating is unlikely to interfere with the satisfactory operation of a high proportion of cars currently on the roads, and would make it possible for countries which at present permit a high level of lead additives by a modest amount.

Major reductions in lead additives, on the other hand, would require either the lowering of engine compression ratios or the addition of higher octane blending components to motor fuel. The first strategy has been adopted in the United States and Japan where compression ratios of engines are being reduced to permit satisfactory operation on 91-93 RON gasoline. This strategy presents a minimum of inconvenience in the case of high horsepower cars where a reduction in compression ratio can be traded off against some of the excess power which they now possess. But for the lower powered cars whose engines are matched more closely to their weight (the power-to-weight ratio of a typical U.S. car is one and a half times greater than that of a typical European car), a return to lower compression ratios would mean increases in the size of engines. This, in turn, would lead to larger and heavier cars and greater specific fuel consumption. In countries where the price of fuel is already high, a substantial loss in fuel economy would be unpopular. Car manufacturers who lowered compression ratios of their cars voluntarily might find themselves at a disadvantage vis-a-vis those who did not take such a step, thus making it unlikely that any individual manufacturer would do this in the absence of specific incentives or regulations.

The alternative strategy of maintaining current octane characteristics in fuel by replacing lead additives with higher octane blending components is technically achievable but could be implemented only over a longer period of time. This is because some of the needed high octane petroleum fractions are not currently available in all countries and new facilities, requiring long lead-time, would have to be built to supply them. The investment in the needed refinery additions might be high; however, the costs would vary widely from country to country, depending on the capacity and technical makeup of existing plant, the amount of unused capacity for distilling high octane components, and the degree of lead removal desired. Special provisions might have to be made for small refiners, for whom making the necessary refinery modifications would represent a heavier burden than for large producers.

Concern has been voiced that a limitation of lead additives could lead to large increases in the use of aromatics as a research octane booster and thereby cause possible increases in the emissions of phenols and of photochemically reactive and polynuclear aromatic (carcinogenic) hydrocarbons. The extent to which an increase in the aromatic content of fuel would be necessary will depend on the desired octane quality, the degree of limitation placed on the use of lead as additive and the refining process chosen. There are many processes available for the production of high octane quality components, some of which do not produce significant quantities of aromatics. For example, it has been estimated that, in order to keep octane ratings at 98 RON, an increase of no more than 3–5 percentage points in aromatics would be needed to compensate for the reduction in lead content to 0.15 g/litre. Other estimates, however, point to an increase in aromatics that could be as high as ten percentage points.

Future emission controls might reduce the significance of such possible aromatic increases from the standpoint of exhaust emissions. However, a fuel which has had its lead content replaced with aromatics in order to maintain the same research octane number (RON) will not necessarily continue to be satisfactory for all engines. For example, if already high aromatic levels were to increase by 8 to 10 percentage points, the resulting fuel blend might not provide enough resistance to detonation under conditions of high speed and eventually lead to failure in the case of some high-compression engines. For this reason, when judging the adequacy of fuel quality, entire reliance should not be placed solely on the RON rating but must be considered in conjunction with the motor octane rating (MON).

CONSULTATIONS ON MOTOR FUEL QUALITY

Concern has been voiced about the fact that divergent national policies toward lead in motor fuel could adversely affect international trade in motor vehicles and might inhibit the freedom to drive freely within Europe. In considering these assertions one must be careful to distinguish between two distinct phenomena: reduction in the fuel's lead content *per se*, and any consequential effects of such reduction on the quality of the fuel.

Clearly, a reduction of lead could not, of itself, have any negative repercussions. But different results might follow if restrictions placed on lead additives caused the quality of the fuel to be modified in such a manner as to adversely affect the driving performance of some vehicles. There is no consensus that reductions in the lead content would indeed have such consequences. If, however, for any reason, a significant degradation of the overall quality of motor fuel did occur, its effect could be to oblige some manufacturers to modify the design of their engines (in itself not a discriminatory effect) and to limit the ability of some motorists to go where they please.

To guard against this, some countries believe that there should be agreement on uniform legislation concerning fuel additives. The object of such consultations would be (1) to seek mutual assurances that fuel of adequate overall quality and satisfying prevailing emission control requirements will be available throughout Europe as long as is necessary; (2) to reach agreement as to what constitutes "adequate" fuel quality in terms of research and motor octane numbers (RON and MON), volatility and such other indices as together might accurately describe overall fuel performance.

The group also considered the question of the necessity to assure an adequate supply of lead-free motor fuel. In view of the fact that the appearance of cars equipped with catalytic converters is likely to be, if at all, the result of European-wide collaboration, the Group does not believe that this issue warrants at this time special consideration. Nor is the immediate unavailability of lead-free fuel likely to affect sufficiently large numbers of American tourists who wish to purchase and drive cars abroad before shipping them home to have a significant effect on international trade. In the longer run, however, the feasibility of promoting world-wide harmonization of national requirements concerning fuel composition and quality should be explored.

REGULATORY OR ECONOMIC CONTROLS

The most direct method of controlling pollutants and noise at source—and one which has been favoured by countries to date—is to set and enforce maximum limits on emissions from new motor vehicles. The major alternative which has been proposed to this approach is a system of economic penalties or emission charges. This involves the establishment of a schedule of fees related to the emission characteristics of different vehicles. Since the amount of the tax would be proportional to the emission of pollutants and noise, car manufacturers would have a financial incentive to reduce emissions from their vehicles.

Advocates of emission charges argue that this approach offers a number of advantages over regulatory forms of control. In the first place, emission charges, they say, would encourage the most efficient means of controlling automotive air pollution and noise; car makers would see to it that it occurred in the cheapest possible way. Another advantage claimed for charges is that they would require far less government interference in the internal operations of the auto industry; once the charge has been established, technology and economics would be factors debated within the firm and not between car manufacturers and government agencies. Finally, the argument is made that manufacturers would be under no compulsion to modify every vehicle type, but would have the freedom to set the emission characteristics of all their models in such a way as to minimize their overall control costs.

The use of economic incentives and penalties has been advocated primarily by economists who see them as a way of achieving the optimal economic balance between the costs of pollution damage and the costs of pollution control. The system, in effect, utilizes the workings of the market economy to promote in principle the least-cost solution to environmental clean-up.

While the concept of emission charges (particularly if they were to be used in conjunction with, rather than as a substitute for, regulatory controls) may have merit from a strictly economic standpoint, the arguments in favour of them are not entirely convincing for they seem to overlook one crucial point: namely that, in the field of public policy, the tests of practicality and administrative feasibility are of overriding importance. Applying these criteria to the field of automotive emission control it is felt that the economic approach is deficient in several major respects.

1. With a charge scheme it would be impossible to fix clearly defined targets and deadlines for reaching emission reductions or attaining an ambient air quality objective. This is because the response of the automobile industry to a given emission charge schedule could not be anticipated in advance. Manufacturers could ignore the charge incentive altogether: they could absorb the cost of the emission charge by reducing other expenditures (e.g. through less frequent model changes), or transfer the cost of the charge to consumers in the form of higher car prices—while making little effort to reduce emissions.

2. Charges would introduce a disturbing element of uncertainty into the planning of the motor industry. Manufacturers could no longer plan the introduction of emission control improvements on an orderly basis, as each tried to anticipate and match the moves of his competitors.

3. Governments could come to rely upon emission charges as a source of revenue and be tempted to continue them long after the underlying public health justification for them had ceased to exist.

4. A main rationale for the use of charges is that they would reflect more truly the amount of actual damage caused by the pollution and noise. However, measurement of environmental damage and its expression in quantitative terms has so far eluded the professional community. In practice, therefore, any emission charge scheme would probably be based on criteria derived from the same considerations as those currently used in setting standards.

5. With charge schemes it would be more difficult to achieve international uniformity in relation to automotive emission control requirements.

6. The use of charges could be criticised on grounds of equity and fairness. Some motorists might choose to pay for the continued right to pollute and make noise rather than accept the lowered driving performance that accompanies very stringent emission control requirements. It is unlikely that public opinion would tolerate a situation in which individuals have the freedom to accept or reject at will measures designed to protect the public health.

7. The lack of experience with the use of charges to control vehicle emissions means that public acceptance of them and their administrative feasibility are unproven. Their use in other fields of environmental management, such as control of effluents into rivers, provides little indication of how they would operate when applied to millions of moving sources.

In summary, the choice is one between the certain (both for governments and industry) and proven regulatory approach on the one hand, and the uncertain and untested approach with emission charges. In view of the growing seriousness and urgency of the air pollution and noise problem, the ability to achieve specific levels of emission control by certain deadlines is an important objective of any environmental control policy. On these grounds alone the weight of evidence would seem to favour the continued use of standards.

CONTROLLING AIR QUALITY AND NOISE BY REDUCING THE USE OF CARS

The use of cars is likely to be increasing at such a rapid rate in the near future that levels of emissions and noise will continue to grow in some urban areas despite the use of the best available control technology. In these circumstances measures that involve indirect controls may be required as a supplementary strategy for reducing air pollution and noise.

One of the most widely considered alternatives is the use of measures to reduce and divert motor vehicle traffic. Action has already been taken in some cities in this direction, although the motives have generally been to relieve congestion. Instruments used for this purpose include parking restrictions, selective prohibitions on heavy traffic, ringroads and bypasses, the creation of car-free zones, priority lanes for buses, and other improvements in the quality and convenience of public transport.

Most of the available techniques for reducing or diverting traffic could be used to abate emissions and noise. Already important shopping streets have been closed to cars in Japan primarily to reduce the exposure of shoppers to vehicle emission and noise. The practicality in political terms of such actions will depend on whether or not parallel efforts are made to improve the speed, frequency and availability of public transport. Only if public transportation service markedly better than that available today is provided, is there a prospect of persuading people to switch from cars to public transport.

Variations in climate, population densities, the form of cities, the level of car ownership, in the availability of public transport and in environmental conditions will make the case for reducing the use of cars for purposes of air pollution and noise abatement differ in strength from city to city. In some places there may be no need for it at all, but in metropolitan areas where air pollution and noise are widespread and a cause of serious public concern, both the limitation of emissions at source and some measures to restrain motor vehicle traffic are likely to be necessary.

Local Measures

Experience makes it clear that schemes to exclude cars from city streets can be highly effective in abating local concentrations of pollution and noise. Experiments in a number of countries—New York, Tokyo, Marseilles and Vienna, for example—have shown that carbon monoxide concentrations can be reduced by as much as two thirds, and noise levels by 5 decibels (dBA) in individual streets.

By now there is firm evidence, based on a large number of cases, that cars can be barred from limited areas without adversely affecting retail trade within them. Increased congestion and pollution in adjoining areas has not become evident, though the creation of traffic-free areas is usually linked with some measures to increase parking capacity on the periphery of the area. The long-range consequences of traffic restraints cannot be predicted with certainty; however, a study of the OECD Sector Group on the Urban Environment concluded on the basis of available evidence that they do not interfere with the healthy development of the inner city.

Perhaps the most convincing evidence of the practicality and public acceptance of traffic-free zones is their recent growth, both in extent and in number. According to the study just cited, over 100 cities have banned traffic from portions of their central districts. Certain of them such as Essen, Leeds, The Hague, Oldenburg and Munich have each extended an original nucleus of a few streets into large pedestrian areas. Other cities which have created vehicle-free zones include Athens, Bologna, Brussels, Florence, Ravenna, Rouen and Verona.

The huge scale of the central areas of cities such as New York and London makes it necessary to adopt somewhat different approaches, and to combine the exclusion of private cars with the continued operation of public transport. Ambient pollution and noise levels are likely to decrease considerably in such transitways as a result of excluding private motor traffic, though peak noise levels would be attenuated only if special action were taken to modify public transit vehicles.

Nevertheless, the potential of traffic restrictions as a tool of pollution abatement should not be exaggerated. While exclusion of vehicles can effectively serve to reduce the street-level concentrations of pollutants, particularly of carbon monoxide, it will contribute little to improving air quality on a city-wide scale; the latter is a function of emissions from thousands of stationary as well as moving sources spread over an entire urban area and is relatively insensitive to small-scale reductions.

In most cities, however, climatic and topographic conditions ensure a relatively rapid dispersion of pollutants, so that they do not accumulate as an envelope over the urban area. The main problem is one of local conditions in particular streets where carbon monoxide concentrations often tend to build up to levels dangerous to human health. In such circumstances, car-free zones and traffic bans deserve close consideration as vehicular pollution abatement measures.

Measures with More Wide-Ranging Effect

Air Pollution Where pollutants tend to build up into envelopes blanketing entire metropolitan regions, strategies of a different kind will be called for. In such cases it may be necessary to introduce mixes of measures which, while individually small in effect, can together be quite far reaching. Car-free zones in both metropolitan and suburban centres are likely to be part of such a mix even though their influence will be only local. Parking restrictions can provide additional relief. In Marseilles an experimental ban on parking in the core of the city resulted in a lowering of carbon monoxide concentrations by 10 to 60 per cent depending on the time of day. The average reduction, based on 502 samples at four locations, was 40 per cent.

Suggestions have been made that tolls, parking fees or other charges designed to increase the cost of entering and parking in inner cities could be used to reduce travel by car. The effect of pricing on damping down demand in other fields is well established but doubts remain about its effectiveness in limiting automobile use, at least at the levels of charges that might be acceptable politically. In the case of commuters, charges might discourage poorer car owners only to attract additional motorists with higher incomes, or whose costs were offset by their employers. On top of this there would be administrative difficulties.

Among other possibilities are the diversion and rerouting of through traffic, particularly heavy trucks, incentives to encourage the sharing of cars by commuters, public automobile service and staggered working hours. All are the object of experiments in various countries. In San Francisco cars carrying three or more passengers in rush hour traffic are offered a toll-free passage across the Bay Bridge into the city, though so far this device has been employed to increase the number of people carried rather than to reduce the number of cars.

In Montpellier a self-drive taxi service has been introduced thus enabling participants in the experiment to have the use of a car in the city centre even though they do not use one to get there. And in Göteborg, traffic has been stopped from crossing the city centre by dividing it into quadrants and barring access to them except from a ring road.

The practice of allowing the employees of large industrial, commercial and governmental organisations to stagger their working hours is spreading in both Europe and North America. It is prompted by a desire to reduce congestion and to give employees greater freedom in planning their own lives. Its contribution to alleviating air pollution and noise is likely to be through smoothing out traffic flows during rush hours.

Noise Many of the controls over car use just described could help to reduce noise as well as air pollution. Routing heavy trucks away from residential and shopping areas would be of particular value. Measures to improve the flow of traffic are also likely to be beneficial through reducing the extent to which vehicles have to move in low gear. Car-free zones will reduce the impact of noise on workers, residents and visitors within them, but may lead to increases in the levels experienced in adjacent streets, and if so may need to be combined with additional abatement measures.

Noise is also a serious problem in suburbs where traffic densities may be too low to pollute the air. Combined programmes of land use planning controls and traffic regulations such as have been implemented in certain inner residential districts in Stockholm and London can be employed in those circumstances.

THE LONGER TERM

In the longer term, levels of automotive pollution and noise could also be lowered through rational urban and land use planning. The need to travel could be decreased by minimizing distances between home and work, home and school and home and shops. Reliance on cars for moving about in cities could be reduced by promoting development around existing railway and bus lines and by extending such networks in order to place public transport within the reach of a greater proportion of metropolitan residents.

The design of vehicles is itself capable of considerable evolution. Advances in technology could lead to propulsion means that are virtually pollution-free and noiseless. Growing concern about diminishing resources and possible increases in energy costs may add impetus to such developments. Other pressures on vehicle manufacturers to provide increased comfort and increased protection in collisions could at the same time push car design in conflicting directions. A possible result of these circumstances could be the emergence of specialisation in the use of vehicles. Compact, light, low-polluting vehicles might gain favour among commuters and those who use cars primarily in metropolitan areas. Larger, heavier and more comfortable cars could come to serve mainly the needs of intercity highway travel.

Outside events are themselves pushing the car-buying public towards such differentiation. Parking difficulties and traffic congestion already encourage many urban residents to purchase smaller cars. Higher incomes will widen the market for second cars specially adapted to urban conditions. The expansion and development of car-carrying train services and increases in the practice of renting cars to complete plane and train trips could reduce the reliance of long distance travellers on cars.

Such a process of differentiation in the design and use of cars could be accelerated if given a deliberate impetus. Government support for research and development into unconventional forms of propulsion might be a necessary part of such a strategy. Transportation and fiscal policies could be harnessed to provide additional incentives. International cooperation could also play a role. An example of what may be done already exists in the Experimental Safety Vehicle (ESV) programme. Enlarged to cover a wider range of vehicle characteristics, such cooperation could lead to the emergence of an international market for cars specially tailored for use in the cities.

Basic changes in the design and use of cars and efforts to shape urban development so as to make people less dependent on cars could be expected to affect life styles and the entire economic fabric of cities profoundly. The extent to which such changes are deliberately encouraged will therefore depend on the political climate within individual countries. The question they raise is whether present systems of personal transportation and present trends in urban development represent the greatest good for the greatest number. This issue is beyond the scope of this report but it deserves priority in the deliberations of all nations.

URBAN GROWTH, MOTORIZATION AND ENVIRONMENTAL QUALITY

We are experiencing a period of substained urban growth. The biggest cities are growing fastest and becoming the homes of increasing proportions of people. These trends, coupled with rising car ownership, mean that growing numbers of people are living in places where traffic congestion is endemic.

Between 1960 and 1970 the proportion of people living in cities increased from 58 to 64 per cent and growth was fastest in the bigger metropolitan areas. Cities with over a million residents now accommodate more than a quarter of the population of OECD nations, and between 1960 and 2000 they are forecast to increase in number from 47 to 60. Over the same period, cities with half a million to a million residents are expected to increase in number from 14 to 24. ("Urban Growth in OECD Countries," Organisation for Economic Cooperation and Development, 1971.)

The dominant role of the car is characteristic of all these cities. Figures from four German cities, for example, show a doubling in car registrations between 1960 and 1970. A survey of traffic forecasts in nine cities indicates that car usage will double in these cities over periods of about twenty years. (OECD Urban Transportation Questionnaire, 1970.)

Within the metropolitan area the troughs between rush-hour peaks of traffic will tend to fill in, partly due to congestion which will make rush hours last longer, and partly due to the growing use of cars for school, business and personal trips of all kinds throughout the day. These trends are already well apparent in some cities where traffic is at an almost continuously high level during the length of the working day. Increased use of cars for trips connected with shopping, entertainment and social life will, in turn, tend to fill up the roads during weekends to weekday levels.

The scope for traffic to increase will be generally greater in suburbs than in already crowded central areas. This is because downtown traffic is already running close to saturation, while in outlying districts it still has plenty of room to move in and thus to increase. This trend is evident in every large city. In Chicago and Paris, for example, traffic per square mile is expected to show almost no increase in the central area between 1972 and 1980. In most cities congestion will therefore tend to spread out rather than to grow in intensity.

The principal result of this evolution will be to increase the area of a metropolitan region experiencing high pollution concentrations and noise levels and to extend their influence through time. More people will be spending more hours living, commuting and working in places where the quality of air is substandard and where noise levels are annoyingly high. This will be expecially so if people are forced to move further and further from their work in order to find satisfactory homes and to spend longer and longer hours driving to and from work.

Variations in the rate of growth of car usage, in weather, in topography and in demographic trends in different cities make it impossible to predict at what rate individual metropolitan areas will experience this evolution. Most cities, particularly in Western Europe, may be sufficiently windy to avoid the frequent occurrence of area-wide blankets of pollution. Incidents of photochemical smog are nevertheless likely to increase in frequency, particularly in places prone to prolonged atmospheric inversions and exposed to sunlight, unless countermeasures are taken. In the micro-environment of the inner city, concentrations of carbon monoxide and levels of noise already appear to be substantially similar in most major cities. Congestion is the ultimate regulator of levels of automotive pollutants and noise in the street environment, and once traffic has reached near-saturation levels there is little scope for conditions to deteriorate any further except under the influence of exceptional climatic events.

HEALTH AND WELFARE EFFECTS OF AUTOMOTIVE POLLUTANTS

Pollution of the atmosphere due to motor vehicles is viewed as a rising threat to the public health and to the welfare of the population, although there is no agreement on how serious and immediate this threat has already become. What follows is a brief summary of the effects of automotive pollutants. Detailed discussion of such effects is beyond the scope of this report, but may be found in a number of specialized studies such as, for example, the 1969 Report of a W.H.O. Expert Committee.[1] In particular, the summary below does not purport to assess the relative toxicity of automotive pollutants under normal conditions such as may be encountered in a typical urban area most of the time. The actual effects which these pollutants may cause will depend on the length of exposure and the ambient concentrations of them in the atmosphere. The latter, in turn, will be a function of traffic density, climate and topography. Moreover, due to the variety of pollution sources in urban areas and the complexity of the chemical and photochemical reactions taking place in the atmosphere it is difficult to single out and define the degree of responsibility of each of the pollutants.

Carbon monoxide is well known for its poisonous effects at high concentrations. It is absorbed through the lungs and reacts primarily with the hemoglobin of red blood cells. As an air pollutant, carbon monoxide represents a potential danger to human health and safety. It decreases the oxygen-carrying capacity of the blood and thus reduces the availability of oxygen transported to vital tissues. Carbon monoxide concentrations of 10 ppm have been known to produce blood carboxyhemoglobin levels of 2 per cent in nonsmokers. This level has been associated with impaired time interval discrimination and visual acuity. Carbon monoxide concentrations of 30 ppm for 8 to 12 hours have been associated with impaired psychomotor performance and reduced visual acuity in normal subjects and with increased physiological stress to patients with heart disease.

Hydrocarbons represent the major class of reactive organic matter in the atmosphere that is responsible for photochemical smog. Through their reaction intermediates and photochemical oxidation products, they are directly responsible for the eye irritation associated with photochemical smog and much of the characteristic vegetation damage. Hydrocarbon oxidation products are also believed to be important contributors to the atmospheric aerosols responsible for the reduced visibility associated with smog. In addition, ethylene, a specific hydrocarbon, is directly responsible for certain forms of plant injury.

Nitrogen oxides in the atmosphere are essential to the photo-oxidation of hydrocarbons and the development of photochemical oxidants. Sufficiently reduced levels of either NO_x or HC alone in the air tend to alleviate the formation of photochemical oxidants, but the exact relationships are extremely complex. Nitrogen dioxide, a type of oxide of nitrogen, is also a specific air pollutant associated with increased incidences of acute bronchitis in infants and school children and acute respiratory disease in the entire family group. Nitrogen dioxide has also been associated with damage to vegetation and corrosion of electronic components. Increases in the incidence of respiratory disease have been associated with nitrogen dioxide levels ranging from 0.06 ppm to 0.08 ppm over a 6-month period.

Photochemical oxidants result from a complex series of atmospheric reactions initiated by sunlight. When reactive organic substances and nitrogen oxides accumulate in the atmosphere and are exposed to the ultraviolent components of sunlight, the formation of new compounds, including ozone and peroxyacyl nitrates, takes place. Photochemical oxidants may adversely affect vegetation, human health, animals and certain manmade materials. They can cause injury to many important species of plants such as beans, tobacco, peanut and pine trees. Injury to sensitive species has occurred after exposure of 4 hours to 0.05 ppm of photochemical oxidants. Rubber, fabrics and dyes are particularly sensitive to oxidants as evidenced by rubber cracking, reduced strength in cellulose fabrics and the fading of certain dyes after exposure. The principal human effects associated with photochemical oxidants include impairment of athletic performance and an increase in attacks among asthmatics, the latter effect having been observed when oxidant levels reached 0.15 ppm for one hour. Eye irritation has been associated with oxidant levels of 0.10 ppm.

Lead is a biologically nonessential metallic element which is clearly toxic under conditions of prolonged and excessive exposure (e.g., ingestion of paint containing lead). Furthermore, lead accumulates in persons exposed to high atmospheric concentrations. Lead is abosrbed primarily through the gastrointestinal and respiratory tracts.

Other substances. The significance of the following automotive emissions on human health or welfare has not been established. Medical and biological investigations are, however, underway:

- Particulate materials, both organic and inorganic
- Aldehydes and other carbonyl compounds
- Nitrogen compounds other than nitrogen oxides, such as amonia
- Miscellaneous organic materials, such as polynuclear aromatic hydrocarbons.

AMBIENT AIR QUALITY STANDARDS

The growth of awareness of the potential harm to health of air pollutants has led some countries to define standards of air quality. The best known are those established in the United States. The U.S. legislation distinguishes between "primary" ambient air quality standards which reflect public health considerations, and "secondary" standards which are aimed at protecting the public welfare and are to be achieved by the earliest date commensurate with technological and economic feasibility.

In Japan an ambient air quality standard for carbon monoxide was established in February 1970. It stipulates that the average hourly level should not exceed 20 ppm over an 8-hour period and 10 ppm over a 24-hour period. The regulation states that the standard should be applicable to any areas or localities where the general public permanently live or engage in activities. It is not applicable to places, such as roadways, which are exclusively reserved for vehicular traffic or parking. An ambient standard for particulate matter was published in January 1972. It provides that the mean value of suspended particulate matter should be less than 100 mg/m^3 over a 24-hour period and that its one-hour value should be less than 200 mg/m^3 except in special industrial areas.[2]

CONTRIBUTION OF MOTOR VEHICLES TO TOTAL ATMOSPHERIC EMISSIONS

Although motor vehicles are suspcected to be the greatest single source of air pollution in most countries, studies that substantiate this belief are generally lacking. Nationwide emissions inventories are available for the United States, Germany and Sweden. A U.S. inventory, carried out in 1969, showed motor vehicles accounting for 51 percent of all emissions by weight, while stationary sources were responsible for 16 per cent, industrial processes for 14 percent and solid waste disposal for 4 percent. (Miscellaneous sources, 17 percent. "Summary of Nationwide Emission Estimates of Air Pollutants, 1969," U.S. Environmental Protection Agency, 1971.) A German inventory, made in 1969, revealed transportation (motor vehicles plus rail and air transport) to be responsible for 50 percent of total emissions by weight, stationary sources for 39 percent and other sources for 11 percent. ("Unweltprogramm der Bundesregierung," October 1971, p. 57.)

The part played by motor vehicles is shown to be even more significant when only carbon monoxide, hydrocarbons and the oxides of nitrogen, the three principal vehicular emissions, are considered. United States and Swedish estimates are shown in table 26.1.

Table 26.1 Estimated Nationwide Emissions of Air Pollutants

	Percentage of National Emissions		
	CO	HC	NO_x
United States (1969)			
Motor vehicles	64.7	45.7	36.6
Other forms of transport	9.0	7.2	10.5
Fuel combustion in stationary sources	1.2	2.4	42.0
Industrial processes	7.9	14.7	0.8
Solid waste disposal	5.2	5.3	1.7
Miscellaneous	12.0	24.7	8.4
Total	100.0	100.0	100.0
Sweden (1970)			
Traffic	89	81	22
Industry	5	17	28
Heating	3	–	39
Power generation	1	–	1
Solid waste incineration	1	1	–
Total	100	100	100

Source: "Sweden's National Report to the United Nations Conference on the Human Environment," May 1971, p. 35. "Development of Systems to Attain Established Motor Vehicle and Engine Emission Standards," Report of the Administrator of EPA to the Congress, July 1971, p. 4-1.

Of particular interest is the contribution of motor vehicles to *urban* air pollution. Table 26.2 shows how vehicles are the dominant source of emissions in metropolitan areas.

In nine out of the twelve cities vehicles are responsible for at least 90 percent of the carbon monoxide emitted, and for at least 70 percent of the hydrocarbons. The proportion of nitrogen oxide emissions is relatively smaller: only in Los Angeles, Stockholm and Ankara is it estimated to exceed 50 percent.

TRENDS IN AUTOMOTIVE EMISSIONS IN METROPOLITAN AREAS

Whether air quality in the cities is deteriorating can be determined only by actual measurements of pollution concentrations over time. However, a rough measure of the *rate* at which the urban air pollution problem is growing can be obtained by comparing the estimates of past, present and future daily vehicle emissions for a given city.

Table 26.2 Percentage Contribution of Motor Vehicles to Total Atmospheric Emissions in Twelve Metropolitan Areas

	Percentage of Total Emissions		
	CO	HC	NO_x
Madrid	95	90	35
Stockholm	99	93	53
Tokyo	99	95	33
Osaka	99	95	25
Toronto	98	69	19
Chicago	94	81	35
Los Angeles	98	66	72
Philadelphia	70	47	27
Washington, D.C.	96	86	44
Pittsburgh	80	70	29
New York City	97	63	31
Ankara	75	57	52

Source: National replies to OECD questionnaire. Published in *Environmental Implications of Options in Urban Mobility*, OECD, 1973.

Techniques for Estimating Gross Vehicle Emissions

Gross vehicle emissions for a metropolitan area can be calculated in two ways. Fuel consumption data may be used to calculate total number of kilometers travelled by all vehicles. Gross emissions can then be obtained by multiplying the total distance travelled by a figure representing the average vehicle emissions per kilometer. This approach depends on the availability of local fuel consumption data, and is prone to errors because fuel purchased within an area may be used outside it and vice versa. Although the two phenomena may balance each other out in very large metropolitan areas, this assumption may be less valid in smaller cities.

Limitations in the fuel consumption approach make it preferable to base estimates of vehicle emissions on travel figures obtained directly from local traffic departments. Many large cities maintain reasonably accurate traffic counts from which such data can be derived. A number of local traffic authorities also have rough estimates of future traffic growth. Once these figures have been obtained, rough estimates of total vehicles emissions for each pollutant can be obtained by multiplying the overall distance travelled by the average vehicle emission rates for each pollutant (representing a weighted average of model year emission factors).

Basic emission rates are obtained by testing cars in laboratories. Such rates will, of course, be influenced by the test procedure used. Direct comparisons between emission rates obtained from a European test procedure and from a United States test procedure cannot be made both because the driving cycle and the method of measuring emissions are different. However, actual emission rates for United States, European and Japanese vehicles measured according to both test procedures are available. These may then be adjusted to reflect the age composition of the fleet, any increase in emissions and any decrease in annual distance driven as vehicles increase in age, and the existence of any uncontrolled cars. The result will be a set of emission factors that can be used to obtain a more faithful estimate of emissions from the vehicle fleet actually on the road.

Growth in Total Emissions

Exhibit 26.1 shows past and projected growth of daily vehicle emissions in selected metropolitan areas. Table 26.3 gives a normalized view of these trends. Because they may have been calculated in different ways, comparisons between the cities should be avoided. One thing, nevertheless, emerges clearly: emissions have been growing at a substantially similar rate in all the cities for which data has been furnished. In the absence of controls they have been doubling roughly every 10 years.

Emission Densities and Population Exposure

Gross emissions are only a crude indicator of the levels of urban air pollution because they fail to take account of the size of the area over which the emissions are distributed and of the numbers of people exposed to them. A more meaningful measure of the degree of pollution might be the amount of pollutants by weight that are on the average emitted daily in each square kilometer of a city (average emission density). A still more sensitive indicator might be average emission density multiplied by the number of people exposed to it. Table 26.4 shows the average daily emissions (in tons) of vehicular pollutants per square kilometer in 11 metropolitan areas.

Assessing trends in air pollution in these terms is somewhat misleading because two important considerations are ignored: that metropolitan areas grow in size as well as in population, with a result that emissions get distributed over an increasing surface area and get diffused in an increasing volume of air; and that growth in traffic, and therefore in vehicular emissions, get distributed over an increasing surface area and get diffused in an increasing volume of air; and that growth in traffic, and therefore in vehicular emissions, will be distributed unevenly

Exhibit 26.1 Past and Projected Gross Daily Vehicular Emissions in Twelve Metropolitan Areas (tons/day)

Oslo	**1969**	**1971**	**1973**				
CO	124	139	168				
HC	19	21	26				
NO_x	3	4	5				
Madrid	**1969**	**1970**	**1971**	**1973**	**1974**		
CO	622	715	767	870	914		
Stockholm	**1950**	**1960**	**1970**	**1980**	**1980**[a]		
CO	88	330	620	1,100	440		
HC	16	62	115	200	78		
NO_x	7	26	49	87	69		
Tokyo	**1961**	**1965**	**1970**	**1975**	**1975**[a]	**1980**	**1980**[a]
CO	2,100	3,700	4.200	4,600	2,700	5,800	2,200
HC	300	550	700	930	380	1,160	100
NO_x	140	250	420	580	390	730	110
Berlin	**1955**	**1960**	**1965**	**1970**			
CO	122	227	371	521			
HC	5	9	13	19			
Frankfurt	**1955**	**1960**	**1965**	**1970**			
CO	77	121	176	231			
HC	3	5	7	9			
Cologne	**1955**	**1960**	**1965**	**1970**			
CO	78	128	196	245			
HC	3	5	8	9			
Munich	**1955**	**1960**	**1965**	**1970**			
CO	133	198	307	413			
HC	5	8	12	15			
Wiesbaden	**1955**	**1960**	**1965**	**1970**			
CO	30	45	72	94			
HC	1	2	3	4			
Los Angeles	**1960**	**1967**	**1970**	**1975**	**1980**		
HC	2,600	2,500	2,000	1,300	600		
Chicago	**1960**	**1967**	**1970**	**1975**	**1980**		
CO	6,400	7,000	5,600	3,700	1,800		
Lisbon	**1970**	**1975**					
CO	255	425					
HC	39	64					
NO_x	8	13					

Source: National replies to OECD questionnaire. Published in *Environ-Mental Implications of Options in Urban Mobility*, OECD, 1973.
Note: Because of different calculation techniques used, comparison between cities should be avoided.
[a]With controls.

Table 26.3 Baseline Estimates of Gross Daily Vehicular Emissions in Ten Metropolitan Areas (1970 = 100)

Pollutants by city	1960	1965	1970	1973–1975	1980
Oslo					
CO			100	129	
HC			100	130	
NO_x			100	140	
Madrid					
CO			100	128	
Stockholm					
CO	53		100		177/71*
HC	54		100		174/68*
NO_x	53		100		178/141*
Tokyo					
CO	50	88	100	110/64*	138/52*
HC	43	79	100	133/54*	166/14*
NO_x	33	60	100	138/93*	174/26*
Berlin					
CO	44	71	100		
HC	47	68	100		
Munich					
CO	48	74	100		
HC	53	80	100		
Los Angeles					
HC	130	130	100	65	30
Chicago					
CO	114	121	100	66	32
Paris					
CO	118		100		83/72*
HC	123		100		83/72*
NO_x			100		112
Lisbon					
CO			100	167	
HC			100	164	
NO_x			100	163	

Source: National replies to OECD questionnaire. Published in *Environmental Implications of Options in Urban Mobility*, OECD, 1973.
*Without/with controls.

Table 26.4 Vehicle Emission Densities in Eleven Metropolitan Areas (tons/day/km^2)

Pollutants by city	Emission density	Pollutants by city	Emission density
Oslo		**Madrid**	
CO	0.14	CO	1.20
Stockholm		**Los Angeles**	
CO	1.12	CO	0.12
HC	0.21	HC	0.03
NO_x	0.09	NO_x	0.01
Tokyo		**Philadelphia**	
CO	2.10	CO	0.43
HC	0.35	HC	0.05
NO_x	0.21	NO_x	0.03
Berlin		**Washington**	
CO	1.09	CO	0.56
HC	0.04	HC	0.12
		NO_x	0.03
Munich			
CO	1.33	**Pittsburgh**	
HC		CO	0.25
		HC	0.02
Lisbon		NO_x	0.03
CO	0.09		
HC	0.01		

Source: Gross emissions data from national replies to OECD questionnaite; data on size of metropolitan areas taken from "International Statistical Yearbook of Large Towns," International Statistical Institute, 1962–1970; emission densities for U.S. cities taken from "Air Quality Criteria for Carbon Monoxide," National Air Pollution Control Administration, Publication No. AP-62, pp. 4–5. Published in *Environmental Implications of Options in Urban mobility*, OECD, 1973.

over a metropolitan area. Emission densities and population exposure would seem nevertheless to constitute more sensitive indicators of pollution harm than gross emissions. And where meteorological conditions are assumed to be similar, they allow meaningful comparisons to be drawn between different cities.

Trends in the Distribution of Emission Densities

It must also be pointed out that all area-wide average emission indicators mirror the conditions at specific points in a metropolitan area imperfectly. They tend to understate the level of emissions in congested streets of the central districts and to exaggerate their severity in more loose-knit, less heavily travelled suburbs. However, as traffic growth will occur mainly in the suburbs rather than in the already traffic-saturated central cores, a trend toward progressive equalization of downtown and suburban levels of emissions can be expected.

THE QUALITY OF AMBIENT AIR IN METROPOLITAN AREAS

The preceding part of this chapter dealt with levels and patterns of emissions. Yet depending on the background levels of pollution and on wind, temperature, sun, topography and other diffusion characteristics, the same levels and patterns of vehicular emissions can lead to wide differences in ambient air quality in different cities or even at different points in the same city. This is why actual measurements of pollution concentrations are in the end the only accurate means of assessing the scope of a city's air pollution problem. To this end, annual mean concentrations of vehicular air pollutants in a number of OECD metropolitan areas are given in the appendix to this chapter. The data was obtained from national replies to the OECD questionnaire.

Trends in Air Quality

From the figures it appears as if particulate matter is either decreasing or remaining at the same levels in the cities for which data was furnished. The results for NO and NO_2 are far less clear. The concentrations of these pollutants appear to be increasing in Tokyo, but in the United States the trends vary from city to city, increasing in some, decreasing in others. The same is generally true of CO concentrations. In Chicago and Tokyo, over the same period, the emission trends as shown in table 26.3 are much steeper than the concentration trends. The more steeply sloped trends in emissions tend to confirm what was remarked before: that pollution in metropolitan areas due to carbon monoxide may be extending in space rather than growing in concentration. However, weather patterns vary greatly from year to year, making the interpretation of any trends somewhat questionable.

Comparison between Cities

Comparisons between cities are risky on the strength of the figures available. Different methods have been used in taking measurements. The character of the sites chosen for measurement also varies from city to city. Both are capable of working a heavy influence on any results.

The location of sampling equipment could also influence the results obtained. Pollution concentrations are greatest at the centre of the roadway and decline with increasing distance from the centre. The vertical concentration profile shows a similar decline. It is almost impossible to know if the cities were sampled at comparable sites. Nor are there firm guidelines on how to monitor and sample urban air quality. One suggestion is that each city should sample its most polluted place, but this would not produce a representative index. A more complicated and expensive approach would involve taking measurements at a standard number of points set out on a standard grid. Presented in the form of a map such measurements would reveal the contours of pollution. Averaged out they would depict the general level of air quality throughout the area under survey.

Agreed rules for the timing of sample measurements are also lacking. Pollution concentrations can vary considerably in and out of rush hours, and can be much lower during the height of holiday seasons than a typical working week. So long as data from one city exists for only part of a day or year it cannot be considered comparable with figures from another averaged over a complete day or year.

Governments are only now beginning to discuss the desirability of standardizing their methods of measuring and monitoring air qulaity. Internationally agreed guidelines to this effect would enormously facilitate a meaningful comparison of air pollution conditions and of changes in air quality over time in metropolitan areas throughout the world.

EFFECTS OF NOISE ON MAN

In many countries urban noise is beginning to cause as much concern as air pollution. Public opinion in a number of countries believes that noise, no less than some of the more tangible forms of pollution, represents a real threat to the quality of the environment and to the well-being of people.

While the sources of annoying sounds in a city are plentiful, few have been extending their influence as rapidly and relentlessly as the noise of motor vehicles. It is the rapidly growing spatial and temporal progression of traffic noise, affecting as it does the lives of an ever growing proportion of the population for an ever longer number of hours each day and night, that distinguishes the problem of traffic noise from most other noise-generating activities. This is also the reason why traffic noise is such a frequent source of complaints. Surveys and opinion polls in a number of countries show that people consider traffic noise as the most annoying kind of unwanted sound. Physical measurements confirm that the intensity of traffic noise exceeds that of any other source over the greater part of urban areas. A survey carried out in London in 1961 showed that noise from motor vehicles predominated at 84 per cent of the locations where measurements were taken. ("London Noise Survey," H. M. Stationary Office, 1968.)

Detailed discussion of the effects of noise on man is beyond the scope of this report but can be found in a number of specialized reports such as the Wilson Report in the United Kingdom, the OECD Report, and a report of the U.S. Environmental Protection Agency to the President and Congress.[3] What follows is a brief review of the effects of noise on man.

Traffic noise at the levels and durations normally encountered in urban areas is unlikely to harm health. However, it can be extremely disturbing and produce secondary adverse effects. Among them are interference with sleep, distraction of concentration and interruption of conversation. General psychological distress produced by noise can also add to the overall stress of life and may increase non-auditory disease, though to what extent remains to be determined. One important—but still unproven—consequence of traffic noise is that it may speed up the process of age-induced deafness (presbycosis). The auditory acuity of persons subjected to traffic noise during many years may decrease more rapidly than that of persons living in quiet areas. This is a hypothesis put forward by scientists who regard noise as a decisive factor in age-induced loss of hearing at high frequencies in industrialized and developed societies. Further research, however, seems necessary before this opinion can be confirmed or refuted.

Disturbance of sleep. Sociological surveys show that the disturbance of sleep is often given as the main reason for annoyance. People are disturbed by traffic noise if this interrupts their sleep or prevents them from falling asleep. The increase in traffic noise during the night may change the sleeping habits of city dwellers, and the cumulative physiological effect of not enough sleep is likely to be harmful for most people.

Effects on activities. In addition to producing a subjective impression of annoyance and causing sleep disturbance, traffic noise can interfere with people's daytime activities, especially in their homes. In dwellings exposed to traffic noise the activities most easily interfered with are conversation, listening to radio and television, intellectual work and various delicate tasks. The effects of noise on task performance have been extensively studied, but mainly with reference to noise of higher intensity and frequencies than that due to urban traffic. One of the most obvious effects of traffic noise is interference with communication. Noise can interrupt verbal communication or adversely affect the hearing of music or television. The extent of the interference will depend on the conditions under which the communication takes place, the intensity level and the noise spectrum, and the time distribution of the noise.

DESIRABLE AMBIENT NOISE LEVELS

In many countries efforts have been made to define acceptable or desirable environmental noise standards. The best known are those proposed in 1963 by the Wilson Committee.

Studies carried out in France, Sweden and the United States support these recommendations. In France, a mean sound level of 60 dBA in front of residental property has been recommended as a limit not to be exceeded (with windows closed the corresponding level indoors would be around 40 dBA); in Sweden, a daytime outdoor level of 55 dBA has been suggested by the Board of Urban Planning as an acceptable limit in areas for outdoor living spaces such as balconies. A maximum indoor sound level of 35 dBA during daytime and 25 dBA during nighttime has also been suggested. Investigations in the United States indicate that areas where outdoor median noise levels exceed 56 to 60 dBA in daytime are not suitable for detached houses. The same studies show that when daytime noise levels are greater than 66 dBA relaxed conversation is not possible unless the building is air-conditioned.[4]

THE NOISE ENVIRONMENT IN METROPOLITAN AREAS

Table 26.6 shows noise levels recorded in London during a 1961 investigation. As can be seen, the levels of noise vary widely according to the location, type of street, density of traffic and time of day. In a more recent Paris inquiry median noise levels (levels exceeded 50 percent of the time) were measured in front of buildings in 43 different locations.[5] As can be seen in figure 26.1, it is noise at night that varies the most from place to place: a 15 dBA difference was observed between the noisiest 5 per cent and the quietest 5 per cent of the streets. Noise levels have also been measured in many other places, for example in Austria, Canada, Germany, Japan, Belgium and the United States. Some of the results of these surveys are shown in figures 26.2 and 26.3. In Brussels measurements have shown that noise levels decrease with distance away from the center. This appears to be true for daytime, evening as well as nighttime background noise levels. Ambient noise levels increase not only as a function of proximity to the center of the city but also as a function of city size. A Canadian study has shown that background noise levels rise rapidly as a small town expands, but tend to level off as the city reaches a certain radius.

Table 26.5 Sound Levels Which Must Not Be Exceeded for More Than 10 Per Cent of the Time in Bedrooms and Living Rooms

Location	By day	By night
Rural areas	40 dBA	30 dBA
Suburbs (remote from main roads)	45 dBA	35 dBA
Busy urban areas	50 dBA	35 dBA

Table 26.6 Ambient Noise Levels in London in 1961

	Noise climate in dBA[a]	
Location by group	Day (8 A.M.–6 P.M.)	Night (1 A.M.–6 P.M.)
A Arterial roads with many heavy vehicles and buses (kerbside)	80–68	70–50
B Major roads with heavy traffic and buses	75–63	61–49
Side roads within 15–20 yds.of roads in groups A or B1 aove		
C[b] Main residential roads	70–60	55–44
Side roads within 20–50 yds of heavy traffic routes		
Courtyards of blocks of flats, screened from direct view of heavy traffic		
D Residential roads with local traffic only	65-56	53–45
E Minor roads	60-51	49–43
Gardens of houses with traffic routes more than 100 yds. distant		
F Parks, courtyards, gardens in residential areas well away from traffic routes	55-50	46–41
G Places of few local noises and only very distant traffic noise	50–47	43–40

Source "Noise," Final Report of the Wilson Committee on the Problem of noise, 1963.

[a]By noise climate is meant the range of noise level recorded for 80 per cent of the time. For 10 percent of the time the noise was louder than the upper figure of the range and in the case of Group A attained peak levels of about 90 dBA; for 10 percent of the time the noise was less than the lower figure in the range.

[b]In Groups C to F, noise from other sources, such as trains or children's voices, predominated over road traffic noise at particular times, but traffic was the most frequent noise source.

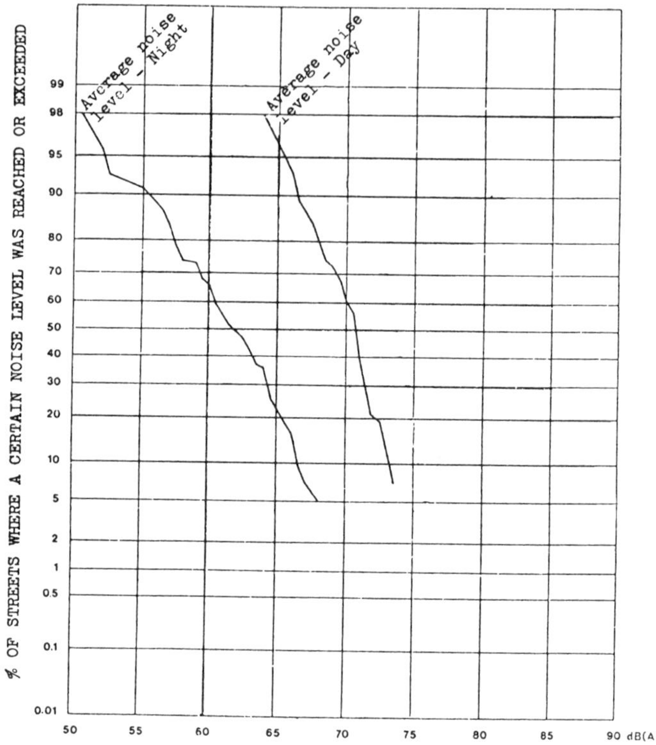

26.1 Average urban noise levels during the day and night, Paris. (Source: "Etude de la gêne due au trafic automobile urbain," Centre Scientifique et Technique du Bâtiment, Paris, 1971).

PAST AND ANTICIPATED EVOLUTION IN THE URBAN NOISE CLIMATE

Past Trends

Although it is difficult to document any dramatic rise in traffic noise levels over the past two decades in terms of sound intensity, there is ample evidence that noise has become more pervasive in space and in time: more urban residents are exposed for longer hours to annoying levels of traffic noise today than they have been in the past. This phenomenon has been illustrated by the results of two inquiries carried out in London, the first in 1948, the second 14 years later, in 1962. Whereas more than half the citizens made no reference to noise in 1948, 90 per cent were conscious of it in 1962, during this same period, persons annoyed by noise doubled in number, rising from 23 per cent to 50 per cent.

Studies in the United States support these findings. An investigation by the Environmental Protection Agency shows that the land area in the United States seriously affected by noise from highway traffic increased from 75 square miles in 1960 to 545 square miles in 1970. At the same time, the number of people affected rose from approximately half a million to 2.7 million. These figures do not take into account the increase in noise resulting from motor traffic on secondary roads.

26.2 Urban noise levels, Vienna, 1964. In the residential zone, the nighttime outdoor levels are very similar to the daytime indoor levels with the window open. (Source: "Technical Backgound for Noise Abatement in HUD's Operating Program," Bolt, Beranek and Newman, Report No. 2005, 1970.)

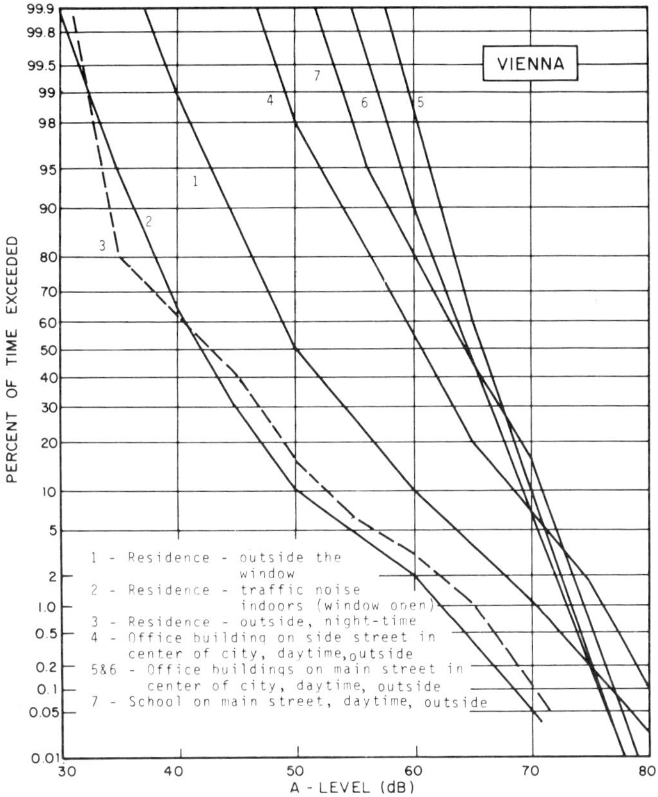

26.3 Urban daytime noise levels, Tokyo, 1966. (Source: "Technical Background for Noise Abatement in HUD's Operating Program," Bolt, Beranek and Newman, Report No. 2005, 1970.)

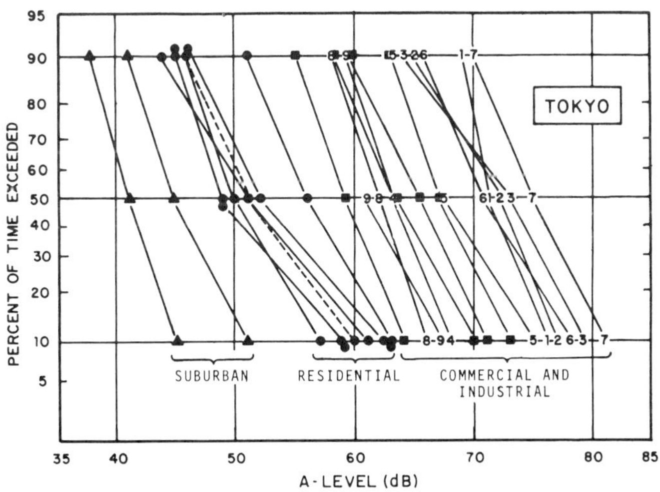

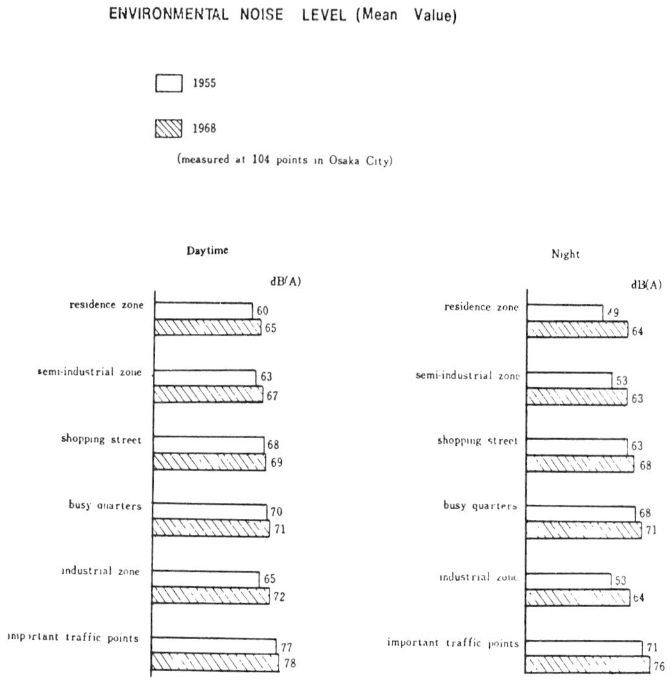

26.4 Changes in the Noise Climate of Osaka, 1955–1968. (Source: "Environmental Pollution in Osaka Prefecture," April 1970.)

The 1961–1962 London inquiry showed that the duration of an average "noise night," i.e. the time during which sound level is lower, was surprisingly short: only 5½ hours (from midnight to 5 or 6 A.M.). This period of quiet began before midnight only in about 25 per cent of places where measurements were made and extended beyond 6.30 A.M. only in 11 per cent of cases. ("London Noise Survey," HMSO, London, 1968.) What is even more significant, subsequent measurements have shown this quiet period to have grown even shorter, since 1962. ("Tackling the Problems of Big City Vehicle Noise," C. H. Vulkan, *Municipal Engineering*, August 1969.)

These findings are confirmed by the results of a study in Osaka, Japan which compared daytime and nighttime noise measurements made in 1955 and 1968. ("Environmental Pollution in Osaka Prefecture," Environmental Pollution Control Office, Osaka Prefectural Government, April 1970.) The results revealed only a slight increase in daytime noise levels, particularly in the center of the city, but very substantial increases at night, especially in residential areas. This can be seen in figure 26.4.

Future Trends in the Urban Noise Climate

Future trends in urban traffic noise levels are extremely difficult to predict. They hinge on a number of factors such as the rate and magnitude of growth of urban areas, the increase in

Table 26.7 Average Daytime and Nighttime Outdoor Noise Levels in the United States

	Average daytime 7 A.M.–7 P.M.		Average nighttime 10 P.M.–7 A.M.		Difference between day and night (dBA)
	Range (dBA)	Mean (dBA)	Range (dBA)	Mean (dBA)	
Residual Noise Level (L_{90})					
City (4 locations)	61–77	69.1	51–69	60.8	8.3
Suburban and urban detached housing Residential (11 locations)	38–53	45.6	35–46	39.8	5.8
Median Noise Level (L_{50})					
City (4 locations)	64–80	73.0	55–75	65.5	7.5
Suburban and urban detached housing Residential (11 locations)	44–59	50.9	38–50	44.2	6.7

Source: "Report to the President and Congress on Noise," Environmental Protection Agency, December 31, 1971, Table 2-1.

vehicular traffic, as well as the rate of progress in the technology of noise reduction and trends in noise abatement legislation.

In France, the increase in traffic noise levels has been estimated at 2 to 3 dBA between 1970 and 1985. Studies in the United States indicate that, in the absence of controls, outdoor noise levels in residential areas of urban communities would rise from a typical level of 47 dBA to 50 dBA by the year 2000. The land area seriously affected by noise would increase from 545 square miles in 1970 to over 2000 square miles thirty years later and the number of people affected would jump from 2.7 million to 10 million. However, in view of the fact that local conditions play a decisive role, it is clear that estimates of this kind are of limited value for decision-making. In the saturated centres of big cities the evolution of the noise problem is likely to be quite different than in the suburban areas.

Table 26.8 Trend in Population Numbers Exposed to Excessive Noise Levels in the United Kingdom

	Urban population exposed to an L_{10} ($\leqslant$ 70 dBA $\geqslant$ 65 dBA)	
Situation	1970	1980
Vehicle noise unchanged	8.5–21 million (19–46%)	14–29 million (30–61%)
Increase of 5 dBA by 1975, then no further change	8.5–21 million (19–46%)	29–45 million (61–93%)
Reduction of 5 dBA by 1980	8.5–21 million (19–46%)	4.3–14 million (9 to 30%)
Reduction of 10 dBA by 1980	8.5–21 million (19–46%)	0.9–4.3 million (2–9%)

Source: Published in *Environmental Implications of Options in Urban Mobility*, OECD, 1973.

Other forecasts have been made in the United Kingdom by the Working Group on Road Traffic Noise ("A Review of Road Traffic Noise," Road Research Laboratory, Report LR 347, United Kingdom, 1970), showing the trend between 1970 and 1980 in urban population exposed to excessive noise level (sound level L_{10} = 65–70 dBA), according to whether noise emitted by vehicles remains unchanged, rises or diminishes (see table 26.8).

APPENDIX A: TYPICAL 1971 MOTOR VEHICLE EMISSION RATES

TABLE 26.9 (Emissions from 1971-Model Vehicles in the United States

	Emissions (grams/mile)		
Manufacturer	HC	CO	NO_x
U.S. manufacturer A	2.17	34.12	3.90
U.S. manufacturer B	2.12	35.94	4.23
U.S. manufacturer C	1.93	26.86	4.02
U.S. manufacturer D	1.75	19.41	3.51
Major import	2.15	35.79	2.66
Applicable emission standard	2.20	23.00	–

Source: U.S. Environmental Protection Agency staff. The test used was the now obsolete 7-mode/7-cycle test, against which 1971 model year cars were certified.
Note: The above data is derived from a study by the U.S. Environmental Protection Agency of 1971 model cars in use during 1971. Emission tests were made at sabilized mileage points (4000–7000 miles).

Table 26.10 Results of Measurements of Exhaust Emissions of Carbon Monoxide and Hydrocarbons from Uncontrolled Vehicles, Performed at the Vehicle Emissions Laboratory in Studsvik

Vehicle model/ year	Service weight (kg)	Number of tested vehicles	ECE test (g/km)		U.S. test in accordance with federal law, 1970 (g/km)	
			CO	HC[a]	CO	HC[b]
Volvo Amazon 1962-1968	1140-1250	33	46	2.2	27	4.0 (2.2)
Saab V4 1967-1968	960	10	36	1.6	24	2.0 (1.1)
Volkswagen 1100 & 1600 1962-1966	820-1050	27	53	3.0	30	4.4 (2.4)
Ford 17M 1961-1965	1010-1060	10	40	1.7	20	2.9 (1.6)
Opel Record 1700 1962-1966	1020-1120	12	46	2.1	23	3.0 (1.7)
Fiat 850S 1968	760	3	36	1.7	16	2.8 (1.6)
Average values						
For vehicles with a service weight of 760 to 1250 kg (as above)	1050	95	46	2.3	26	3.6 (2.0)
For vehicles with a service weight of 1460 1460 to 2560 kg	1800	10	114	3.0	48	3.6 (2.0)

Source: "Air Pollution from Motor Vehicles," Report by the Swedish Ministry of Communications Group, Stencil K 1971:1, page 79. Published in *Environmental Implications of Options in Urban Mobility*, OECD, 1973.
[a]Hydrocarbons as hexane (NDIR).
[b]Hydrocarbons (1.8 × ax hexane NDIR). (Hydrocarbons as hexane are given in brackets.)

Table 26.11 Results of Survey Investigations into Emissions of Carbon Monoxide and Hydrocarbons in Vehicle Exhaust Gases

	Number of vehicles tested	Average emission in ECE test (g/km)	
		Carbon monoxide	Hydrocarbons
Sweden At the Emissions Testing Laboratory in Studsvik, representing "Swedish vehicle stock"	95	46	2.3
Great Britain At MIRA, representing "British vehicle stock"	105	42	2.7
CEC In European vehicle factories	469	50	3.0

Source: Published in *Environmental Implications of Options in Urban Mobility*, OECD, 1973.

Table 26.12 Emissions of Carbon Monoxide and Hydrocarbons from Controlled Vehicles, Results of Tests Performed at the Vehicle Emissions Testing Laboratory in Studsvik

Vehicle model	Number tested	Model	ECE test (g/km) Carbon Monoxide	ECE test (g/km) Hydrocarbons
Volvo Amazon	1	1968	19.3	0.79
Volvo Amazon, automatic	1	1968	23.0	1.01
Volvo 144	1	1968	28.4	1.00
Saab V4	3	1968	27.9	1.20
Volkswagen 1500	3	1968	22.6	0.77
Volkswagen 1600	3	1968	24.8	1.45
Valiant 130, automatic	1	1968	61.8	1.34
Chrysler Satellite, automatic	1	1968	46.8	1.28
Ford Falcon	1	1968	22.6	1.48
Buick 350, automatic	3	1968	110.2	1.51
Buick 400 GS, automatic	1	1968	24.7	1.12
Mercedes 280 SE	1	1968	37.9	2.22

Source: "Air Pollution from Motor Vehicles," Report by the Swedish Ministry of Communications Group, Stencil K 1971:1, pages 125 and 88. Published in *Environmental Implications of Options in Urban Mobility*, OECD, 1973.

Table 26.13 Emissions of Nitrogen Oxides from Controlled Vehicles, Results of Tests Performed at the Vehicle Emissions Testing Laboratory in Studsvik

Vehicle model	Number tested	Nitrogen Oxides (as NO_2), idling warm engine (ppm)	ECE Test (g/km) Carbon monoxide	ECE Test (g/km) Hydrocarbons
Volvo 142	26	105	36.5	1.89
Volvo 164	25	92	53.8	1.80
VW 1600 Electronic	26	112	47.2	1.78
Volvo Amazon 1966 Model	1		50	2.0

Source: "Air Pollution from Motor Vehicles," Report by the Swedish Ministry of Communications Group, Stensil K 1971:1, pages 125 and 88. Published in *Environmental Implications of Options in Urban Mobility*, OECD, 1973.

APPENDIX B: AUTOMOTIVE POLLUTANT CONCENTRATIONS IN SELECTED URBAN AREAS

Table 26.14 Concentrations of Carbon Monoxide in Selected Urban Areas

City	Sampling dates	Mean concentration (Mg/m^3)
Vienna	1969–1971	14
Graz	1969–1971	8–11
Innsbruck	1969–1971	6–11
Toronto	1970	4
Windsor, Ontario	1970	8
Hamilton, Ontario	1970	6
Tampere[a]	1969	1
Paris	1970	19
Munich	b	9
Wiesbaden	Sept. 1965–March 1966	13
Berlin	Spring 1966–Autumn 1967	16
Frankfurt	Jan. 1964–Feb. 1965	13
Cologne	b	14
Tokyo	1969	5
Osaka	1969	4
Madrid	July 1970	11
Stockholm[c]	1969–1970	7
Chicago[d]	1966	16
Cincinnati[d]	1967	7
Denver[d]	1967	10
Philadelphia[d]	1967	8
St. Louis[d]	1967	7
Washington D.C.[d]	1967	6

Source: National replies to OECD questionnaire except as otherwise noted. Published in *Environmental Implications of Options in Urban Mobility*, (ECD, 1973.

Note: All concentrations have been converted into mg/m^3 and have been rounded to the nearest unit digit.

[a]Data was presented for morning, noon, afternoon and night. The average was calculated by weighting the first three time periods by 1 and the last by 3.

[b]Dates of sampling not given.

[c]The median concentrations for each of four stations were given. (The average median concentration was 6 mg/m^3.) Using methodology and approximate standard geometric deviations given in "Air Pollution Control Association Journal," 19 (1), 24–30, Larson, 1969, the medians were converted into means.

[d]Data from "Air Quality Data for 1967," Environmental Protection Agency, Publication No. APTD-0741, August 1971.

Table 26.15 Concentrations of Hydrocarbons in Selected Urban Areas

City	Sampling dates	Mean concentration (mg/m^3)
Toronto	1970	3
Windsor, Ontario	1970	4
Hamilton, Ontario	1970	4
Edmonton, Alberta	1970	5
Tokyo	1969	2
Osaka	1969	1
Chicago[a]	1967	6
Cincinnati[a]	1967	5
Denver[a]	1967	5
Philadelphia[a]	1967	5

Source: Published in *Environmental Implications of Options in Urban Mobility*, OECD, 1973.
Note: All concentrations have been converted into mg/m^3, and have been rounded to the nearest unit digit.
[a]Data from "Air Quality Data for 1967," Environmental Protection Agency, Publication No. APTD-0741, August 1971.

Table 26.16 Concentrations of Nitric Oxide (NO) in Selected Urban Areas

City	Sampling dates	Mean concentration (mg/m^3)
Tokyo	1969	.06
Osaka	1969	.08
Chicago[a]	1967	.10
Cincinnati[a]	1967	.04
Denver[a]	1967	.05
Philadelphia[a]	1967	.08
St. Louis[a]	1967	.05
Washington[a]	1967	.06

Source: National replies to OECD questionnaire except as otherwise noted. Published in *Environmental Implications of Options in Urban Mobility*, OECD, 1973.
Note: All concentrations have been converted into mg/m^3 and have been rounded to the nearest hundredth of a milligram.
[a]Data from "Air Quality Data for 1967," Environmental Protection Agency, Publication No. APTD-0741, August 1971.

Table 26.17 Concentrations of Nitrogen Dioxide (NO_2) in Selected Urban Areas

City	Sampling dates	Mean concentration (mg/m^3)
Vienna[a]	1969–1971 (7h–19 h)	.05–.08
Tampere[b]	1969	.04
Paris	1969	.03
Berlin	Spring 1966–Autumn 1967	.05
Cologne	c	.23
Tokyo	1969	.08
Osaka	1969	.04
Stockholm[d]	1969–1970	.10–.14
Chicago[e]	1967	.10
Cincinnati[e]	1967	.06
Denver[e]	1967	.08
Philadelphia[e]	1967	.09
St. Louis[e]	1967	.05
Washington[e]	1967	.09

Source: Published in *Environmental Implications of Options in Urban Mobility*, OECD, 1973.
Note: All concentrations have been converted into mg/m^3 and have been rounded to the hearest hundredth of a milligram.
[a]The measurements provided were given in terms of normalized cubic meters (adjusted to $0^\circ C$), but practically speaking, they have the same value when expressed in terms of cubic meters (adjusted to $25^\circ C$).
[b]Data was presented for morning, noon, afternoon and night. The average was calculated by weighting the first three time periods by 1 and the last by 3.
[c]Sampling period not specified.
[d]The range of median concentrations was given.
Using methodology and approximate standard geometric deviations given in "Air Pollution Control Association Journal," 19 (1), 24–30, Larson, 1969, the medians were converted into means.
[e]Data from "Air Quality Data for 1967," Environmental Protection Agency, Publication No. APTD-0741, August 1971.

Table 26.18 Concentrations of Particulate Matter in Selected Urban Areas

City	Sampling dates	Mean concentration
Vienna	1969-1971 (7h-19h)	.2-.0.6 mg/m^3
Linz	1969-1961 (7h-19h)	.35 mg/m^3
Toronto	1970	.111 mg/m^3
Windsor, Ontario	1970	.172 mg/m^3
Sarnia, Ontario	1970	.110 mg/m^3
Hamilton, Ontario	1970	.150 mg/m^3
Calgary, Alberta	1968	.111 mg/m^3
Edmonton, Alberta	1969	.098 mg/m^3
Tokyo	1969	.20 mg/m^3
Osaka	1969	.119 mg/m^3
Madrid	1970	197 ppm[a]
Stockholm	1968-1969	.096-.465 mg/m^3
Chicago[b]	1967	.143 mg/m^3
Cincinnati[b]	1967	.107 mg/m^3
Denver[b]	1967	.106 mg/m^3
Philadelphia[b]	1967	.162 mg/m^3
St. Louis[b]	1967	.124 mg/m^3
Washington DC[b]	1967	.104 mg/m^3

Source: Published in *Environmental Implications of Options in Urban Mobility*, OECD, 1973.

[a]Because of the variable weight of particulate matter, parts per million cannot, in general, be converted into mg/m^3.

[b]Data from "Air Quality for 1967," Environmental Protection Agency, Publication No. APTD-0741, August 1971.

APPENDIX C: VEHICLE NOISE EMISSION CHARACTERISTICS

Under urban traffic conditions, the noise from a heavy truck or bus is equivalent to that from 10 to 15 private cars together. It is well to point out that the unit of sound measurement—the decibel (dB)—is a logarithmic unit: an increase of 3 dB is equivalent to an increase of sound intensity by a factor of 2, 10 dB by a factor of 10; and 20 dB by a factor of 100.

Typical noise emission characteristics of vehicles found in British traffic are given in Table C.1.

For purposes of comparing noise levels emitted by American motor cars and those of the European type, results may be quoted from a 1970 inquiry by the California Highway Patrol.[6] The measurements were made at a distance of 15 metres (50 feet), the general practice in the United States (whereas in Europe measurements are usually made 7.5 metres from the axis followed by the vehicle).

Table 26.19 Noise Emitted by Different Types of Vehicles Traveling at a Constant Speed of 40 to 50 km/h (25 to 30 m.p.h.) (measurements carried out at a distance of 7.5 m)

Vehicle type	Mean level (dBA)	Noise level range for 80% of vehicles (dBA)[a]	Relative sound intensity
Private cars			
Light car, under 1100 cc	70	67–75	1
Medium car, 1100–1600 cc	72	67–75	1
Heavy car, over 1600 cc	72	68–77	1
Light, four-wheeled commercial	73	69–77	
Heavy commercial	81	76–86	10
Buses	83	80–85	16
Motorcycles	77	72–83	4

Source: "Traffic Noise" by R. J. Stephenson and G. H. Vulkan, Journal of Sound and Vibration, 7 (2), 1968.
[a]These figures include neither the noisiest 10 per cent nor the quietest 10 per cent.

Nine thousand vehicles, each weighing less than 6,000 pounds, were checked. In urban surroundings, the average noise level recorded was 68 dBA when the speed was below 55 km/h (35 m.p.h.) which is equivalent to about 74 dBA at a distance of 7.5 metres. Some vehicles emitted less than 63 dBA and others were than 74 dBA, equivalent to 68–80 dBA at 7.5 metres.

As regards private cars, an investigation by the French Institut de Recherche des Transports confirms these different results.[7] At 7.5 metres with the vehicle travelling at 50 km/h—a typical rate of speed in an urban area—the noise varies, depending on engine capacity and the make of car, between 65 and 75 dBA when the vehicle is cruising in third gear, and between 75 and 85 dBA when it is accelerating in second gear.

Other measurements made in the United Kingdom by the Motor Industry Research Association under test conditions recommended by the International Standards Organisation yielded the following results:[8]

Heavy commercial vehicles 88–92 dBA
Light commercial vehicles 79–91 dBA
Small cars 77–83 dBA
Large cars 80–82 dBA
Sports cars 81–91 dBA

From these few results it may be observed that:

• a private car in general emits 10 dBA less than a heavy commercial vehicle or a powerful sports car;
• acceleration can result in 10 to 20 dBA higher noise levels than at urban cruising speed.

Factors Influencing Levels of Traffic Noise

Unlike the noise emitted by a single vehicle, which depends solely on the vehicle's own characteristics (except where due to tyre/road contact), noise due to traffic, in addition to the characteristics of the different kinds of vehicle involved, hinges on a number of independent parameters. These parameters may be classified as follows:

1. *Traffic parameters*
• Speed and flow rate
• Composition
• Traffic fluidity (traffic lights, one-way streets)
• Driver behaviour
2. *Road parameters*
• Road design (tunnels, cuttings, embankments, or at level of adjacent land)
• Gradient and degree of curvature
• Type of road surface
• Width
3. *Environmental parameters*
• Distance and height with respect to the roads, of the point at which the noise is heard
• Existence of natural or artificial screens between road and point of perception
• State of the ground between the road and point of perception (noise absorption)
• Reflection of noise from buildings at each side of the road (problem linked to road width)
4. *Weather parameters*
• Rain, snow or dry conditions
• Wind (direction and speed)
5. *Dwelling parameters*
• Orientation of living areas
• Attenuation of noise through windows
• Size of window openings

These different parameters are not of equal importance. Some significantly affect the observed noise, others only negligibly.

27 CONTROL OF VEHICLES IN USE

WHY CONTROL USED VEHICLES?

There are two reasons for controlling vehicles that are in actual use. For one thing, emissions tend to increase with mileage and age. There is also a time lag before vehicles with better emission controls become sufficiently numerous to influence overall levels of pollution.

Figure 27.1 shows how exhaust emissions increase with mileage. The figures are based on some 5000 tests of California cars and weighted to reflect the actual distribution of vehicle types and makes in that State. Both carbon monoxide and hydrocarbon emissions can be seen increasing with mileage.

Emission rates also increase with vehicle age. This is not surprising since the age of a vehicle and its mileage are closely related. Hydrocarbon emissions from cars on American roads have been calculated to increase on the average by 6 per cent after one year and by 25 per cent after five years; carbon monoxide emissions show a somewhat smaller rate of increase (see table 27.1).

At present this phenomenon seems to apply to controlled as well as to uncontrolled vehicles. It may be avoided in future as control systems become more durable and reliable, and particularly in countries where standards for new vehicles are coupled with a requirement on manufacturers to guarantee the performance of emission control devices for a specified time or mileage. Eventually, improvements in the durability of control systems may reduce the need for their inspection and repair.

The reduction in total emissions is governed by rates of replacement, which in most countries run at about one tenth of the vehicle fleet every year. In densely developed cities where pollution is likely to be a hazard to health during the next five years, control over vehicles in service can be employed to slow down the rate of growth of emissions until normal attrition eliminates the older, more polluting cars from use.

27.1 Exhaust emissions vs. mileage, California automobile population. (Source: "Control Techniques for CO, NO and HC Emissions from Mobile Sources," National Air Pollution Control Administration, Publication No. AP-66, p. 3. Published in *Environmental Implications of Options in Urban Mobility*, OECD, 1973.

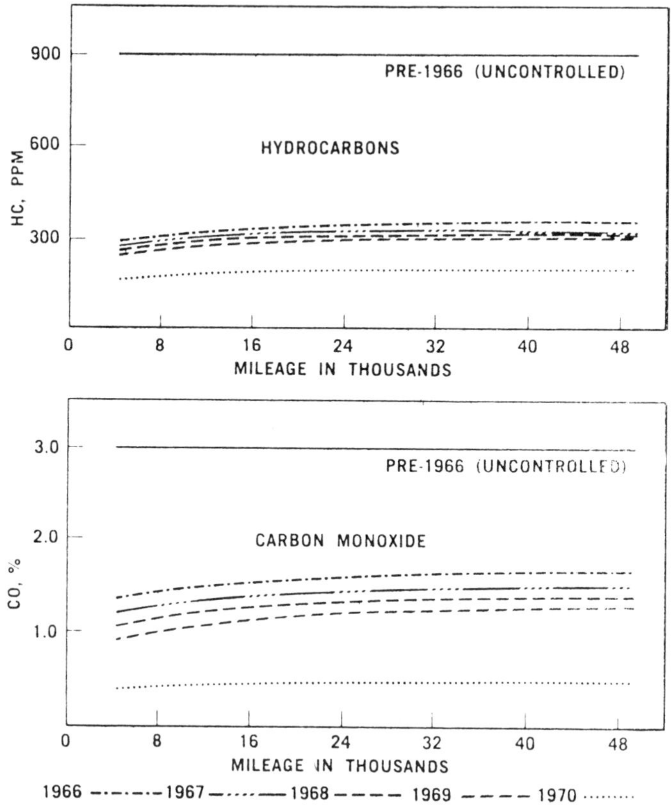

Table 27.1 Deterioration of Hydrocarbon and Carbon Monoxide Emissions in Used Vehicles

Vehicle age (years)	Deterioration factor	
	HC	CO
New	100.0	100.0
1	106.3	105.5
5	125.2	119.5
10	129.6	120.1
15	130.2	120.3

Source: California Air Resources Board. Published in *Environmental Implications of Options in Urban Mobility*, OECD, 1973.

ALTERNATIVE WAYS OF CONTROLLING POLLUTANTS AND NOISE FROM VEHICLES IN USE

The control of exhaust emissions from used vehicles can be approached in a number of different ways. Owners can be obliged to install emission control devices; they can be made to submit their vehicles to regular maintenance; or they can be obliged to have them regularly inspected and brought back into order.

The control of noise fits this framework only at some points. In used cars it is confined to efforts to promote more effective and longer-life silencers, to provide for their regular maintenance and to develop improved techniques for conducting noise checks.

Mandatory Modification

One possibility is to oblige owners of older cars to install devices that would bring exhaust emissions and noise down to or towards the level of new cars. Where vehicle registration is required annually, retrofitting of older cars with the necessary control devices could be made a condition of renewing the annual licence; it could also be required when a car changed hands.

The technical and economic feasibility of retrofit has not been demonstrated fully to the satisfaction of all. Nor is voluntary installation of retrofit devices likely to occur to any great extent. Available evidence suggests that the cost and the effects on fuel consumption and driving performance may discourage car owners from installing the equipment. They include roughness at low speed, rough idling and sluggish acceleration; average fuel consumption is increased by 2 per cent.

Mandatory Maintenance

The regular maintenance of emission control systems is another possibility. Old and faulty silencers may be replaced. Adjustment may be made to ignition timing and fuel mixture; exhaust, crankcase and evaporative emission control systems may be serviced or replaced. These and other minor maintenance measures can all produce reductions in emissions and noise. Vehicle owners could be obliged to have such maintenance performed at regular intervals and to show proof of it on applying to renew their annual licence where annual registration of vehicles is required.

Studies of the effectiveness and costs of using various maintenance regimes to reduce exhaust emissions have been carried out in a number of countries. Their effectiveness is summarized in table 27.2. The results of more recent tests in the United States are given in table 27.3.

In view of the estimated 1972 cost (Retrofit alone, $20; Tune-up alone, $55; Tune-up and Retrofit combined, $70) it

Table 27.2 Reductions in Emissions of Carbon Monoxide and Hydrocarbons from Used Cars due to Engine Adjustments

	CO		HC,
	At idle	Drive	drive
United Kingdom[a]	30%	20%	8%
Sweden (1968)[b]	40%	20%	7%
Sweden (1971)[b]	24–48%	10–15%	
United States [c]		9–29%	5–24%

Note: Caution must be used when interpreting these results since the emission reductions in all the studies summarized here were based upon tailpipe concentration measurements only, and some of the studies refer to maintained vehicles only.

[a]"Exhaust Emission of Vehicles Tested to the Draft European Procedure," C. D. Haynes and M. Southall, November 1968.

[b]"Air Pollution from Motor Vehicle Exhausts," Report by the Swedish Ministry of Communication Group, Stencil K 1971¾ 1, pp. 114, 118.

[c]"Cost and Effectiveness for Several tune-Up Inspection and/or Maintenance Approaches," Control Techniques for CO, NO and NC Emissions from Mobile Sources, NAPCA Publication No. AP–66, March 1970, pp. 4–8.

would appear that a requirement for retrofit without tune-up is the most cost effective means of these three alternatives for reducing air pollution from pre-controlled automobiles. Tune-up and retrofit combined gives better results for HC and CO, whereas tune-up alone gives inferior results all around. However, even the retrofit alone requires periodic maintenance to retain all of the emission reductions initially achieved. Furthermore the retrofit kit is designed for pre-controlled automobiles only and the reductions quoted above are for such vehicles; the impact of a retrofit programme on overall vehicle emissions will depend upon the contribution of pre-controlled vehicles to the total emissions.

Mandatory Inspection

A third possibility is periodic checks of cars to see that their emissions are within allowable limits. Such checks could range from simple measurements of the carbon monoxide content of exhausts from engines at idle to sophisticated tests involving dynamometers and the measurement of hydrocarbon and carbon monoxide emissions during simulated drives, and static noise tests. Vehicles that failed to pass could be made to undergo repair within a specified time and to return for another inspection.

Idling tests for carbon monoxide have been introduced in Switzerland, Germany, Sweden, Japan and in certain states in the United States (including California and New Jersey). Japan

Table 27.3 Average Reductions in Emissions due to Engine Adjustments (%)

	Average reduction achieved		
	Retrofit w/o tune-up	Tune-up w/o retrofit	Retrofit and tune-up
HC	26	15	34
CO	16	11	18
NO_x	22	-4[a]	20

[a]Not statistically significant.

has a large number of inspection stations, backed up by 6,000 commercial garages which are licensed to carry out full-scale inspections. The interval between compulsory inspections is one year for trucks and two years for cars. In Switzerland all taxis, rental vehicles and buses are subject to a compulsory annual inspection, while all other vehicles are inspected every three years. The inspection includes a smoke check as well as a carbon monoxide idle test. The United Kingdom obliges heavy trucks to undergo an annual inspection which includes a visual smoke check.

THE CHOICE OF INSPECTION PROCEDURES

As has been made clear, governments have a range of inspection and maintenance procedures to choose from, and one that is suitable for one country may not be suitable for another. Inspection regimes should be chosen only after taking into account the number of cars to be covered and their average age, the availability of inspection stations, the reduction in aggregate emissions that is needed to achieve the desired air quality and the ability of motorists to pay.

Visual Checks for Smoke

This is the least expensive of the available alternatives. In the United Kingdom it led to the detection of 6,600 excessively smoking vehicles in annual inspections and to some 2,300 more in spot checks in 1970. The cost of this campaign was estimated at £35,000 for the inspections and £4,000 for the spot checks. The possibility of introducing an instrumented smoke check is being considered.

Although smoke control has no value for reducing emissions of carbon monoxide, hydrocarbons and oxides of nitrogen, it is useful for controlling emissions of particulates and for curring down haze due to smoke.

Spot Checks for Carbon Monoxide Emissions

Experience gained in Tokyo suggests that spot checks using portable testing equipment can reduce the number of cars

Table 27.4 Cars Spot-Checked in Tokyo and Found to be Exceeding Permitted Limits for CO Emissions at Idle

	Number of cars tested	Warning and maintenance order issued	Percentage found exceeding limits
1970			
August	12,807	5,821	45
September	13,683	5,635	41
October	30,746	10,753	35
November	24,864	8,695	35
December	19,477	6,232	32
1971			
January	17,027	6,096	36
February	15,568	5,182	34
March	18,169	5,046	28
April	15,751	4,404	27
May	9,328	2,647	27
June	9,566	2,943	31
July	13,219	3,640	27
August	9,010	2,615	29
September	11,286	2,724	24
Total	220,501	72,438	32

Source: Tokyo Capital Police Bureau, Traffic Control Enforcement Division.
Note: The number of cars tested over the 14-month period represents approximately 10 per cent of all those registered in the Tokyo metropolitan area.

with excessive carbon monoxide emissions. During a period of 14 months, ending in September 1971, it was found possible to check 200,000 cars or about 10 per cent of those registered in the city. About one third were found to be exceeding the permitted limits for carbon monoxide at idling speeds and obliged to go for maintenance. When spot checking was introduced, 45 per cent of all cars stopped were found to be in need of adjustment. Fourteen months later this percentage was down to 24, having fallen off steadily throughout the entire period. Table 27.4 shows the results in detail.

In Sweden 533,000 cars were tested for carbon monoxide emissions at idle during the first quarter of 1971. Of these, 30 per cent needed adjustment. After adjustment all but 2 per cent of the corrected vehicles met the carbon monoxide emis-

Table 27.5 Result of Carburetor Adjustments on Vehicles Which Were Exceeding the Stipulated Limit at the First Measurement in Sweden

Carbon monoxide content before adjustment (%)	Carbon monoxide content after adjustment -5.5%	5.6–7.0%	7.1+%	Total
Number				
5.6–7.0	36,283	148	–	36,431
7.1-	95,110	1,542	682	97,334
Total	131,393	1,690	682	133,765
Per cent				
5.6–7.0	27.1	0.1	–	27.2
7.1-	71.1	1.2	0.5	72.8
Total	98.2	1.3	0.5	100.0

Source: "Periodic Inspections of Private Cars During the First Quarter of 1971," p. 9-10, Stockholm, 1971.

sion requirement of 5.5 per cent by volume. Detailed results are shown in table 27.5.

Available evidence indicates that an idle test for carbon monoxide has no significant correlation with a full emission test. However, in city traffic, where cars spend a significant proportion of time in idling condition at traffic lights and in congested streets, reductions in idling emissions can bring substantial improvements in air quality, particularly at street level. Hence, carbon monoxide idle tests when followed by appropriate adjustments, can be an effective though limited measure.

Noise Checks

The main cause of avoidable noise from cars is not age but bad maintenance and breakdown of exhaust silencers through corrosion. In trucks, badly maintained engines, brakes, transmission and bodywork can be additional sources of excessive noise. Periodic inspections offer the best opportunity to check and correct these faults. There is general agreement that a quick, stationary test is necessary for noise checks to be fair and effective. The United Kingdom is currently considering the introduction of such checks for heavy trucks, to be combined with the annual safety inspection. Some relief in ambient noise levels may be expected from this approach, although in view of the inherent noisiness of heavy commercial vehicles, the overall benefits are likely to be small.

Roadside spot checks for noise involve a number of practical problems, the main one being the difficulty of singling out the noise of individual vehicles in heavy traffic. Neverthe-

Table 27.6 Cost and Effectiveness for Several Tune-Up Inspection and Maintenance Approaches Using Dynamometers

Type of inspection and maintenance	Reference	Av. total annual out-of-pocket cost per car, repairs	Immediately after service: Exhaust HC reduction, percent	Immediately after service: Exhaust CO reduction, percent	Cost-benefit ratios: $ per car/% HC removed	Cost-benefit ratios: $ per car/% CO removed
New Jersey—ACID identifies high emitters, corrective action taken	12[a]	$4-7.50[b]	13-19	7-14	0.31-0.39	0.57-0.54
Clayton—Key Mode diagnosis, corrective action taken	12[a]	$13.85[b]	24	23	0.58	0.60
ACID type cycle used for diagnosis—indicated adjustments and repairs made	5[c]	$10.30	6	13	1.72	0.79
Dynamometer and oscilloscope used for diagnosis—indicated adjustments and repairs made—all carburetors rebuilt or replaced—idle set 75 rpm above manufacturers' specifications	6[d]	$50-60.00[e]	28.9	13.4	1.73-2.08	3.73-4.48

Source: "Control Techniques for CO, NO and HC Emissions from Mobile Sources," National Air Pollution Control Administration, Publication No. AP-66, March 1970, p. 4-10.

Note: Data in this table is based on the now obsolete 7-mode/7-cycle reference emission test. However, preliminary data using the CVS emission test indicated that the reductions achievable by the various schemes are of the same relative order of magnitude as those expressed in the table.

[a]Based on 22% equipped, 78% not equipped with exhaust control systems

[b]Estimated—large-scale, full-utilization basis.

[c]All vehicles studied equipped with exhaust control systems.

[d]None equipped with exhaust control systems.

[e]Reference 6 indicates that these costs should be reduced by $8.80 to account for gasoline savings resulting from tune-up, and by $7.31 to account for what owners normally spend voluntarily on major tune-ups. Since all other methods of maintenance and/or inspection have some uncertain dollar value (other than emission control) average, total, annual, out-of-pockets are reported for the years in which studies were done.

less, this method is being used in some countries and is under consideration in others, notably in the United Kingdom and the Netherlands. Switzerland has the widest experience where anti-noise brigades have been set up in Lausanne and other cities. Roadside detectors have also been used in California where six noise abatement teams have checked 600,000 vehicles and issued 2,000 citations in 1970.

Full-Scale Inspection

Several studies of the practicality and effectiveness of more representative types of emission inspection are underway. Some concern the measurement of carbon monoxide from idling engines, others concern tests that involve dynamometers and the measurement of HC and CO from loaded engines. Good results are reported from the United States with tests designed to produce diagnostic information of direct value in repairing defective parts of engines (see Table 27.6).

PRACTICALITY AND EFFECTIVENESS OF IN-USE VEHICLE CONTROLS

The practicality of large-scale emission testing and inspection programmes for vehicles in service rests on three main planks.

1. A test is needed that is cheap and quick, that can be applied to warmed-up cars, and that gets results that correlate well with those from a full driving cycle. Both the European and U.S. full tests are too time consuming and expensive to use for routine purposes. (The Swedish Government estimates the cost of an ECE-type test at 700 to 1,000 Kr. ($140–$200) per car.)

2. Test procedures that reveal what is wrong with cars that fail the inspection and that indicate what needs repairing are essential. So are simple manufacturer's maintenance instructions and manuals. Few if any mechanics today would know how to reduce excessive emissions.

3. Assuming that a short test capable of diagnosing troubles becomes available, it will be necessary to determine what reduction in emissions it is capable of bringing about, how much it would cost governments and motorists, and whether the resulting air quality improvements are worth the cost.

The potential emission reductions from inspection and maintenance cannot be determined precisely until more information is available about such factors as the rate at which emission control devices deteriorate in the interval between maintenance. According to best estimates such reductions are not expected to amount to more than 10 to 25 per cent, with a more likely figure in the lower range of this estimate.

28 TRANSPORTATION AND LAND USE CONTROL STRATEGIES

THE RATIONALE FOR TRANSPORTATION CONTROLS

Air pollution is present in metropolitan regions in two forms: in high but local concentrations in streets and near major highways, and as a more widespread phenomenon consisting of a mixture of vehicle exhaust and pollutants from domestic and industrial sources which, in certain circumstances, tends to build up into blankets covering an entire region.

In most cities airborne pollutants mix with the air and disperse rapidly and do not, as a rule, build up and hang over the whole city. The main problem is to reduce local concentrations arising mainly from car exhausts. Measures that reduce the use of cars in urban cores therefore offer considerable promise as a tool of air pollution control.

Noise is also present in cities in two forms: in concentrations along main highways where heavy vehicles and buses appear as small peaks amidst a high intensity background, and on other streets where all vehicles, be they cars or trucks, form higher peaks against lower levels of background noise.

Current trends are for noise in both these circumstances to get worse as the number of vehicles in use in metropolitan regions increases, as new highways are built and suburbs spread out beside them. Noise is thus a growing problem in suburban as well as central districts.

Furthermore, the tendency for heavy commercial vehicles, which emit up to 10 dBA, or up to twice the perceived sound intensity of the average car, to increase in number faster than all other lighter goods vehicles, creates problems that have no parallel in air pollution.[1]

Measures that reduce traffic flows in densely inhabited city centres and eliminate congestion generally therefore promise valuable relief to city dwellers burdened by noise. Additional relief is likely to flow from acts that are specifically aimed at diverting heavy vehicles from places thronged by people by day and from where they sleep at night.

A wide choice of measures can be employed to deal with both air pollution and noise. They can be roughly grouped into:

- traffic and parking restraints
- making more intensive use of cars
- staggering of work hours
- improving the flow of traffic
- design and location of highways
- control of urban development

Experience suggests, however, that none of these policies are likely to be pursued solely to abate noise and improve air quality. They will have to be justified on complementary grounds such as the relief of congestion, savings of time and money to commuters and, more widely, a wish to enhance the commercial and aesthetic appeal of inner cities.

TRAFFIC AND PARKING RESTRAINTS

Traffic Bans and Car-Free Zones

Total bans on vehicles have the potential of greatly improving air quality in the streets directly affected. In New York City the temporary closure of Madison Avenue to traffic in the Spring of 1971 was found to have resulted in a lowering of carbon monoxide concentrations from 22 to 7 parts per million

Table 28.1 Effects of Banning Vehicles on Air Pollution Levels in Four Shopping Districts in Tokyo

	CO (ppm)		
Sampling station	Before	After	Remarks
Ginza			
Okura Building	14.2	2.9	average of 5 hourly
Victor Building[a]	5.5	2.4	readings
Gas Hall	5.4	2.3	(1 P.M.–6 P.M.)
Shinjuku			
Koma Theatre	2.2	1.2	average of 8 hourly
Yamaichi Sec. Bldg.	9.8	2.3	readings
Electro-board	11.3	2.3	(11 A.M.–7 P.M.)
Ikebukuro			
Parco	9.5	3.5	average of 7 hourly
Seibu Dept. Store	6.7	3.0	readings
Sumitomo Bank	5.7	4.2	(12 noon–7 p.M.)
Asakusa			
Rokku	1.7	1.9	average of 9 hourly
Ward Office Branch	3.2	2.2	readings
			(10 A.M.–7 P.M.)

Before: July 26, 1970 (wind velocity: 3.9 miles per second)
After: August 2, 1970 (wind velocity: 3.3 miles per second)

Source: Traffic Division, Metropolitan Police Department, Tokyo, Japan (courtesy of Mr. Shinji Hishida, Chief, Traffic Regulations Section). Published in *Environmental Implications of Options in Urban Mobility*, OECD, 1973.
[a]Survey by automatic recorder; in all other locations, hourly sampling.

(ppm),[2] although the significance of this data cannot be fully evaluated because information about the manner in which the measurements were carried out is lacking. In Tokyo, however, detailed air quality measurements are available which show that exclusion of cars can produce substantial reductions in the levels of carbon monoxide. These are shown in table 28.1.

Further evidence of the effectiveness of car exclusion in reducing air pollution is available from the cities of Marseilles and Vienna.

During the period of October 7-27, 1971, Marseilles conducted a series of experiments in traffic restrictions of unprecedented scope. A total ban on parking was imposed in the central area of the city, covering some 24 hectares (about 0.25 km^2). Nine kilometers of exclusive bus lanes were added to the existing bus lanes in order to improve public transport services.

During the first phase of the experiment (October 7-16) all vehicles except taxis and buses were excluded from the central area. Air quality was monitored throughout the experiment, the results of which are shown in table 28.2.

In Vienna a vehicle-free zone was established in the inner city in November 1971. All traffic except buses and service vehicles are forbidden entry between 10:30 A.M. and 7 P.M. on weekdays. Deliveries are allowed between 7 A.M. and 10:30 A.M. Carbon monoxide concentrations were monitored continously at three locations: two within the exclusion area and a third in an adjoining street open to deliveries. The results from one station, located in the center of the exclusion area, are shown in table 28.3.

Traffic bans have also demonstrated an impressive capacity to reduce noise. The closing of Madison Avenue in New York City lowered levels from the range of 70-78 dBA to 64-66 dBA. ("Madison Mall," *op. cit.*, p. 12.) In Tokyo median noise levels during the traffic ban in the Ginza went down by 5-7 dBA. This is shown in table 28.4.

In Vienna measurements at ten locations showed that noise levels in the vehicle-free zone decreased roughly 3-6 dBA. In streets adjoining the zone noise levels increased somewhat during delivery hours, while at other times they have remained the same as before the creation of the traffic-free zone.

By now there is firm evidence, based on a large number of cases, that cars can be barred from limited areas without adversely affecting the retail trade within them. Increased con-

Table 28.2 Effects of Banning Cars on Air Pollution Levels in Central Marseilles

	CO (ppm)		
	Before[a]	After[b]	Remarks
Sampling station			
Banque Italienne	19.3	3.9	Average of 7 readings/day at each location (8 A.M.–6 P.M.)
Dames de France	19.4	2.8	
Magasin Général	17.5	3.8	
Belle Jardinière	18.9	4.0	
Mean value	18.8	3.6	
Time			
8 A.M.	20.2	5.5	Average of readings at four locations
10 A.M.	19.8	3.3	
12 noon	14.7	3.6	
2 P.M.	14.2	2.7	
4 P.M.	19.8	3.3	
5 P.M.	20.3	3.9	
6 P.M.	22.3	4.1	

Source: Association pour la Prévention de la pollution atmosphérique, Comité Marseille-Provence. Published in *Environmental Implications of Options in Urban Mobility*, OECD, 1973.
[a]Before: September 13–October 6, 1971. Total of 1,138 samples taken at 2-hour intervals.
[b]After: October 7–16, 1971 (ban on private cars; buses and taxis allowed). Total of 496 samples taken at 2-hour intervals.

gestion and pollution in adjoining areas has not become evident, though the creation of traffic-free areas is usually linked with some measures to increase parking capacity on the periphery of the area. The long-range consequences of traffic re restraints cannot be predicted with certainty; however, a study concludes on the basis of available evidence that they do not interfere with the healthy development of the inner city.[3]

Perhaps the most convincing evidence of the practicality and public acceptance of traffic-free zones is their recent growth, both in extent and in number. According to the study just cited, over 100 cities have banned traffic from portions of their central districts. Certain of them, such as Essen, Leeds, The Hague, Oldenburg and Munich have each extended an original nucleus of a few streets into large pedestrian areas.

The huge scale of the central areas of cities such as New York and London makes it necessary to adopt somewhat different approaches, and to combine the exclusion of private cars with the continued operation of public transport. Ambient

Table 28.3 Effects of Banning Cars on Air Pollution Levels in Central Vienna

Time	CO (ppm) Before	After	Percentage reduction
Arithmetic mean value			
24-hour reading	7.4	4.1	45
9 A.M.–7 P.M.	10.0	4.0	60
2 P.M.–6:30 P.M.	10.8	3.8	65
Maximum 24-hour value			
24-hour reading	11.3	7.8	31
9 A.M.–7 P.M.	15.9	6.2	61
2 P.M.–6:30 P.M.	16.2	3.8	77

Source: "Fussgängerzonenexperiment-Wien 1971," R. Baumann, K. Irlweck, J. Schedling, Umweltschutz (Bohmann Verlag K.G.) 9, (6) 127, 1972.
Note: All results are based on half-hourly average readings during weekdays. "Before" measurements were made during 10 days; "after" measurements, during 23 days. Sampling point was 2.6 meters from the ground.

pollution and noise levels are likely to decrease considerably in such transitways as a result of excluding private motor traffic, though peak noise levels would be attenuated only if special action were taken to modify public transit vehicles.

Partial Traffic Restrictions

In many cities banning all vehicles from large areas may prove to be politically or practically impossible. For this reason schemes that discourage rather than prohibit traffic are receiving increasing attention. The cities of Göteborg and Bremen have pioneered this approach. By dividing the city centre into quadrants and by erecting physical barriers between them, traffic across the centre has been made impossible except for trams and buses, which are given special privileges. Each quadrant is in effect a precinct with only local circulation allowed. All other traffic must use a ring road, leaving and entering each quadrant at designated locations. Göteborg's experience has shown that this traffic management scheme can reduce internal circulation by 17 per cent. Total number of accidents has been reduced by 7 per cent, although accidents involving personal injury have gone up slightly. Concentrations of carbon monoxide in the central business district are claimed to have been lowered in places from 30 ppm to less than 5 ppm; noise levels have been reported down from 75 dBA to 72 dBA.[4]

Table 28.4 Effects of Banning Cars on Noise Levels in the Ginza, Tokyo

Time	Median value (dBA)	90% range (dBA)
11 A.M.–12 noon[a]	75	68–83
1 P.M.–2 P.M.	69	67–72
2 P.M.– 3 P.M.	70	68–74
3 P.M.–4 P.M.	68	66–71
4 P.M.–5 P.M.	70	68–73
5 P.M.–6 P.M.	71	69–75

Source: Department of Noise Pollution, Environmental Pollution Research Institute, Tokyo Capital Government.

Selective Restrictions

The practice of applying selective restrictions on heavy trucks is growing. In most German cities with pedestrian precincts, goods vehicles are banned from making deliveries between 10 in the morning and 8 at night. In Hamburg all vehicles over 6 tons are banned from about 1 square kilometer of the central area between 10 A.M. and 6 P.M. on weekdays. Many towns require heavy through traffic to use by-passes or specified routes, and residential neighbourhoods in a number of cities exclude commercial vehicles altogether. As trucks and buses are amongst the noisiest classes of vehicles, their exclusion from a traffic flow can bring considerable noise abatement benefits.

Studies made in Sweden show the following reductions in noise levels as a result of eliminating heavy vehicles from the traffic stream:

Original percentage of trucks in the total traffic volume	20%	15%	10%	5%
Reduction of effective dBA level after prohibition of trucks	5%	4%	3%	1%

Another insight into the benefits of thinning out traffic can be obtained from looking at the reductions in noise on a traffic route when a by-pass for it is opened. Results of before-and-after studies made by the Greater London Council are shown in table 28.5.

Parking Prohibitions, Restrictions and Charges

Parking restrictions on private cars are the most commonly practised way of limiting traffic and promoting its flow. The effects of parking restraints on levels of congestion and therefore on air quality and noise have not been widely explored. Some evidence comes from the Marseilles experiments referred

Table 28.5 Noise Levels on a Traffic Route Before and After a By-Pass Is Opened (dBA)

	Before	After
Day	80	73
Evening	75	66
Night	69	57

Source: "Noise from Urban Motorways," G. H. Vulkan, Greater London Council Scientific Branch, 1971, p. 224.

to above. During the second phase of that experiment (October 17–27, 1971) a total ban on parking in the core of the city was imposed, although traffic was allowed to move freely within and through it. Carbon monoxide concentrations went down by 10 to 60 per cent, depending on the time of day; the average level, based on 502 samples at four locations, decreased by 40 per cent. This is shown in table 28.6.

The total elimination of parking, like total bans on traffic, is feasible only in limited areas, though municipalities might be prepared to extend them wider and wider as experience with parking prohibitions increases. On the scale of city centres, however, it is necessary to use parking charges or parking restrictions to reduce the use of cars. Restrictions and variable charges that discouraged parking in areas and at times prone to high pollution concentrations might be particularly appropriate.

The use of parking measures as an instrument of traffic restraint and air pollution and noise control is subject to three limitations. Through traffic, which in many cities constitutes a relatively large proportion of total traffic even during peak hours, will not be deterred by restrictions on parking, and may indeed increase as a result of them. Moreover, parking restrictions may cause some commuters to be driven to work by other family members and dropped off, with the reverse sequence repeated at the end of the day. Parking restrictions, therefore, could conceivably result in doubling the use of some cars. The sharp rise in the level of traffic at peak hours on rainy days is often cited in support of this phenomenon.

Finally, parking charges in downtown areas might have little deterrent effect on the regular commuter: the cost of parking is likely to be assumed by his employer who will view it as an increased cost of doing business in the city centre.

Pricing Policies

The experience of many cities suggests that the closing of streets and parking controls are inadequate by themselves to achieve major reductions in vehicular noise and air pollution.

Table 28.6 Reductions in Air Pollution through a Ban on Parking in Central Marseilles

	CO (ppm)		
	Before[a]	After[b]	Remarks
Sampling station			
Banque Italienne	19.3	12.9	Average of 7 readings/day at each location (8 A.M.–6 P.M.)
Dames de France	19.4	10.9	
Magasin Général	17.5	10.4	
Belle Jardinière	18.9	12.1	
Mean Value	18.8	11.6	
Time			
8 A.M.	20.2	18.9	Average of readings at four locations
10 A.M.	19.8	15.7	
12 noon	14.7	9.7	
2 P.M.	14.2	9.4	
4 P.M.	19.8	12.8	
5 P.M.	20.3	7.4	
6 P.M.	22.3	7.0	

Source: Association pour la prévention de la pollution atmosphérique, Comité Marseille-Provence.
[a]Before: September 13–October 6, 1971. Total of 1,138 samples taken at 2-hour intervals.
[b]October 17–27 (total ban on parking; no ban on traffic). Total of 502 samples taken at 2-hour intervals.

Economic measures such as congestion tolls, permits to drive in certain areas and discriminatory taxes on fuel have therefore been suggested in some quarters to back them up. The objective of all such economic measures is to discourage drivers from entering congested central districts without banning their freedom to drive outright.

Congestion tolls. Theoretically, the most effective method of restraining the use of cars, short of an outright ban, is by means of road pricing in one form or another, that is using a system whereby the car owner is charged a licence fee according to the mileage driven in a particular area. In an ideal system such charges would reflect the time as well as the place of driving and would be higher during peak hours when the social cost of driving is the highest. There are, however, a number of technical, administrative and political problems in instituting such a system.

Variable road-user charges of this kind presuppose the continuous monitoring of all cars in a designated area. Systems designed to do so are being worked on, but they are not expected to be ready for widespread use before the end of this decade.[5] Even if the necessary technology becomes available, large costs and practical difficulties are likely to be encountered in the administration of the system and in collection of the charges. In many countries, moreover, it is doubtful whether the public would accept direct economic restraints on driving. The existing freedom of the motorist to come and go at will, means that anything which inhibits this ability and which involves a tax to boot, is bound to be viewed with dislike. Only if a political constituency can be found which would have something to gain from road pricing, does it seem likely that such charges will be introduced.

An alternative would be to assess motorists on the basis of the mileage they travel, but in such simplified form it is doubtful whether the costs of establishing and collecting the charges would justify the theoretical benefits. It is also questionable whether the necessary odometers could be made sufficiently tamper-proof to prevent widespread tax evasion.

Entry permits for high-pollution areas. A more straightforward approach would be to charge motorists fees based on the places where they use their cars. A series of concentric control zones, delineated according to the traffic dnesity or air quality prevailing within them, would have to be established. In a metropolitan area the central business district might form an inner zone, the inner suburbs could form an intermediate one and the outer suburbs a third. Vehicle owners would apply for distinctive stickers licencing them to drive and park in a specific zone. The cost of a sticker for inner city driving would be higher than one for the outer suburbs. A licence permitting the use of a vehicle in a higher pollution area would, of course, permit it to be driven in zones of less stringent control, but the reverse would not apply. Any vehicle found without the appropriate sticker would be fined or towed away.

Permits of this kind would discourage the use of cars in pollution-prone areas without banning them outright. They are administratively feasible, although the cost of enforcement might run high. A drawback is their failure to discriminate between heavy and light polluters—in other words, between cars used intensively or only occasionally in any zone.

Taxes on Fuel

Suggestions have also been made that taxes on fuel could be used to discourage the use of cars and so promote clean air.

This would involve calculating the reduction in mileage necessary to achieve desired air quality objectives and then forecasting the level of taxation necessary to bring it about. It seems extremely unlikely that such calculations would be of more than academic interest. Even if they were, the tax would be evaded unless it was applied over a very wide area, in which case it would be inefficient, inequitable or, more likely, both.

Effectiveness of Pricing Mechanisms

The underlying theory upon which pricing schemes are based is rooted in classical economic theory. According to this theory the individual's decision to use the auto for a given trip will be influenced by the cost of driving relative to the cost of using an alternative means of transport or of not making the trip at all. Increasing the cost of driving (or parking), so goes the argument, will cause some people to switch to public transport and others to forego non-essential trips altogether. In practice, however, comfort, convenience and reliability are likely to weigh much more heavily in most people's judgement than the added expense of driving—particularly when this cost is assumed, as it is likely to be for many people, by the center city employers and merchants bent on retaining their employees and customers. Even if an increase in the cost of driving did cause some people to shift to buses and trains, the resulting reduction in congestion may well attract other auto users, thus bringing traffic volume quickly up to its original level.

LONG-TERM EFFECTS OF LIMITING THE USE OF CARS IN CITY CENTRES

Fears that limiting the use of cars in city centres will lead to their dispersal of activities to the suburbs are widely held and have deterred many municipalities from taking firm action to reduce traffic. Available evidence, however, tends to allay these fears. In 1960 the city of Bremen imposed severe restrictions upon car movement within the historical central area by dividing the district into four precincts and forbidding traffic between them, much as Göteborg has done in more recent days. Part of each precinct was turned into a pedestrian area.

In 1970, a full decade after the introduction of the scheme, the city centre remained Bremen's principal focus of activity and attraction. Business and commerce have also prospered. Whether this has happened because or in spite of the traffic measures is hard to assess. The failure of severe restrictions on cars over ten years to damage the center city's development or to promote the dispersal of activities into the suburbs is nevertheless notable.

Table 28.7 Occupancy Rates of Cars in London

Trip purpose	Average occupancy
Work	1.16
Shopping	1.54
Social	1.62
Entertainment	1.82

Source: Greater London Development Plan Inquiry Proof E12/1, November 1970, p. 44.

Even if some changes in land use patterns were to occur as a result of restricting the use of cars, it does not necessarily follow that such shifts would be undesirable. The activities most affected by such restrictions would be those of large commercial firms whose relocation to more outlying places in the metropolitan area could be beneficial.

For many employees this could lead to shorter journeys to work and to a reduction in commuting. A new life could open up for the inner cities under these conditions. It could be based on a mixture of entertainment, education, shopping and specialized commercial activities.

MAKING MORE INTENSIVE USE OF CARS

The sharing of cars is another way by which congestion, noise and air pollution might be reduced. The scope for sharing is indicated by occupancy rates in London. These tend to vary only slightly from city to city.

These figures make it clear that the greatest potential for sharing lies among people going to and from work.

Incentives for Car Sharing

Riders who are not close neighbours at home and at work cannot be expected to share cars. The failure of a campaign to promote voluntary car pools in Los Angeles in October 1971 helped to make this point. Despite considerable publicity and efforts by more than 100 companies to organize computerized carsharing pools, the effort failed, probably because of the highly dispersed pattern of origins and destinations of the commuters in the Los Angeles area.

Even where common travel routes make car sharing feasible, more tangible incentives than the desire to reduce congestion and pollution are likely to be necessary to stimulate commuters to travel together. Where toll bridges or tunnels exist, pools can be encouraged by giving fully occupied vehicles a faster or cheaper passage in peak hours. A scheme of this kind was introduced on the San Francisco-Oakland Bay Bridge in California in December 1971, though such techniques, like improvements to traffic flow, run the risk of attracting additional vehicles and being self-defeating from the standpoint of air quality control.

Where bus lanes have been created on main highways, any space not used by buses can be offered to car pools, though this would presumably necessitate some form of enforcement. The use of bus lanes by taxis in several cities shows this principle already in use.

In Washington, D.C. certain government agencies give preference to car pools in according space in their basement garages. In cities where municipal garages or parking lots exist, cars full of passengers could be charged lower fees.

STAGGERING WORK HOURS

Another device for reducing concentrations of air pollution is the staggering of work hours. It would be effective in two ways. To start with it would thin out traffic in the height of the rush hours, thus smoothing out flows. This would help reduce the total volume of pollutants emitted. On top of this it would spread them over a longer period. Staggering the rush to work would be particularly effective in cities where the wind is slack in the morning and the atmosphere generally stable—conditions that inhibit the dispersal of pollutants.

Experiments in the United States and in Germany indicate that work staggering is feasible. In April 1970 about 50,000 persons employed by 45 firms in lower Manhattan adopted staggered work hours, shifting from the traditional 9 A.M. to 5 P.M. schedule principally to a new 8:30 A.M. to 4:30 P.M. one. By April 1971 the number of employees involved had grown to about 60,000. Public response to the scheme was extremely favourable.

In Germany some 2000 firms have taken the same course. They include Messerschmitt-Bolkow-Blohm with 7000 employees in Munich, Lufthansa in Cologne and the Federal Ministry of Transport in Bonn. Participating firms are open for business from 7 A.M. to 7 P.M., but employees are free to pick any time they like between 7 A.M. and 9:30 A.M. to start work and any time after 3:30 P.M. to go home, so long as they work a 42-hour, 5-day week. The response of employees to "Gleitende Arbeitszeit", as staggered hours are called in Germany, has been overwhelmingly favourable.

Table 28.8 Percentage of Pollutants Emitted per Mile for Different Conditions of Engine Operation

	Gross hydrocarbons (%)	Carbon monoxide (%)	Oxides of nitrogen (%)
Idle	5.9	7.5	0.03
Cruise	14.1	14.3	21.4
Acceleration	56.2	62.2	78.5
Deceleration	23.7	16.1	0.17

Source: "Variables Affecting Traffic and Vehicular Operating Conditions in Urban Areas," *Air Quality Standards*, 1970.

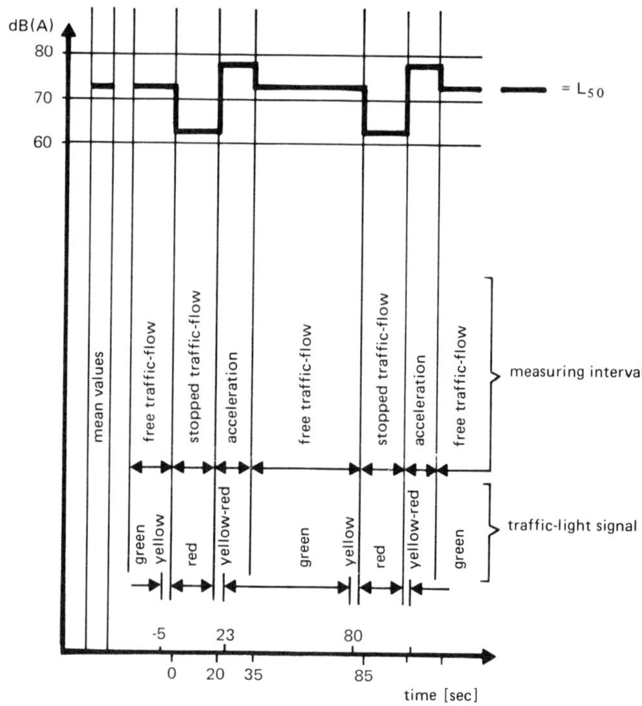

28.1 Effects on noise of stopping traffic at a signal. (Source: Prepared by the Federal German Institute for Roads and Motorways, Cologne (BAST). Published in *Environmental Implications of Options in Urban Mobility*, OECD, 1973.)

28.2 Vehicular emissions vs. speed (pounds per vehicle mile). (Source: Second Report of Secretary of Health, Education and Welfare to U.S. Congress Pursuant to Public Law 88-206 Clean Air Act, 2/19/65, Table 1. Published in *Environmental Implications of Options in Urban Mobility*, OECD, 1973.)

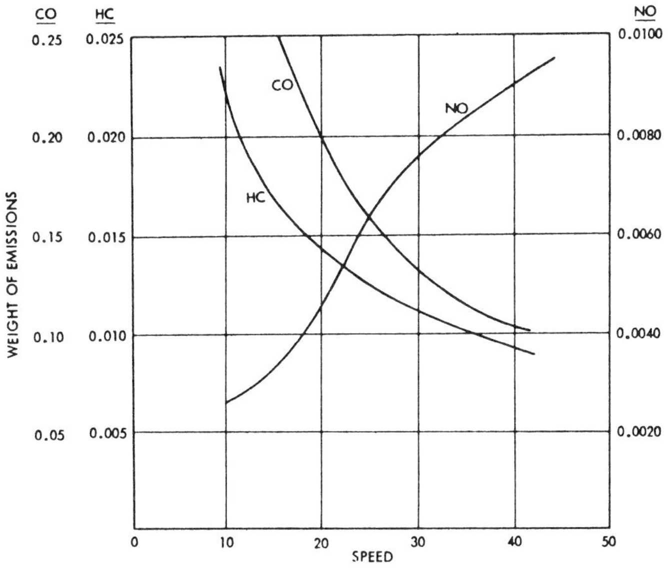

Four-Day Working Weeks

Four-day, forty-hour working weeks go further towards reducing air pollution than staggering hours because they cut weekly commuting trips by a fifth and shift the people involved in them away from a conventional travelling pattern. They thus combine the advantages of work staggering with an additional benefit. On the assumption that four-day weeks would be spread out over six days from Monday to Saturday, with different groups of people working different sets of four days, the number of daily commuting journeys could be cut by as much as a third.

The profound social implications of a large shift to a four-day working week preclude the possibility of its being introduced merely to relieve road congestion and to combat air pollution. Nevertheless the environmental benefits of such a policy should be borne in mind when it is being evaluated.

IMPROVING THE FLOW OF TRAFFIC

Any measure that smooths the flow of traffic by reducing stop-and-go conditions will tend to reduce vehicle exhaust emissions and noise. This is because the rates of emissions of the three main vehicular pollutants and of noise are at their greatest during acceleration. This is shown on table 28.8 and figure 28.1.

Reductions in emissions may also be obtained by increasing traffic speeds. Carbon monoxide and hydrocarbon emissions decline with rising vehicle speeds, at least up to about 60 km/

hour, though nitrogen oxides increase. This relationship is illustrated in figure 28.2.

There is, however, a serious possibility that any reductions in emissions and noise obtained by such means will be short lived. Additional capacity created by improvements in traffic flow will attract additional cars which would offset the gains in speed and add to the traffic volume. Where such improvements are obtained by building additional highways, they will also encourage people to drive further to work and could actually increase emissions of air pollutants. It would therefore appear that most techniques for improving traffic flow, unaccompanied by restraints of one kind or another on car usage, offer little prospect of relief from urban air pollution and noise.

REDUCING EXPOSURE TO VEHICLE EMISSIONS THROUGH THE DESIGN AND LOCATION OF HIGHWAYS AND BUILDINGS

The design and location of highways and the relationship of buildings to them offers another range of possibilities for reducing the exposure of people to vehicle emissions and noise. Roads may be built through industrial areas or along existing railways. They may be sunk in cuttings or even tunnelled to minimize their impact on places where people live and work.

In addition to this, windows may be double-glazed and sealed, and dwellings may be rearranged so as to reduce exposure of bedrooms and living rooms to busy roads. Such measures are likely to be more effective in combatting noise than air pollution. Levels of carbon monoxide inside of downtown office buildings adjacent to heavily travelled roadways tend to be high. One study found that the difference between indoor and outdoor concentrations in New York buildings seldom exceeded 4 parts per million.[6]

LAND USE, URBAN DEVELOPMENT AND TRANSPORT POLICIES

Levels of pollution and noise could also be lowered by reducing reliance on cars for moving about in cities. This could be done in two ways: patterns of land use that minimise distances between home and work, home and school and home and shops could be encouraged; in addition, development around existing railway and bus lines could be promoted and such networks extended in order to place public transport within the reach of a greater proportion of metropolitan residents.

Changes in the structure of the cities and in the provision of public transport cannot, however, be engineered except over long periods. The social and economic consequences of different city forms and of new transport systems are also profound. Justification for a particular construction of them is therefore unlikely to be found in the abatement of air pollution alone. Only when a wide range of urban objectives are taken into account will it be possible to find preferred solutions.

29 LIMITING THE USE OF LEAD IN MOTOR FUEL

THE CASE FOR LIMITING THE USE OF LEAD IN MOTOR FUEL

A growing number of countries have decided to limit the content of lead in motor fuel. Their plans range from modest reductions to near-total elimination of lead additives; in some cases they have proposed to introduce one grade of lead-free low octane fuel.

Two factors have prompted Member countries to limit the lead content of fuel: firstly, the effect of airborne lead as a pollutant and its impact on public health; secondly, the adverse effect of lead on the performance of catalytic converters. There is disagreement whether present levels of airborne lead constitute a danger to public health. However, many governments believe it is desirable that present levels of airborne lead should not be exceeded and should, if possible, be reduced. On the other hand, there is evidence that lead, even in minute quantities, can poison some catalysts. It is not clear at present, however, whether catalytic devices will be required for the conditions encountered in Europe and many other areas of the world. This will depend on the policies adopted in the future for vehicle exhaust emissions.

Arguments for the elimination of lead have also been made on the grounds that lead adversely affects engine performance. There are conflicting claims as to whether or not lead additives contribute to the build-up of deposits in engine combustion chambers and result in higher hydrocarbon emissions.[1] It is generally agreed, however, that lead can have an adverse effect on the life of exhaust systems, spark plugs and crankcase oil. In any case, however, none of these effects are the determining factors in the decision to eliminate lead from fuel.

A number of factors, essentially of an economic nature, have been advanced against major reductions of the lead content of fuel. The magnitude of the costs involved would depend on the extent of reduction and whether the removal of lead is accomplished with or without lowering octane levels. Some countries believe that problems may be encountered in maintaining the overall quality of fuel required by existing vehicles, but there is no unanimity on this point.

RATIONALE FOR LEAD ADDITIVES

The purpose of lead additives is to raise the octane rating of motor fuel. It is the cheapest and easiest method of doing this. Lead also provides the refiner with great flexibility in maintaining octane levels if, for some reason, alteration in fuel composition is made necessary at some stage in the refining process.

Octane number is a measure of the resistance of a fuel to a particular uncontrolled combustion phenomenon, known as detonation or "knocking," which can result in considerable power loss, severe engine damage and increased nitrogen oxide emissions. Knock can occur at either high speed or low speed depending on the octane number of the different components used in the fuel. The total fuel blend must be designed to avoid both of these conditions. There are several ways to express the octane rating. Research Octane Number (RON) is the most widely known index and is determined in a single cylinder (CFR) engine under special laboratory conditions. The Motor Octane Number (MON) is determined in the same type of engine but under more rigorous conditions. The Road Octane Number, determined from actual road trials, reflects the actual performance of a fuel better than the other two indices. In general, the Road Octane Number falls between the RON and MON. Practically, all emphasis is placed on the RON at present despite the importance of all three parameters.

The octane number necessary for good combustion depends primarily on the engine's compression ratio. Generally, the higher the compression ratio, the higher the octane requirement, but the greater the efficiency and therefore the better the fuel economy. Considerably different engine designs at a given compression ratio can require fuels with different octane ratings. The octane number depends not only upon the compression ratio but also upon combustion chamber geometry, valve configuration, ignition timing, mixture strength, cylinder head temperature, coolant temperature, type of coolant, humidity of the air, the altitude at which the engine is working and the restrictiveness of its exhaust system. Differences in detonation susceptibility can also occur in apparently similar engines due to manufacturing tolerances and different rates of deterioration in engine components.

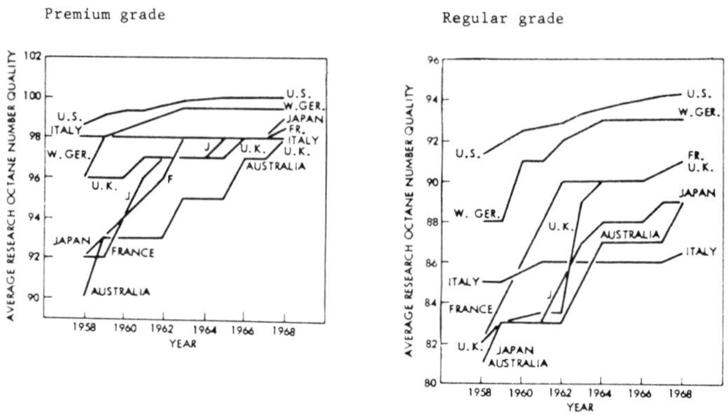

29.1 Trends in research octane number quality, 1958-1968. (Source: "Fuel Requirements of Passenger Cars Throughout the World," F. B. Fitch and R. H. Thena, Society of Automotive Engineers, New York, January 1969.)

The average octane requirements of new cars have risen considerably since 1958, but in the last few years octane requirements have remained relatively constant. During the period 1958-1968 premium grade fuel has increased up to eight octane numbers. A similar trend has occurred in regular grade fuel. This can be seen in figure 29.1. Today, premium grade fuel stands at fairly uniform 98 to 100 Research Octane Number (RON) in most countries with a lead content ranging from 0.1 to 0.9 grams per litre. Regular grade fuel varies more widely in octane rating, with a low of 85 (Spain, Portugal) and a high of 96 (United Kingdom), and a lead content from zero to 0.8 grams per litre.

Table 29.1 shows one estimate of the relationship between RON requirements and satisfactory driving performance. Table 29.2 gives the percentages of premium fuel sold in several nations. It is suspected that there is some wasteful overbuying of premium fuel—partly voluntary and partly forced because of the octane ratings of present fuels. In the United Kingdom for example some 40 per cent of all cars in 1971 needed premium fuel, but 1971 sales of premium fuel represented as much as 62.4 per cent of the total. This is not due to nonavailability of lower grade fuels, because four grades are available. Some of this apparent "overbuying" can be attributed to the use of premium grade fuel in blending pumps which allow delivery of the correct grade of fuel to meet particular engine requirements. The use of a higher grade fuel than the engine requires does not lead to better performance but increases the emission of lead into the air, as lead content is generally higher with increased octane levels. Less emphasis on "high octane" in fuel advertisements might reduce the voluntary overbuying. The introduction of a third grade of fuel, the introduction of

Table 29.1 Relationship between RON Requirement and Satisfactory Driving Performance for Different Compression Ratios

	Percentage of cars satisfied[a]			
RON	10:1 CR	9:1 CR	8:1 CR	Avg. 1970 cars
90		25	68	32
91		30	75	45
92	10	45	80	59
93	20	55	90	65
94	30	60	93	70
95	40	80	97	83
96	50	90	98	90
97	70	92	99	93
98	75	95	100	95
99	85	98	100	97
100	90	100	100	99
101	95	100	100	100

Source: Derived from data provided by individual car manufacturers, l'Argus de l'Automobile, July 1971 and "An Economic Analysis of Proposed Schedules for Removal of Lead Additives from Gasoline," Bonner and Moore, June 1971.

Note: Average octane satisfaction has been derived using the following assumptions as to distribution of new vehicles on the basis of their engine compression ratio: 37% of 1971 cars have a compression ratio of 7.5–8.5 (avg. 8:1), 49% have a ratio of 8.5–9.5 (avg. 9:1), and 6% have a ratio of 9.5–10.5 (avg. 10:1). These assumptions are based on incomplete figures obtained from individual manufacturers.

[a]"Satisfied" means operating without developing a knock perceptible to a trained driver. General consumer satisfaction might be somewhat higher.

blending pumps, or a slight increase in the octane-number of low octane fuel in some countries would enable a greater proportion of the motorists to dispense with high octane fuel.

THE IMPLICATIONS OF REDUCING LEAD CONTENT OF FUEL

A modest reduction in octane ratings would be unlikely to interfere with the satisfactory operation of a high proportion of cars currently on the road, and would make it possible for countries which at present permit a high level of lead in gasoline to lower their permitted levels by a modest amount. More drastic reductions in lead would require changes that might take longer to implement. Such major reductions would require either (a) the modification of car engines to operate on low lead, low-octane fuel; or (b) the modification of fuels so that they could retain their present octane ratings. Each alternative has its own set of consequences.

Table 29.2 Consumption of Premium Gasoline as a Percentage of Total 1971 Gasoline Consumption

	Premium as percentage of total
Austria	64.8
Belgium and Luxemburg	87.0
Denmark	57.0
Finland	55.6
France	74.1
Germany	52.9
Italy	78.2
Netherlands	83.6
Norway	52.9
Portugal	67.7
Spain	64.7
Sweden	64.5
Switzerland	82.8
United Kingdom	62.4
Canada	38.0
United States	39.0
Japan	16.0
Australia	83.0

Source: World-Wide Survey of Motor Gasoline Quality, The Associated Octel Company, Ltd., May 1972.

THE CONSEQUENCES OF LOWERING ENGINE COMPRESSION RATIOS

Within limits the thermal efficiency of an engine increases with compression ratio. For a given engine design capacity, the higher the compression ratio, the greater the power output, or alternatively, the lower the fuel consumption for a specific horsepower output. Conversely, lowering the compression ratio reduces the power output or raises the fuel consumption. To compensate for the power loss one must increase the size of the engine and its fuel consumption. One effect of this alternative would be some increase in demand for crude oil.

There has been a continuing world-wide trend from 1960–1970 towards higher compression ratios, although since 1971 this trend has been reversed in the United States. The small high performance engine continues to be widely used in European cars because of the demand for smaller cars and the need for fuel economy. In high horsepower cars a reduction in the compression ratio can be traded off against some of the excess power which they now possess. But for the lower powered cars, which do not have power to spare, a return to lower compression ratios might mean increases in the size (and weight) of engines.[2] This, in turn, would mean larger and heavier cars and, more importantly, a greater specific fuel consumption, leading to higher lead, HC and CO emission rates per vehicle/km if not otherwise controlled.[3] The overall result might be less of a decrease in lead emissions than might have been expected.

In countries where the price of fuel is already high, a substantial loss in fuel economy would be unpopular (even though the demand for fuel has been shown to be relatively inelastic in the face of price changes in highly developed countries). Car manufacturers who lowered compression ratios of their cars voluntarily to permit operation on lead-free fuel might find themselves at a disadvantage vis-a-vis those who did not take such a step, thus making it unlikely that any individual manufacturer would do this in the absence of specific incentives or regulations. The penalties to the consumer in terms of increased fuel consumption and the longer run consequences for the shape and design of the car make the lowering of engine compression ratios an unacceptable strategy for Europe.

Moreover, an engineering change in new cars would not allow an immediate reduction in fuel octane, as the needs of many older cars must be satisfied with fuels of current octane levels. And, as long as there exists a choice between a leaded high-octane fuel and a low-lead low-octane fuel, there is no guarantee that owners of new cars would choose the low-leaded fuel unless the price differential were large. This choice will not be available in the United States where design guidelines for filler system components will prevent owners of new cars from buying leaded fuels.

THE CONSEQUENCES OF MODIFYING FUEL

Addition of lead is the cheapest and perhaps the easiest, but not the only way, of raising octane numbers. Substantially similar results are technically achieveable through the blending of higher octane quality components. Some of the necessary blending components, however, are not currently available in all countries.

Because of limited refinery capacity, it would be impossible to produce sufficient quantities of high octane blending components to eliminate or drastically reduce lead in motor fuel *immediately*, and still maintain present octane levels of 98–100 RON in premium grade fuel. At nearly all refineries new facilities, perhaps including isomerization or alkylation units, would have to be constructed to produce high-octane fractions. The demands on the refinery construction industry limit the speed at which such facilities can be brought into operation. The speed at which high octane fractions could be made available also depends upon the current refinery set-up, the type of crude oil imported, and other market demands upon refineries.

A grade of lead-free fuel could be placed on the market now by diverting high octane components from premium and regular blends to a lead-free blend. But unless this were part of a much wider programme of conversion, the net effect would be to increase total lead emissions from car exhausts. This is because extra lead would have to be added to the other blends to compensate for the shifting of the high octane petroleum fractions to the lead-free blend. The more lead is added, the less effective it is in raising octane. Thus, the net effect would be to add more lead to other blends than has been saved on the lead-free fuel. However, any large scale shift to lead-free fuel would result in refining industry process changes which would ultimately lead to lower overall lead consumption in all grades of fuel marketed.

In Europe it seems likely that the *total* emissions of lead can be reduced only by longer term changes that raise the overall average octane level of the fuel coming from the refineries. The time and the huge sums required to construct new refining facilities mean that this must be a gradual process.

Cost of Additional Refining Capacity

The cost of the refinery additions needed to produce higher octane blending components will vary widely from country to country depending on the capacity and technical makeup of existing plant, the amount of unused capacity for distilling high octane components and the degree of lead removal desired. In those plants where new units need to be built, the cost of producing lead-free fuel will increase exponentially with the octane index of the fuel.[4]

Impact on Small Refiners

The economic burden of making refinery changes would fall most heavily on small refiners who would be required to make much higher than average investment per ton of present capacity. Small refiners have operated at a cost disadvantage for many years. The trend of increasing gasoline octane has accentuated this disadvantage, and the production of lead-free fuels would further increase this burden. According to United States data, the added cost in cents/gallon of producing lead-free gasoline would be 25–30 per cent higher for a refinery with an output of 30,000 barrels/day than for a 100,000 barrels/day refinery.[5] Consequently, some small refiners might not be able to continue to compete in the production of unleaded fuels without buying isoparaffins from larger producers.

Cost of Distribution and Marketing

The distribution and marketing sector of the oil industry would be affected, particularly if a new, unleaded grade of gasoline were to be added to the existing two-grade structure. The U.S. Environmental Protection Agency, on the basis of data reports for service stations which have converted for commercial reasons, estimates that the cost of conversion to a three-pump system ranges from $2,780 to $4,200. Other sources have estimated the cost to be $8,000.[6] Substantial extra costs would also be involved in segregating leaded from unleaded gasoline at all stages of manufacture and distribution and in preventing the delivery of leaded fuel to vehicles requiring lead-free gasoline.

Effects on Fuel Price

The added cost of refining, marketing and distributing unleaded fuel would have a direct impact on its price at the pump. Studies in the United States show wide variations in the estimates of fuel price increases. One computation ranges from 0.2 to 0.9 cents per U.S. gallon (0.06 to 0.2 cents/litre), depending on how soon industry would have to provide the unleaded fuel and whether it would have to be for all grades.[7] The schedule of lead reduction proposed by the Environmental Protection Agency leads to price increases of about 0.25 cents per gallon.[8] This would represent an increase in the purchase cost of motor fuel of about 1 per cent. Additional calculations for a range of lead reduction schedules estimated the total increase in pump price at 1.5 to 2 cents/U.S. gallon.[9] Assuming a fuel economy penalty of 12 per cent, the overall cost to the consumer might amount to 5.5 to 7 cents/U.S. gallon.

In Germany, a reduction in lead content down to 0.15 g/litre has been estimated to result in a price increase of fuel from 1 to 2 D pf/litre. Swedish calculations show a 2 per cent rise in price due to a total removal of lead from fuel. U.K. estimates are 0.3 per cent for 0.64 g/litre, 1 per cent for 0.5 g/litre and 3 per cent for 0.15 g/litre. These estimates, however, do not have a common basis—e.g. unleaded or equal RON fuel. For the Common Market area as a whole, CONCAWE estimates have put the increase at 2 cents (U.S.)/litre.

Savings in vehicle maintenance costs accompanying the use of lead-free fuel would, at least partially, offset increased fuel costs. Such savings, principally due to longer spark plug life, less oil contamination and greater exhaust system life, have been estimated at 2 cents (U.S.)/gallon.[10]

Effects on Motor Vehicle Emissions

Concern has been voiced that the limitation of lead additives could lead to large increases in the use of aromatics as a research octane booster, and thereby cause possible increases in the emissions of photochemically reactive and polynuclear aromatic (carcinogenic) hydrocarbons and of phenols, and adversely affect engine performance because of inadequate boost in motor octane number.

The extent to which an increase in aromatic content of fuel will be necessary would depend on the refining process chosen, the desired octane quality, and the degree of limitation of lead usage. There are many processes available for the production of high octane quality components, some of which do not produce significant quantities of aromatics. However, their economic application depends on the refining/product pattern of the area under consideration. A fuel which has its lead content replaced with aromatics to give it the same research octane number (RON) will not necessarily continue to provide the same resistance to detonation under normal driving conditions. For this reason, entire reliance should not be placed solely on the RON rating method, but should be used in conjunction with the alternative motor method of octane rating. Estimates of the aromatic increase needed to compensate for the reduction in lead content to 0.15 g/litre if fuel is to maintain octane rating of 98 RON vary from 3–5 percentage points (CONCAWE), to 3–9 percentage points (*Lead Content of Gaolines*, Technical Committee on the Use of Petroleum Products, Paris, June 1972), to 8–10 percentage points (Report of a working group of the BTC Fuel Commission, U.K.).[11] If already high aromatic levels were to increase by 8–10 percentage points, the resulting fuel would probably be unsatisfactory since the motor octane number would be low and this could lead to engine failure at high outputs.

With current motors there appears to be a practical limit on the amount of aromatics that can be put in fuels. Beyond a certain point high compression engines, under conditions of

Table 29.3 Aromatic Content of Motor Fuel Produced in Selected Countries in 1970–1971

	Percentage by volume	
	Premium	Regular
France	23–39	5–33
Germany	40–45	22–30
Italy	21–39	14–21
Netherlands	32–42	21–33
Scandinavia	25–46	19–36
Switzerland	36–39	25–28
United Kingdom	32–44	19–41
United States	20–25	20–25

Source: Octel Group Gasoline Survey of Europe, October 1971.

long periods of high-speed, high-load driving, have shown signs of "high-speed knock." (In general, however, elevated olefin levels are more important in avoiding this phenomenon.) Current aromatic contents of fuel produced in selected countries are shown in table 29.3.

Higher aromatic fuel content may not necessarily lead to increased reactive hydrocarbon emissions. Higher aromatic levels are usually accompanied by lower olefin levels which tend to cancel out reactive hydrocarbon effects. Moreover, HC emission controls will reduce the significance of increased aromatic content. For automobiles with no emission controls, however, results from American tests indicate that an increase in fuel aromatics will lead to increased polynuclear aromatic hydrocarbon and phenol emissions, and to a decrease in olefin emissions. Unless an extremely rigid fuel lead reduction schedule were to be adopted, the concern over aromatic increases is probably not justified.

Effects on Automobile Performance

The German Government has indicated that in order to maintain satisfactory automobile performance reducing the lead content of fuel to 0.4 g/litre has necessitated either an increase of 0.5–1.0 in the research octane number of "clear" gasoline or a retardation of 1–1.5 crankshaft degrees in the ignition timing. They believe the further reduction of lead content to 0.15 g/litre will necessitate a 2.0–2.5 increase in the research octane number or a 3–5 crankshaft degree retardation of the ignition timing. In the latter case, there could be a 2–3 per cent drop in fuel economy.

On the other hand, the British Technical Council of the Motor and Petroleum Industries is concerned that fuel with low or no lead will adversely affect the operation of current automobiles in addition to the reduction of RON that is to be expected. This is on account of the phenomenon of high speed knock. A measure of high speed knock tendency is the Motor Octane Number (MON) but it has also been demonstrated that as lead content is reduced the propensity for a gasoline to knock in the high engine speed range increases beyond that indicated by MON, i.e., a "lead bonus" exists in relation to high speed knock. Compared with the majority of premium fuels available in Europe at present, the BTC believes that fuels with lead content reduced to 0.15 g/litre could have an overall penalty of 4 to 5 MON in terms of road performance. In existing engines this would be equivalent to a spark retard approaching 10° crankshaft angle. In many cars a retardation of this magnitude could result in a 5 per cent increase in fuel comsumption as well as penalising the acceleration response. BTC further believes that the loss in cooling efficiency could also cause overheating and so defeat the objective of compensating for lower octane number. To allow for such quality reductions in new engines a compression ratio reduction approaching 1 ratio would, in its opinion, be necessary, resulting in approximately 5 per cent increase in fuel consumption. To re-establish the performance level obtained in existing engines would require an increase in engine size and a further increase in fuel consumption.

Lead helps to prevent recession of valve seats caused by the abrasive action of iron oxide deposits. Motor manufacturers fear that complete absence of lead in gasoline may lead to valve damage, though very low concentrations of lead (of the order of 0.10 g/litre) will prevent this trouble. The valve recession problem can be avoided in *new* cars by using valve inserts made of more resistant alloys.[12]

Prevailing opinion holds that if catalytic converters are installed to control motor vehicle emissions, then unleaded fuel must be available for them. Even a small quantity of leaded fuel can destroy the efficacy of these devices, and their replacement would be costly. Research on the development of a catalytic converter which operates satisfactorily in the presence of lead is continuing. An oxidizing catalyst which operates satisfactorily in the laboratory with 0.10–1.15 g Pb/1 has been reported, but no reducing catalyst which operates satisfactorily under the same conditions has yet been found.[13] Lead filters do not appear to offer a satisfactory solution to this problem as they cannot prevent all the lead from reaching the catalysts. (There is some promise for the use of lead filters to reduce atmospheric emissions, but further testing under actual driving conditions is needed to investigate their effectiveness.)

There is some concern that divergences in national fuel policies could affect the freedom of some motorists to drive freely within Europe. This could happen if restrictions placed on lead additives in one country lowered the octane rating of fuel sufficiently to adversely affect the driving performance of some vehicles manufactured in other countries. Large stocks of existing cars with high compression ratios make it extremely unlikely that fuel satisfying these cars will cease to be sold in the 1970's but, if for any reason, octane levels or overall fuel quality were to be reduced sufficiently to affect driving performance, the fear of those who envisage adverse effects on international trade may be well-founded.

APPENDIX: THE HEALTH EFFECTS OF AIRBORNE LEAD

The health effects of the use of lead in automobile fuel have been examined often, but there is no unanimity of opinion concerning the degree of risk due to its presence there. This appendix attempts to summarize the information presently available about the health effects of lead from fuel.

The Presence of Lead in the Atmosphere

Most of the lead in the urban air appears to come from motor vehicle fuel. Some fuel is evaporated before being utilized, which leads to the escape of alkyl lead compounds to the environment. According to tests made with American automobiles about 70–80 per cent of the lead from fuel consumed leaves the exhaust system of the automobile and reaches the environment (7). In a national study of lead emission sources in the United States in 1968, inorganic lead emissions from motor vehicle fuel accounted for about 98 per cent of the atmospheric emissions of lead (25). This study, however, did not include lead sources from the attrition of lead-containing materials. The length of time the lead particle remains airborne depends upon several factors: the physical size and shape of the lead particle, its chemical formulation, and the meteorological conditions present. Studies performed in several countries have found a relationship between atmospheric lead content and traffic density and/or motor fuel sales (4, 15, 16, 34). Occasionally these studies have found a low correlation between traffic density and lead concentrations in the atmosphere: This is probably because lead concentration is dependent upon meteorological and geographical factors in addition to the amount of lead discharged. Hourly atmospheric lead levels have been found to correlate well with diurnal traffic patterns, the highest levels occurring at rush hours (33). Another type of study by Chow and Earl (4) further suggests the importance of lead in fuel on the lead existing in the atmosphere. They measured the isotopic composition of the lead compounds present in the air and in gasoline, and found very similar compositions in the lead present in air and in automobile fuel for several places (Boston, Denver, San Diego, Switzerland). They also note the radically different isotopic composition of lead found in surface soils. They, however, fail to examine the isotopic composition of lead from possible attritive sources. A recent study by Lawther et al (18), however, suggests that there may be important urban sources of atmospheric lead other than motor vehicle fuels. They found higher concentrations of lead relative to other vehicle exhaust pollutants than one would expect on a roof-level sampling station.

Table 29.4 presents the atmospheric concentrations of lead for selected areas. Different sampling techniques make specific comparisons impossible, but it should be noted that the more urban areas report higher lead concentrations.

Estimates of the amount of air breathed by the average adult (children are considered separately below) vary from 10-23 cubic meters per day. The amount of lead which enters the respiratory tract, the amount deposited and retained there, and the quantity finally absorbed, is dependent upon the size, shape and chemical composition of the particles. Several estimates of particle size for exhaust lead have been made, and there is general agreement that most particles are less than 1 micron in diameter (2, 7, 8, 27). Estimates obtained from human experiments of the amount of lead absorbed generally vary from 25 to 50 per cent with the maximum rate of absorption for particles of median size 0.6–1.0 μ (2, 6, 7, 16, 27). Most of these experiments, however, were carried out with experimental aerosols rather than motor vehicle exhausts. Lawther et al (18) have reported some preliminary work on lead levels in the air of London, and have described the electron microscopic appearance of some particles found in streets and in the exhausts of motor cars. They found that small electron-dense particles, apparently lead, and of the order of

Table 29.4 Atmospheric Concentrations of Lead in Selected Areas

	Sampling date	Mean concentration ($\mu g/m^3$ g)	Reference
Vienna	1969-1971	2.8[a]	Austrian reply to questionnaire
Graz, Austria	1969-1971	5.8[a]	Austrian reply to questionnaire
Toronto	1961-1962	1.64	28
Berlin	Spring 1966- Autumn 1967	3.8	German reply to questionnaire
Frankfurt (light traffic)	Jan. 1964- Feb. 1965	2.3[b]	German reply to questionnaire
Cologne	c	4.0	German reply to questionnaire
Tokyo	1969	1.04	Japanese reply to questionnaire
Osaka	1969	1.94	Japanese reply to questionnaire
Kawasaki	Apr. 1968- Mar. 1969	0.89	16
Stockholm	1968-1969	2.0-3.5	30
London (Fleet Street)	Apr.-Aug. 1971	5.5	18
Zurich (in urban streets)	1970	3.8	15
Composite Rural U.S.		0.5	11
Remote California Mountains		0.12	11
Greenwich Village, N.Y.	1970-1971	2.13	32
Los Angeles, Ca.	1968-1969	3.6	31
Philadelphia, Pa.	1968-1969	1.7	31
Cincinnati, Oh.	1968-1969	1.2	31
Los Alamos, N. M.	1968-1969	0.17	31
Houston, Tx.	1970-1971	0.86	32
Pasadena, Ca.	1968-1969	3.39	31
Chicago, Il.	1970-1971	1.63	32

[a]Concentrations are actually presented in terms of normal cubic meters.
[b]The median concentration is presented.
[c]Sampling date unspecified.

0.02 μ in diameter were either attached to, or included in, carbon particles and aggregates thereof; some lead may be in solid solution in the carbon particles. They believe that Brownian motion is therefore more important than sedimentation in determining the deposition of particles in the respiratory tract. In such a case they suggest that an absorption rate of about ten per cent might be more realistic.

Food and drink are important sources of lead. For example, it is estimated that Americans on the average ingest from 120 to 350 μg of lead daily (7), of which some five to ten per cent is absorbed. In Japan, it is estimated that the average normal healthy adult ingests 230 to 320 μg of lead daily through food and the average German is believed to ingest 518 μg/day (16).

The contribution of atmospheric lead from motor vehicle exhaust to lead content in food is thought to be small by comparison with other sources of lead. It appears that plants have very little capacity for taking up lead from soils and that most of the lead taken up by plants is concentrated in the roots (17). Moreover, it appears that plants grown under intensive agriculture take up less lead than poorly nourished plants (17). On the other hand, lead dust settles on plants and on water surfaces. Dairy cattle fed grass growing along a Swiss highway showed higher lead content in their milk (1). Most crops are grown away from major highways, however, and it is believed that pesticides, fertilizers, and the natural lead content of soils are much more important sources of lead to agricultural products.

Acute Human Response to Atmospheric Lead

Human physiological response to atmospheric levels of lead has been demonstrated in several cases. Foremost of these studies have been those of occupational exposure. The blood of occupational groups exposed to automobile exhaust demonstrated higher concentrations of lead than the blood of groups less exposed. Results from occupational studies from Switzerland, Japan, Germany, and the Untied States are presented in table 29.5. The lead concentration in the blood of all the occupationally exposed groups are greater than the average blood lead concentration of those living in the rural United States, and, except for the concentrations in the Los Angeles policemen, are greater than the average level of urban United States residents. It must be emphasized that the differences in the blood-lead levels among the groups are not always statistically significant and that comparability of measurement among

Table 29.5 Occupational Studies of Lead Exposure

Place and Group	Number Sampled	Mean blood level (μg/100 g blood)	Standard deviation	Range	Reference
Rural United States		16			3
Urban United States		21			3
Cincinnati					
Service station attendants 1956	130	28	7	15-55[a]	33
Parking attendants 1956	48	34	6	25-65[a]	33
Garage mechanics 1956	152	38	9	25-65[a]	33
Police traffic officers 1963	40	30	9	15-65[a]	33
All police 1963	123	25	7	15-65[a]	33
City Health Department employees	36	21	5	15-35[a]	33
Los Angeles					
Male aircraft employees 1960	272	19	10	5-75[a]	33
Los Angeles policemen 1960	155	21	6	5-45[a]	33
Switzerland					
Medical students 1956	19	13.5		10-19	15
Garage workers 1956	33	22		10-35	15
Japan					
Tokyo policemen 1970	312	23			b
Germany					
Garbage collectors	77	22	12.1	0-55[c]	19
Road maintenance workers	75	27	12.2	3-55[c]	19

Source: Published in *Environmental Implications of Options in Urban Mobility*, OECD, 1973.
[a]Midpoints of interval data.
[b]Personal communication from Dr. Takeo Suzuki, Director, Department of Environmental Pollution, National Institute of Public Health, Japan.
[c]Approximate, read from graph.

studies can be questioned. Within each study location, those professions with stronger automobile exhaust exposure have higher blood-lead levels. Hence Cincinnati garage mechanics working in enclosed quarters have higher lead-blood contents than service station attendants, and Swiss garge workers have higher levels than medical students, etc.

Other studies have investigated the levels of lead in the blood for groups of persons with different degrees of exposure to automobile exhaust, but not occupationally exposed to lead. Table 29.6 presents a summary of the results from these studies. The results vary. Some groups more highly exposed to lead demonstrate higher levels of lead in the blood than less exposed groups. In other studies there appears to be no difference in blood-lead levels between the higher and lesser ex-

Table 29.6 Results of Studies Relating Atmospheric Lead Exposure to Lead-Blood Levels

Location	Air lead level ($\mu g/m^3$)	Number Sampled	Mean blood-lead level (μg/100 g blood)	Standard deviation	Reference
Rural United States	0.5		16		3
Urban United States	1.0		21		3
Remote California Mountain residents (male)	0.12		12		11
Cincinnati, Oh.					33
Post Office employees' home address					
Industrial area		20	22	5	
Commercial area		14	26	11	
Residential area		68	24	7	
Rural area		17	23	5	
Philadelphia, Pa.					33
Males living and working in center city	2	66	24	8	
Males commuting to center from suburbs		43	19	9	
Males spending most time in suburbs		23	13	5	
Selected U.S. neighborhoods					
Greenwich, Village, N.Y.	2.13	142	16.5		32
Port Washington, N.Y.	1.14	203	15.4		32
Bridgeport, Chicago	1.78	148	17.6		32
Houston, Tx.	0.86	206	12.9		32
Los Alamos, N.M.	0.17	204	15.4		31
Pasadena, Ca.	3.39	209	17.6		31
Ritterhouse Square	1.67	137	20.6		32
Osaka, Japan					
Persons facing heavy traffic	5.08–10.1	26	23.5		16
Persons living away from heavy traffic	0.51–14.3	28	15.7		16

Source: Published in *Environmental Implications of Options in Urban Mobility*, OECD, 1973.

posed groups. The inconsistent results of these studies is disturbing. One explanation for their lack of consistency is that other factors influence the lead-levels of blood. Smokers and males appear to have higher levels of lead in the blood than non-smokers and females. Differences of lead in diet will also influence blood-levels. Of the studies reported, only one, the study of Philadelphia residents reported in table 29.6 (33), appears to adjust for smoking and sex. (It also adjusts for age which turns out to be statistically non-significant.) An analysis of variance of the data from this study indicates that place of residence and work is statistically significant in determining blood-lead levels. (It should be noted that the analysis of variance included a group of 113 policemen who worked in

Table 29.7 Comparison of Mean Concentrations of Lead in Blood, Philadelphia Study (μg/100 g blood)

	Smokers	Non-smokers	Average atmospheric lead exposure
Living and working in center of city			
Police	26	24	2 $\mu g/m^3$
Males	25	22	
Females	18	18	
Commuting to center from suburbs			
Males	21	13	
Females	18	12	
Suburban workers living in suburbs			
Males	15	11	1 $\mu g/m^3$
Females	16	11	

Source: Published in Working Group on Lead Contamination (1965) "Survey of Lead in the Atmosphere of Three Urban Communities," U.S. Dept. of Health, Education, and Welfare, Cincinnati, Ohio.

the downtown area of the city. Their mean blood concentration of lead is not very different from that of the downtown males; hence given the highly statistically significant F-value in the analysis of variance, one can consider the geographical differences in lead-blood levels to be significant.) Table 29.7 gives the mean blood levels of lead for each group. The effects of smoking and atmospheric exposure to lead are clearly visible. There is also a statistically significant interaction between smoking and environmental exposure, with cigarette-smoking contributing less to blood-lead levels when the ambient atmospheric concentrations of lead are high.

Other studies, some parallel to the blood-level studies above, measure lead concentrations in the urine. The results are similar to those of the studies using lead concentrations in the blood. However, the concentration of lead in the urine is not considered to be as reliable as an indicator of lead insult as the concentration of lead in the blood. In one study (9) three groups were examined: a group exposed to lead with no symptoms, an exposed group with mild symptoms of lead poisoning and/or mild anemia (itself a sympton), and an exposed group with severe symptoms. Levels of lead in the urine were a poor discriminator among the three groups. Blood-lead levels appeared to be a better discriminator, although they reflect only current and recent absorption of lead. The best discriminator presently appears to be delta aminolevulinic acid dehydrase activity (29). This indicator will be discussed later.

However, blood-lead is the indicator for which the most information exists.

The significance of the blood-lead levels found in the various studies needs to be considered. A World Health Organisation survey defines the "normal" amount of lead in a person's blood to be between 14–33 μg/100 g blood (6). E.P.A. investigators in the United States believe 40 μg/100 g to be a more realistic upper limit for this mean, pointing out additionally that many, both occupationally and nonoccupationally exposed people who have not been surveyed for excessive lead exposure show blood lead levels above normal. The mean blood-lead levels for all of the non-occupationally exposed groups and for many of the occupationally exposed groups reported above fall within the normal range. Moreover, no occupationally exposed group has a mean blood-lead level which exceeds the normal limit by more than 16 per cent. This result suggests that even if ambient exposures of atmospheric lead were to approach occupational exposures, the mean blood-lead levels of persons exposed would remain in the normal range.

Two important questions remain: what is the individual variation from these means, and what blood levels of lead are associated with adverse health effects? Together the answers to these questions should provide an estimate of the risk of lead poisoning from atmospheric environmental exposure to lead. No overt symptoms of clinical lead poisoning have been reported in adults with blood-lead levels below 80 μg/100 g or in children with blood-lead levels below 50 μg/100 g (7, 8, 25). However, U.S.E.P.A. investigators state that some clinical and biochemical (see below) symptoms have been reported in both children and adults at blood-lead levels in the 40 μg/100 g range. The study of three U.S. urban communities (Philadelphia, Cincinnati, and Los Angeles) (33) measured the blood-lead content of 2,342 adults. Of this number only eleven persons had concentrations of lead above 60 μg/100 g blood, none of whom indicated any symptoms, and all of whom were occupationally exposed to lead. Some thirty additional adults, most of whom were occupationally exposed to lead, had concentrations of lead in the blood between 50 and 60 μg/100 g. Many more persons showed blood-lead levels between 40 and 50 μg/100 g. The upper limits of the ranges of German studies in table A.2 also indicate that some occupationally exposed adults there have blood levels in the 50 μg/100 g range. However, there have been no recorded cases of lead poisoning from ambient atmospheric concentrations. It must be kept in mind, however, that the mildest symptoms of lead poisoning are very general and non-specific (20) and hence may elude detection as lead-poisoning.

The Lead Problem in Children

The situation with respect to children is less clear. There have been numerous cases of lead poisoning among children, but they are apparently the results of ingestion of lead through the mouth, particularly of lead-based paint chips. The fact that practically all of the persons in the three-community study with blood levels greater than 50 μg/100 g blood were occupationally exposed suggests that atmospheric lead can at most only contribute marginally to lead levels in the body producing health effects. One often-quoted study cities a difference in blood-lead levels between normal and mentally retarded children, but before any conclusions can be drawn, it should be noted that mentally retarded children often tend to have a high prevalence of pica (the tendency to ingest dirt, paint, plaster, etc.).

Of concern, however, is the high lead-content of dust in city streets and parks (12, 22, 25). It is a potentially important source of lead to children through ingestion, and a source which is not readily removed.

Chronic Responses to Atmospheric Lead

More difficult to consider is the problem of chronic lead poisoning. Lead is known to accumulate in the body, particularly in bone tissue (3, 16). Moreover, there is some concern that under certain conditions the lead deposited in bone tissue can be released into the blood stream (5, 21). However, the only evidence of ambient levels of lead affecting the human body is the decrease in delta aminolevulinic acid dehydrase activity. Lead appears to inhibit the activity of enzymes which are dependent upon free sulf-hydrol (SH) groups for their activity. There is evidence of this inhibitory effect in the biosynthesis of heme, an ingredient in hemoglobin, as well as an essential constituent of the cytochromes, which are important in energy metabolism. (In fact, cytochrome inhibition may be more important than heme inhibition.) The production of heme in the body is a complicated process, but at least two steps in the process are inhibited by the presence of lead. One of the earlier steps in the biosynthesis of heme is the metabolism of delta-aminolevulinic acid (ALA), which is dependent upon free sulfhydrol groups for its activity. It has been shown that lead, even in the smallest amounts, inhibits the ALA-dehydrase activity in laboratory preparations of red blood cells (13, 14). The fact that this enzyme activity is inhibited at very low blood concentrations of lead suggests that there is no threshold concentration of lead in the blood which does not interfere with this metabolism. The substrate of the delta-aminolevuline dehydrase enzyme, ALA, however, does not appear to increase in the urine until lead concentrations in the blood begin to exceed 45 μg/100 g (14). This apparent discrepancy could be explained by the fact that at lower concentrations of lead in the blood there remains enough residual enzyme to exhaust all of the ALA substrate. Lead then at blood levels less than 45 μg/100 g does not appear to produce consequential metabolic changes in humans. There are no acute responses to these metabolic changes at lower blood-lead levels, and no chronic effects have been demonstrated. There remains some concern about chronic effects, particularly in heme synthesis in other organs. In one experiment (23) with high levels of lead concentration, ALA-dehydrase activity was inhibited at equal rates in the brain tissue and in the blood of laboratory rats. However, at blood levels of 30 μg/100 g blood, there was no significantly diminished ALA-dehydrase activity in the brains of rats.

The contribution of atmospheric lead to blood-lead and ALA-dehydrase activity is not clear. Goldberg (10) has found a closer correlation between individual δ-aminolevulinic acid dehydrase levels and drinking water lead content than with the levels of lead in the ambient air. Some work of Schlipköter (27, 34) suggests that a larger percentage of breathed lead may be absorbed at high concentrations, as some lead tends to impair the ability of cells in the lungs to rid the lungs of lead particules. This work has, however, been carried out with experimentally prepared lead aerosols, and its applicability to lead from motor vehicle exhausts needs to be verified.

Other work has shown that elevated levels of lead exposure have lead to kidney malfunction (3), chronic nephritis (3), chromosome breakage (7, 24, 26), anemia (3), and infertility (24), cancer (23), and teratogenic effects (24). However, all of these effects have been demonstrated only for high levels of exposure (with blood-lead levels in excess of 55 μg/100 g in the case of chromosome breakage and in excess of 80 μg/100 g for the other effects). Further research would be desirable, however, to ensure the absence of any of these effects with blood-lead levels associated with ambient exposure.

Of interest in the consideration of the health risks of leaded fuel are the special risk groups. Children, pregnant women, persons occupationally exposed, and segments of the population exhibiting certain diseases such as anemia and glucose—6—phosphate dehydrogenase deficiency have been cited as high risk groups. The acute effects of atmospheric lead on children have been considered above. Pregnant women and new mothers tend to show reduced ALA-dehydrase activity (12), but this could also be related to mineral deficiencies at the end of pregnancy. Among occupationally exposed groups, present measurements of blood-lead levels suggest the most vulnerable groups are those working in enclosed spaces, and that other occupationally exposed groups exhibit mean blood-lead levels in the normal range. Also, at present lead levels in ambient air, 3–5 per cent of adult urban citizens show blood levels of 40 μg/100 g and above (35), which are compatible with biochemical lead changes. It is doubted that lead-containing paints are the sole lead source for the 20–25 per cent of U.S. urban children who show blood concentrations above this level (35). Lead present in dirt and, in some cases, air, as well as in diet is also a contributing source (35). There is concern about sensitive population segments, but to date no research has demonstrated any increased health hazard from atmospheric lead, but the research in this area is limited.

There remain two important considerations in any examination of the health effects of atmospheric lead. The trends in the increased number of motor vehicles and in increased gasoline octane ratings (6) suggest that if no restrictions are placed upon the lead content in gasoline, we will be subjected to higher levels of lead in the atmosphere than at present. Indeed, in the period from 1961 to 1968–1969, the seven-city study in the United States found that urban atmospheric lead levels generally increased, and by as much as 64 per cent (31). (Other urban areas showed no increase.) Any consequences of even more elevated levels of lead upon the general population are unknown. Although future ambient levels in all probability will remain below current occupationally occurring levels, the absence of effects in self-selected occupational groups does not imply an absence of effects in the general population. Secondly, although no chronic effects have yet been demonstrated in response to levels of lead currently present, the amount of atmospheric lead has increased greatly in the past decade and it may be too early to detect any delayed effects; moreover, the causes of any effects may be confounded in a multiplicy of urban factors.

REFERENCES TO APPENDIX

1. Bovay, E. (1971) "Les dépôts de plomb sur la vegetation le long des autoroutes—Essai d'affouragement de vaches laitières avec du foin souillé par le plomb," in *Das Problem der Benzinverbleiung*, Untersuchungen der Eidgenössischen Kommision für Lufthygiene, Beilage B Nr. 3/1971, 169–185.

2. Bruaux, P. (1972) "Métabolisme du Plomb Chez les Animaux: Possibilités d'utilisation des resultats pour l'Homme," presented at the Séminaire sur le Métabolisme du Plomb dans le Corp Humain, Commission des Communautés Européennes, Luxemburg, 25–27 April, 1972.

3. Chisolm, J. J. (1971) "Lead Poisoning," *Scientific American*, 224, 15–23.

4. Chow, T. J., and Earl, J. L. (1970) "Lead Aerosols in the Atmosphere—Increasing Concentration," *Science, 1969*, 577–580.

5. Craig, P. P., and Berlin, E. (1971) "The Air of Poverty," *Environment, 13(5)*, 2–9.

6. Danielson, L. (1970) "Gasoline Containing Lead," Ecological Research Committee, Bulletin No. 6, Swedish Natural Science Research Council.

7. Engle, R. E., Hammer, D. I., Horton, R. J. M., Lane, N. M., and Plumlee, L. A. (1971) *Environmental Lead and Public Health*, Environmental Protection Agency, Research Triangle Park, North Carolina.

8. Environmental Protection Agency (1972) *Health Hazards of Lead.*

9. Gibson, S. L. M., Mackenzie, J. C., and Goldberg, A. (1968) "The Diagnosis of Industrial Lead Poisoning," *British Journal of Industrial Medecine*, 25, 40–51.

10. Goldberg, A. (1972) personal communication.

11. Goldsmith, J. R., and Hexter, A. C. (1967) "Respiratory Exposure to Lead: Epidemiological and Experimental Dose-Response Relationships," *Science, 158*, 132–134.

12. Haas, T., Wieck, A. G., Schaller, K. H., Mache, K., and Valentin, H. (1972) "Die usuelle Bleibelastung bei Neugeborenen und ihren Muttern," *Zentralblatt für Bakteriologie Hygiene, I. Abt. Orig. B 155*, 341–349.

13. Hernberg, S. and Nikkanen, J. (1970) "Enzyme Inhibition by Lead Under Normal Urban Conditions," *The Lancet, 1*, 63–64.

14. Hernberg, S., Nikkanen, J., Mellin, G., and Lillius, H. (1970) "δ-Aminolevulinic Acid Dehydrase as a Measure of Lead Exposure," *Archives of Environmental Health, 21*, 140–144.

15. Högger, D. (1971) "Le Problème de l'essence au plomb" in *Das Problem der Benzinverbleiung*, Untersuchungen der Eidgenössischen Kommission für Lufthygiene, Beilage B Nr. 3/1971, 88–104.

16. Horiuchi, K. (1970) "Lead in the Environment and its Effect on Man in Japan," *Osaka City Medical Journal, 16*, 1–28.

17. Jones, L. H. P., and Clement, C. R. (1972) "Lead Uptake by Plants and its Significance for Animals," presented at a conference on "Lead in the Environment," 27 January 1972, London.

18. Lawther, P. J., Commins, B. T., Ellison, J. Mck., and Biles, B. (1972) "Airborne Lead and its Uptake by Inhalation," presented at a conference on "Lead in the Environment," 27 January 1972, London.

19. Lehnart, G., Mastall, H., Szadkowski, D., and Schaller, K. H. (1970) "Berufliche Bleibelastung durch Autoabgase in Grosstadtstrassen," *Deutsche Medizinische Wochenschrift, 20*, 1096–1098.

20. May, C. R. (1971) "Lead-Health Effects" presented to the Air Pollution Control Association, Pacific North West International Section, Calgary, Alberta.

21. McCallum, R. I. (1972) "Lead in the Environment: Possible Health Effects on Adults," presented at a conference on "Lead in the Environment," 27 January 1972, London.

22. Miettinen, J. K. (1972) "Uses, Emissions and Analysis of Lead, and its Level in Foodstuffs and Diet," presented at the 16th Session of the Joint FAO/WHO Committee of Experts on Food Additives; Geneva 4–12 April, 1972.

23. Millar, J. A., Battistini, V., Cumming, R. L. C., Carswell, F., and Goldberg, A. (1970) "Lead and δ-aminolevulinic acid dehydrase levels in mentally retarded children and in lead-poisoned suckling rats," *The Lancet, 2*, 695–698.

24. Muro, L. A., and Goyer, R. A. (1969) "Chromosome Damage in Experimental Lead Poisoning," *Archives of Pathology, 87*, 660-663.

25. National Academy of Sciences (1972) *Airborne Lead in Perspective*, Washington, D.C.

26. Schwanitz, G., Lehnert, G., and Gehart, E. (1970) "Chromosomenschäden bei Beruflicher Bleibelastung," *Deutsche Medisinische Wochenschrift, 32*, 1636–1641.

27. Schlipköter, H. W. (1972) "Dépôt et résorption du plomb inhalé dans les poumons" presented at the Séminaire sur le Métabolisme du Plomb dans le Corp Humain, Commission des Communautés Européennes, Luxemburg, 25–27 April, 1972.

28. Smith, Ralph G. (1969) "Air Quality Standards for Lead," Air Quality Monograph No. 69–11, American Petroleum Institute, New York.

29. Stokinger, H. E. (1970) "Experimental Toxicology of Certain Persistent Air Pollutants—Evaluation of Biomedical Indices of Non-disease Effects," Inter-Regional Symposium on Air Quality Criteria and Guides, Geneva, 5-9 October, 1970.

30. Swedish Ministry of Communications Group concerning Development Work in the Field of Motor Vehicle Exhaust Gases (1971) "Air Pollution from Motor Vehicles," Final Report.

31. Tepper, L. B. (1971) "Seven-City Study of Air and Population Lead Levels—An Interim Report."

32. Tepper, L. B. (1971) "Seven-City Study of Air and Population Lead Levels—Interim Report II."

33. Working Group on Lead Contamination (1965) "Survey of Lead in the Atmosphere of Three Urban Communities," U.S. Department of Health, Education, and Welfare, Cincinnati, Ohio.

34. Reply of the German Government to OECD Questionnaire.

35. Reply of the U.S. Government to OECD Questionnaire.

APPENDIX VI A:
National Vehicle Noise Requirements

Table 1 Maximum Permissible Vehicle Noise Emission Limits (Recommendations of the United Nations Economic Commission for Europe, Working Party No. 29)

Category of vehicle	Values (dBA)
Two-wheeled motor vehicles	
With a two-stroke engine of cylinder capacity	
Over 50 cm^3 but not exceeding 125 cm^3	82
Over 125 cm^3	84
With a four-stroke engine of cylinder capacity	
Over 50 cm^3 but not exceeding 125 cm^3	82
Over 125 cm^3 but not exceeding 500 cm^3	84
Over 500 cm^3	86
Three-wheeled motor vehicles (except public works vehicles, etc.)	
With a cylinder capacity exceeding 50 cm^3	85
Motor vehicles with four or more wheels (except public works vehicles, etc.)	
Private motor cars and conversions of such vehicles	84
Goods transport vehicles of permissible maximum weight	
Not exceeding 3.5 t	85
Over 3.5 t but not exceeding 12 t	89
Over 12 t	
With an engine of 200 h.p. DIN or less	89
With an engine of 200 h.p. DIN	92
Motor buses and motor coaches of permissible maximum weight	
Not exceeding 3.5 t	85
Over 3.5 t	
With an engine of 200 h.p. DIN or less	89
With an engine of more than 200 h.p. DIN	92

Source: Published in *Environmental Implications of Options in Urban Mobility*, OECD, 1973.

Table 2 Directive of the Council of the European Communities Relating to Maximum Permissible Noise Levels (New Vehicles)

Class of vehicles	Acceptable Noise Levels (dBA)
Passenger vehicles with seating capacity for not more than 9 persons including the driver	82
Passenger vehicles with seating capacity for more than 9 persons including the driver and a maximum permissible weight not exceeding 3.5 tons	84
Goods vehicles with a maximum permissible weight not exceeding 3.5 tons	84
Passenger vehicles with seating capacity for more than 9 persons including the driver and a maximum permissible weight of more than 3.5 tons	89
Goods vehicles with a maximum permissible weight of more than 3.5 tons	89
Passenger vehicles with seating capacity for more than 9 persons including the driver and powered by an engine of 200 h.p. DIN or over	91
Goods vehicles powered by an engine of 200 h.p. DIN or over and having a maximum permissible weight of over 12 tons	91

Source: Published in *Environmental Implications of Options in Urban Mobility*, OECD, 1973.
Note: Measurements to be made at a distance of 7.5 meters using ISO procedures. Vehicles which exceed the above limits by not more than 1 dBA can be certified in compliance with the directive.

APPENDIX VI B:
The Technology of Vehicle Emission Control

MODIFICATIONS TO THE CONVENTIONAL INTERNAL COMBUSTION ENGINE

The petrol-fueled internal combustion engine is an inherently high-emission propulsion system. The high emissions are caused, to a major degree, by the fuel itself and the difficulty of bringing about complete combustion. In addition to reducing exhaust emissions by adjusting, servicing and replacing engine parts which affect combustion (e.g. carburetor, ignition system), a measure which could be applied to older as well as to new motor vehicles, several other approaches have been considered for attaining low pollutant emissions from motor vehicle engines. They include techniques for bringing about more complete combustion, either by actions affecting the fuel supply to the engine or the combustion process itself, and techniques which act on the exhaust gases once they have left the engine.

Modification of combustion chamber design: A major source of hydrocarbon emissions from automobiles is unreacted fuel-air mixtures expelled through the tailpipe. This occurs primarily because the very thin layer of gaseous mixture which makes contact with the relatively cool combustion chamber surfaces does not burn. More nearly complete combustion of the full cylinder charge is promoted by modifying the combustion chamber design to reduce the surface-to-volume ratio and by minimizing crevices.

Modification of induction system: Carbon monoxide in the exhaust results from insufficient oxygen in the fuel-air mixture and consequent incomplete combustion. Incomplete combustion is also an important source of hydrocarbons. Leaner air-

This appendix is based primarily on the following sources: "Air Pollution from Motor Vehicles," Final Report by the Swedish Ministry of Communications Group (Stencil K 1971: 1), pp. 112–147, 166–176; Annual Report of the Environmental Protection Agency to the Congress of the United States, July 1971, pp. 3-9 to 3-15, 5-1 to 5-11; and "Véhicules et nuisances—pour un programme d'action," report by a Working Group (M. Frybourg, Chairman), September 1971, pp. 34–35, 38–46 of English version.

fuel mixtures to assure more complete combustion can be achieved by converting more of the liquid gasoline into vapor form and by providing for improved fuel-air mixing and distribution among the cylinders. Air-fuel induction systems can be adapted to provide heated intake air for more uniform carburetor inlet temperatures, thus allowing the use of leaner fuel-air mixtures. More uniform distribution of the fuel-air mixture to the cylinders can be accomplished through intake manifold heating or through design changes. Unfortunately, modifications of induction systems which improve combustion and reduce hydrocarbon and carbon monoxide emissions also raise temperatures and contribute to higher nitrogen oxide emissions.

Carburetor modifications: The carburetor is a key element in effective emission control because of its role in metering the fuel in proper proportion to inlet air. Precise fuel metering, in accordance with changing engine requirements, makes possible operation with lean air-fuel mixtures. Carburetors can be designed with stronger fuel metering signals and closer calibration tolerances to assure better fuel mixing preparation.

Choke modifications: Gasoline in liquid form does not burn. Therefore, when an engine is started cold, an extra amount of gasoline is needed in order to obtain enough vaporized hydrocarbons to mix with air and provide a combustible mixture at the sparkplug. The carburetor choke supplies the added fuel. However, the unvaporized hydrocarbons pass through the engine unburned. By tailoring choke action to car requirements, enrichment during starting and warm-up can be made compatible with satisfactory driveability over a wide temperature range. Changing the fuel to achieve greater vaporization could obviate the need for the choke or drastically reduce its use.

Ignition system modification: Properly modified, ignition systems support improvements in fuel metering and mixture control. Spark retardation can be used to reduce emissions of hydrocarbons and nitrogen oxides. Electronic ignition systems have been developed which will improve control of spark timing at all engine operating conditions, facilitate adjustments of spark timing on vehicles in use, and have greater reliability. Retarding ignition timing results in more fuel being burned during the exhaust phase of the combustion cycle. Accordingly, some loss in power and fuel economy results and demands on the engine cooling system are increased.

Lower compression ratio: The use of high compression ratios improves engine efficiency and results in more power output for a given amount of fuel. Combustion temperatures are high, however, causing high emissions of nitrogen oxides. The octane requirements of high compression ratio engines are high, necessitating the use of lead or fuel modification. The presence of lead in gasoline severely limits the effectiveness of catalytic converters and reduces the life of other emission control system components.

Exhaust gas recirculation: Recirculation of a portion of the exhaust gas into the air-fuel mixture causes a reduction in the peak combustion temperature and a reduction in the formation of oxides of nitrogen. Dilution of the fuel charge with inert gases has the secondary effect of reducing engine octane requirements, but with some loss in power. Extreme dilution causes misfiring and deterioration in driveability. This can be compensated for by increased throttle openings and providing richer carburetion mixtures, but with some loss of fuel economy. Improved systems provide for proportioning the reciruIated exhaust gas to the air flow demanded by the engine.

Afterburner: The afterburner, designed to oxidize unburned hydrocarbons and carbon monoxide in the exhaust gas, includes a precombustion chamber in which secondary air and fuel are spark-ignited to provide thermal energy for the associated reaction chamber. The reduction in CO and HC pollutants only from vehicle exhaust when afterburners are installed is estimated to be some 70 and 80 per cent, respectively. Overtemperature is a serious handicap.

Air injection: After the gases have left the combustion chamber air is injected as close as possible to the exhaust valve where any carbon monoxide and hydrocarbons present are hot enough to ignite immediately, and their oxidation is thus completed. Changes of the cylinder head and exhaust manifold are required. Since control of nitrogen oxides during the combustion process has tended to increase HC and CO emissions, interest in exhaust port air injection is reviving. Air injection pumps are also helpful for effective operation of thermal reactors and catalytic converters.

Thermal reactors: A thermal reactor functions as a combustion chamber outside the engine and normally appears in the form of an oversized exhaust manifold. Thermal reactors receive the hot exhaust gas from the engine, retaining as much heat as possible with insulation. Additional heat is generated by oxidation of carbon monoxide in the exhaust gases. High CO concentrations are obtained by operating the engine with rich fuel mixtures. Such reactors are known as "rich thermal reactors." Supplementary air is required, and it is necessary to create appropriate mixing and provide adequate residence time for the combustibles present to react with the oxygen. When designed for rich fuel-air mixtures to promote nitrogen oxides control, there is a substantial fuel penalty. In a "lean thermal reactor" system, the carburetor is set lean so that the exhaust is inherently oxidizing and a secondary air pump is not required. Emissions are generally higher than from "rich" reactors.

Because of the extremely high temperatures that can be reached, materials of satisfactory durability must be used. Special protective systems would be needed to prevent damage to engine through overheating.

Catalytic converters: Catalytic converters are devices designed to convert exhaust gases into harmless chemicals through chemical oxidation or reduction. The catalyst bed generally consists of an active material deposited in a thin layer on an inert support. Platinum or transition metal oxides are the most common active materials. A two-stage catalytic converter can be employed to reduce oxides of nitrogen and to oxidize hydrocarbons and carbon monoxide. A rich fuel-air mixture can be used in the engine to give an exhaust containing large quantities of carbon monoxide which chemically reduce NO_x in the first stage of the converter. Air can then be added to oxidize CO and HC in the second stage. This device may be expensive to build and maintain, but offers high potential for efficient emission control.

The catalytic converter is a relatively low-temperature device and need not be located in the engine compartment. Underbody modifications may be required, however, for best insulation and adequate ground clearance. It has a high heat capacity, requiring a relatively long time to reach activation temperature, but by the same token it remains warm longer.

A catalyst and thermal reactor combination represents an attractive system for controlling vehicle emissions. The thermal reactor increases the temperature of both the exhaust gas and the catalyst, and provides improved control during cold-start and warm-up. An unresolved problem with most catalysts is deterioration due to cyclic exposure to high temperatures.

OTHER APPROACHES TO REDUCING POLLUTANTS FROM INTERNAL COMBUSTION ENGINES

Stratified charge engine. This gasoline-fueled internal combustion engine differs from the conventional engine mainly in combustion chamber design and the use of fuel injection. A near chemically-correct fuel mixture is used at the sparkplug for easy ignition and a very lean mixture is used within that portion of the combustion chamber most remote from the sparkplug. Exhaust emissions could conceivably be low because combustion occurs with excess air. However, there is a problem in maintaining an effective stratification at any load; if the most remote and lean parts of the charge refuse to burn, unburned buel in the exhaust can be excessive, resulting in reduced thermal efficiency.

Diesel engine. The emissions of carbon monoxide and hydrocarbons from diesel engines are much less than those of gasoline engines, although nitrogen oxides emissions are the same. Diesel engines emit a black or blue smoke if they are poorly designed, tuned or maintained, and the exhaust gases can have an unpleasant odor. However, diesel smoke can be reduced substantially by using additives, by proper engine maintenance, avoidance of overloading, engine derating, and adherence to proper fuel specifications.

Attention has also been given to the use of *gaseous fuels*, such as liquified petroleum gas (LPG), liquified natural gas (LNG) or compressed natural gas (CNG). The most important aspect of the natural gas application is the relative reactivity of exhaust hydrocarbons. Because most of the unburned hydrocarbons consist of methane, reactivity is about 20 per cent of that from the same vehicle operating on gasoline. It is possible to fit motor vehicles with dual gasoline and natural gas equipment with little engine modification. The switch from gasoline to natural gas is reported to reduce CO by 90 per cent, HC by 60 per cent, and NO_x by 50 per cent. Natural gas is cheaper, cuts fuel costs, and lowers oil and maintenance costs. Unfortunately, it is relatively scarce, and vehicle range on bottled gas is limited to 60–80 kms.

Table 1 gives an example of the reduction in pollutant emissions that can be obtained by converting a petrol engine to gas operation.

Table 1 Vehicle Emissions from Gaseous Fuel Operation

Fuel	CO	HC (g/km)	NO_x
Petrol	47	1.12	2.3
Propane/propane	16.5	1.05	2.3
Propane/butane	20.5	1.55	1.4
Propane 95	10	1.70	1.9

Source: "Air Pollution from Motor Vehicles," Final Report by the Swedish Ministry of Communications Group (Stencil K 1971:1), p. 138.
Note: Measurements were made using ECE cycle.

UNCONVENTIONAL POWER SOURCES

In addition to research centered around the conventional internal combustion engine, efforts are being made to develop alternative power systems.

Brayton-cycle (gas turbine) engine: In its simplest form, the gas turbine engine consists of a compressor, a combustion chamber, and a turbine. Air is taken in by the compressor at atmospheric pressure, compressed to a higher pressure and then delivered to the combustion chamber where fuel is sprayed in and burned. The combustion process takes place at essentially constant pressure. The basic thermodynamic cycle on which this engine operates is called the Brayton cycle.

In passenger car applications, the turbine might be a little noisier and higher in fuel consumption. Acceleration lag would probably be perceptible, and under some conditions an exhaust odor might be noticed. Advantages include smoother idling, quicker warm-up, freedom from cooling-system performance problems, and potentially lower maintenance. The gas-turbine exhaust can be substantially cleaner than exhaust from present controlled-emission gasoline engines in regard to unburned HC and CO, but there is a need for reducing the nitrogen oxide emissions. No lead compounds are present in the turbine-exhaust because unleaded fuels are used.

Steam engines: While in the conventional petrol or diesel engine and in the gas turbine it is the combustion gases which do the mechanical work directly on the piston and blades respectively, a special medium or fluid in the stream engine (piston steam engine or steam turbine) develops heat in a boiler or steam generator which then gives off energy in the actual engine. Any heat generating process, i.e. combustion of solid, fluid or gaseous fuels, can in principle be used in a steam engine. The emissions therefore depend on the process selected. Problems include the need for a large condenser, numerous auxiliaries such as air pumps and blowers, controls which would adapt the boiler's operation to demand, and an automatic stopping device for the burner in case of shock in order to avoid fire. Lubricating difficulties and complications due to frost might occur. The long delay in the starting time of the engine would be a handicap, as well as the size and weight.

Stirling-cycle engine: In its most common form the Stirling-cycle engine, sometimes called a regenerative hot-air engine, is an external combustion, closed-cycle, piston-type powerplant that uses a gaseous internal working fluid, usually hydrogen or helium. While the working fluid in a steam engine exists in two different forms, liquid phase and steam phase, during the course of the operation, the Stirling engine operates with a working fluid which is gaseous at all times. As in the steam engine, the heat source may in principle be of any type. The heat supply required to drive the engine need not be generated by combustion in a combustion chamber directly connected to the engine. Very low emission levels of carbon monoxide and hydrocarbons have been reported for Stirling engines, i.e. less than one per cent of those comparable uncontrolled-emission gasoline engines. NO_x emissions are also lower.

Electric vehicles: Possible power sources for electrically driven vehicles include storage batteries, fuel cells and thermoelectric generators. For the electric vehicle to achieve performance broadly comparable to that of today's internal-combustion-powered cars, a low-cost, high-temperature alkali-metal battery might be developed. The advantages of this class of battery are high energy density. Of the alkali metals, only lithium and sodium are serious candidates: lithium because it is the lightest, and sodium because it is the cheapest. Battery driven vehicles do not emit combustion gases. However, since chemical reactions take place in charging and discharging, environmental and safety risks cannot be completely ignored.

Hybrid systems: A hybrid system involves a small engine which operates at a constant speed and drives a generator. The generator feeds electric driving motors and, when there is an excess of energy, a buffer battery. This arrangement allows the engine size to be dimensioned to suit the average output requirement, with the battery providing the excess power requirements. Since the engine operates at a constant power output, pollution is minimized. Another type of hybrid system would replace the battery with a mechanical storage device, the spinning flywheel.

APPENDIX VI C:
The Technology of Vehicle Noise Control

Noise from motor vehicles depends upon:

1. Engine speed,
2. Vehicle speed.

Sources of noise related to engine speed (i.e. independent of vehicle speed) are: intake and exhaust, cooling-fan engine structure and that part of the gearbox which is rotating at engine speed.

Sources of noise related to vehicle speed are: the interaction between tyres and road surface, that part of the transmission affected by engagement of the different gears and aerodynamic noise at high speed. Certain operational factors, moreover, have an influence on the noise emitted, such as vehicle load, age and general condition of the vehicle.

Reduction of the noise emitted by one or two components could be insufficient to abate the overall noise significantly because different parts of the vehicle may be equally noisy. Thus, to obtain a substantial reduction of noise one must consider the motor vehicle as a total system.

Certain noise reduction improvements of evolutionary nature can be adopted in the fairly short run. However, large noise reductions can be obtained only by radical changes in the engine and vehicle design and are possible only over a longer period of time.

EVOLUTIONARY IMPROVEMENTS IN NOISE REDUCTION TECHNOLOGY

Engine Modification

Engine noise can be reduced by modifying the combustion system and the cover's design, by adding close-fitting shields made from highly damped material, and, in the case of diesel engines, by turbocharging (see Tables 2 and 3).

Engine Enclosure

As shown in Table 1, tests conducted in the United Kingdom by the Motor Industry Research Association (MIRA) show that a reduction of 6 to 10 dBA is possible by complete enclosure of a diesel engine.

Table 1 MIRA Results (Diesel Engines)

Vehicle condition	Vehicle 1 (dBA)	Vehicle 2 (dBA)
Normal with additional silencer	90.5	88
Plus undertray	89.5	87
Plus engine enclosure	88.5	85.5
Plus gearboox enclosure	–	85
Plus absorbent lining	80.5	82
Radiator intake blanked off	–	81.5

It should be noted, however, that certain problems are still to be solved: fire risk, poor cooling, increase in weight, poor accessibility for engine servicing and high cost of the material (see Tables 2 and 3). In the case of passenger cars, putting absorbent materials on the walls of the engine compartment seems to be enough to reduce noise by 3-4 dBA without affecting engine cooling.

Exhaust Silencer

No general rules exist for reducing exhaust noises. Every vehicle type poses a specific problem. The best solution seems to be to divide the silencer into several chambers since each chamber can be so designed as to shorten a certain frequency range.

In designing a quieter exhaust system, it is, however, essential to avoid lowering performance and to take account of the possible deterioration resulting from age and type of use, as well as of the compatibility with anti-emission devices.

In cases where exhaust noise predominates over other sources of noise (sport cars and motorcycles), noisy vehicles could be corrected fairly rapidly, since "retrofit" replacements of the exhaust system might be possible at acceptable cost.

Tyres

Like exhaust silencers, the tyres on vehicles in use could be replaced fairly rapidly by quieter ones. However, it should be noted that tyre noise is only important at high speeds and that current efforts are concentrated on improving tyre safety.

Tyres may be classified in 3 categories, from quietest to noisiest: rib, cross-bar, and retreaded. Differences exceeding 10 dBA can be found between rib and retreaded tyres, and differences of some 5 dBA between rib and cross-bar tyres. Since the type of road surface (pavement, concrete or asphalt) is also important, the reduction of tyre noise depends not only

Table 2 The Sources of Noise of Motor Vehicles and Possible Action for Control

Source	Action	Comments
Motor cycles		
Air intake	Silencer	Available space small. Adverse effect on performance.
Exhaust	Improved silencer	Idem.
Cover vibrations, valve, gear case, crankcase	Damping on vibration isolation	Probably not beneficial unless intake and exhaust first dealt with.
Engine cylinder block	Damping of cooling fans	Comments on cover vibrations apply.
Motor cars		
Exhaust	Improved silencer	Space not necessarily a problem. Silencer shape can be designed to fit any space available.
Air intake	Improved silencer	Same as exhaust.
Cooling fan	Location of fan with respect to obstructions. Optimisation of design paramaters to limit top speed. Thermally controlled operation.	Styling of the car front can be important. Also design of grille and air paths to radiator.
Engine cover vibrations	Improved design where necessary, e.g.. damping, isolation	
Tyres		Only a problem at high speed.
Engine vibrations		It is unlikely that engine redesign, shielding or enclosure wil be applied to motor cars. The engine is usually well shielded by the engine compartment but some sound-absorbing material within the engine compartment will be advantageous to minimize reverberations.
Commercial vehicles		
Engine	See Table 3	The main source of noise is the engine. See Table 3.
Air intake	As for motor cars	
Exhaust	As for motor cars	
Cooling fan	As for motor cars	As for motor cars.
Tyres	As for motor cars	
Transmission	Shields enclosure improved improved structure design	See Table 3.

Source: Published in *Environmental Implications of Options in Urban Mobility*, OECD, 1973.

Table 3 Diesel Engine: Summary of Noise Sources and Methods of Control Giving Possible Reductions, Probable Costs and Attendant Problems

Method of control	Reduction (dBA)	Cost	Comments
Combustion	2–3	Nil	Possible effect on emissions and economy of operation
Turbocharging	2–3	Cost of turbo-compressor unit	Present difficulties in emis-control. Has the advantage of increasing power. Also, for the same power rated speed can be reduced giving further noise reduction.
Cover design	2–5 possibly on initially poor designs	Could increase cost of covers by 100% or more representing 1–2% of total engine cost	Research needed on suitable cover designs, particularly development of highly damped sandwich materials and vibration isolation techniques.
Shields	2–3	Estimated 2% of total engine cost	Requires considerable research and development, particularly on suitable materials and methods of fixing.
Enclosure	up to 10	Up to 3% of total vehicle cost	Numerous attendant problems—fire, risk, accessibility, weight, difficulty of maintenance, cooling, etc. Has far greater potential in buses than in trucks.
Operating parameters	e.g. 6 possible by change from long to short stroke design for same engine output	Not necessarily affected	A feature of initial design. Weight, size, torque characteristics, etc., have to be considered.
Structure	up to 10 considered possible	Impossible to assess but need not of necessity be greatly increased.	Considerable amount of research and development required. Comments of cover design and shields apply.

Source: Published in *Envirionmental Implications of Options in Urban Mobility*, OECD, 1973.

on the manufacturer but also on road construction and maintenance.

RADICAL ENGINE DESIGN CHANGES

Large noise reductions, of the order of 10 dBA, could be obtained only by radically changing the design of the basic engine structure. These design modifications have to be incorporated when the engine is still on the drawing board. An engine structure designed specifically for low noise characteristics will have the advantage of being designed together with its running parts, thus offering scope for overall improvement. The engine form will be based on an optimum combination of all possible parameters influencing noise: configuration, operating cycle, speed, degree of pressure charging, number of cylinders, bore, bore/stroke ratio, etc.

Design Principles of a Quiet Engine

When intake and exhaust noise is properly suppressed, noise from an internal combustion engine is transmitted by its basic components, i.e. the engine crankcase and cylinder block. These components, which are usually of cast iron or cast aluminum, are of two kinds:

- main load-bearing members,
- walls, which add little to the structural strength but act mainly to retain oil and water. The main load-carrying members, i.e. end bulkheads and cylinder decks, are relatively rigid and their vibration level is low.

In order to reduce weight, the walls are cast as thin as possible. It is they which amplify the vibrations of the load-bearing members. The highest vibration levels are thus found at the centre of the walls and in particular at the skirt of the crankcase wall to which the oil sump is attached.

An engine design to reduce noise could thus incorporate:

- a load-carrying frame of heavy proportions reducing vibrations to a minimum,
- highly damped covers fixed directly to the main members,
- if it is necessary to cast certain walls, a raised flange could be provided so that an additional cover could be fixed,
- large vibrating cast surfaces, such as crankcase walls, should be avoided. They can be replaced by a lower enclosure also forming the oil sump. In this design axial stiffness previously afforded by the crankcase walls can be provided by a beam below the crankshaft, which will have only a load-bearing function.

Experimental Efforts

Experiments carried out in the United Kingdom on diesel engines show that noise reductions of as much as 10 dBA can be obtained by careful design of the crankcase and the structure of the cyclinder block (see Table 3).

At the Institute of Sound and Vibration Research (U.K.), an experimental engine of entirely new design has been built using conventional materials and construction methods. A crankcase design is employed where the crankcase walls are completely eliminated. The crankshaft is supported by a crankframe which is enclosed by a non-rigid, highly damped sheet metal cover which also forms the oil sump. This is bolted to the lower deck of the cylinder block where the vibration levels are low. The basic principles of this design can be adapted to suit most forms of engine.

UNCONVENTIONAL, LOW-NOISE EMISSION POWER SYSTEMS

Gas Turbines

Since the gas turbine is designed to operate at high speed, the idling speed of the engine is high; the noise from a gas turbine is therefore louder when the vehicle is stationary (at traffic lights, for instance) than noise from a convential engine. But noise from a gas turbine engine is less uneven and comprises less pulses than noise from a reciprocating internal combustion engine. Efforts are still underway to reduce inlet and exhaust noise.

Stirling and Rankine Engines

These two engine types produce much less noise than a convential engine (about 20 dBA less for a Sterling engine) and would therefore both offer definite noise emission advantages.

Electric Motors

Noise from an electric vehicle is virtually equivalent to its rolling noise: the use of electric motors in vehicles would thus substantially reduce noise in towns.

APPENDIX VI D:
Motor Vehicle Indicators

Table 1 Total Vehicle Population, Passenger Cars + Commercial Vehicles (in thousands)

	1955	1960	1965	1970	1975	1980	1985
Austria	271	623	1,116	1,605	2,164	2,995	
Belgium	663	953	1,509	2,321	2,800	3,350	
Canada	3,913	5,222	6,624	8,415	10,800	13,900	
Denmark	324	578	981	1,423	1,740	2,117	2,493
Finland	140	254	543	810	1,035	1,317	
France	4,443	7,033	10,815	14,370	17,600	21,300	25,600
Germany	2,375	5,247	10,253	15,107	18,790	20,690	21,700
Greece	44	96	177	343	618	1,238	
Iceland	16	21	35	44	64	94	
Ireland	168	214	329	438	655	900	
Italy	1,196	2,431	6,137	11,127	14,812	18,600	
Japan	900	2,174	6,984	17,826	30,900	47,800	
Luxemburg	28	46	73	104			
Netherlands	384	685	1,516	2,838	4,176	5,350	
Norway	208	340	600	904	1,310	1,655	1,930
Portugal	140	217	331	560	900	1,338	
Spain	242	451	1,197	3,095	6,820	10,948	
Sweden	755	1,324	1,935	2,444	3,265	3,980	4,600
Switzerland	328	574	1,025	1,530	9,925	2,305	
Turkey	71	114	189	268	371	811	
United Kingdom	4,870	7,240	10,950	14,300	18,800	22,900	25,800
United States	62,689	73,868	90,360	108,436	121,200	134,200	147,100

Source: Published in *Environmental Implications of Options in Urban Mobility*, OECD, 1973.

Table 2 Vehicle Population: Passenger Cars (in thousands)

	1955	1960	1965	1970	1975	1980	1985
Austria	143	404	791	1,197	1,700	2,400	3,300
Belgium	501	753	1,281	2,053	2,500	3,000	
Canada	2,961	4,104	5,279	6,700	8,600	11,200	14,300
Denmark	252	485	863	1,277	1,570	1,920	2,270
Finland	85	183	455	700	900	1,150	1,400
France	3,240	5,430	9,010	12,470	15,600	19,200	23,400
Germany	1,748	4,489	9,267	13,941	17,400	19,100	19,900
Greece	17	49	104	226	428	960	2,100
Iceland	10	15	28	37	56	86	133
Ireland	128	170	281	389	600	840	1,200
Italy	861	1,976	5,473	10,200	13,600	17,000	20,400
Japan	153	456	2,182	8,779	19,100	33,000	50,800
Luxemburg	21	37	62	91			
Netherlands	268	522	1,273	2,500	3,719	4,862	5,500
Norway	122	225	465	747	1,130	1,455	1,705
Portugal	99	164	244	452	755	1,143	
Spain	135	291	807	2,378	5,620	9,248	
Sweden	637	1,194	1,793	2,288	3,100	3,800	4,400
Switzerland	280	509	919	1,388	1,750	2,100	2,400
Turkey	30	46	88	129	214	510	900
United Kingdom	3,610	5,660	9,140	12,400	16,600	20,600	23,300
United States	52,682	61,682	75,251	39,309	100,000	111,000	122,000

Source: Published in *Environmental Implications of Options in Urban Mobility*, OECD, 1973.

Table 3 Vehicle Population: Commercial Vehicles (in thousands)

	1955	1960	1965	1970	1975	1980	1985
Austria	128	219	325	408	464	595	
Belgium	162	200	228	268	300		
Canada	952	1,118	1,345	1,715	2,200	2,700	
Denmark	72	93	118	146	170	197	223
Finland	55	71	88	110	135	167	
France	1,203	1,603	1,805	1,900	2,000	2,100	2,200
Germany	657	754	980	1,166	1,390	1,580	1,800
Greece	26	47	73	117	190	278	
Iceland	6	6	7	7	8	8	
Ireland	40	44	48	49	55	60	
Italy	335	455	664	927	1,212	1,600	
Japan	747	1,718	4,802	9,047	11,800	14,800	
Luxemburg	7	9	11	13			
Netherlands	116	163	243	338			
Norway	86	115	135	157	180	200	225
Portugal	41	53	87	108	145	195	
Spain	107	160	390	717	1,200	1,700	
Sweden	118	130	142	156	165	180	200
Switzerland	48	65	106	142	175	205	
Turkey	41	68	101	139	157	301	
United Kingdom	1,260	1,580	1,810	1,900	2,100	2,400	2,500
United States	10,544	12,186	15,109	19,127	21,200	23,200	25,100

Source: Published in *Environmental Implications of Options in Urban Mobility*, OECD, 1973.

Table 4 Level of Car Ownership (passenger cars/1000 population)

	1955	1960	1965	1970	1975	1980	1985
Austria	10	55	107	162	227	315	
Belgium	56	82	135	213	255	303	
Canada	189	222	270	281	379	457	
Denmark	57	106	181	260	313	377	439
Finland	20	42	99	146	180	221	
France	74	118	185	246	298	352	415
Germany	32	81	157	230	276	297	
Greece	2	6	12	26	46	101	
Iceland	63	75	140	185	280	344	
Ireland	46	61	97	130	182	227	
Italy	18	40	103	188	243	297	348
Japan	2	5	22	85	176	284	421
Luxemburg	71	118	185	268			
Netherlands	25	45	103	192	269	333	364
Norway	36	62	126	191	283	346	388
Portugal	12	18	26	47	75	110	
Spain	5	10	25	71	163	258	
Sweden	87	159	233	286	380	450	500
Switzerland	56	94	155	221	267	310	
Turkey	1	2	3	4	6	12	17
United Kingdom	71	108	168	222	290	352	387
United States	315	341	387	433	460	478	490

Source: Published in *Environmental Implications of Options in Urban Mobility*, OECD, 1973.

Table 5 Households Owning Cars (percentage)

	Owning at least one car				Owning more than one car			
	1955	1960	1965	1970	1955	1960	1965	1970
Belgium		30	42					
Canada	56	67	75	78	5	8	12	17
Denmark			50					
France	25	30	48	53		2	3	8
Germany		26	36	62				5
Italy		20	32	48			3	6
Japan			6	22				
Netherlands		26	38					
Sweden				75				
Switzerland			46					
United Kingdom	19	30	41	51		2	5	7
United States	74	74	77	80	10	18	24	29

Source: Published in *Environmental Implications of Options in Urban Mobility*, OECD, 1973.

Table 6 Road Network (Uncontrolled Access) (in thousands of kms)

	1955	1960	1965	1970
Austria		88	92	94
Belgium	92	92	92	92
Canada	729	735	788	840
Denmark	58	59	61	62
Finland		68	69	72
France		717	783	783
Germany		357	392	410
Greece	30	32	34	35
Iceland			11	11
Ireland	81	83	85	86
Italy	175	191	198	282
Japan	941	962	985	1,006
Luxemburg		5	5	5
Netherlands			46	48
Norway	47	52	66	72
Portugal	27	28	29	30
Spain	121	131	133	138
Sweden		159	167	173
Switzerland	50	56	58	60
Turkey			43	50
United Kingdom	301	310	331	339
United States		5,748	5,937	5,963

Source: Published in *Environmental Implications of Options in Urban Mobility*, OECD, 1973.

Table 7 Road Network (Controlled Access) (in thousands of kms)

	1955	1960	1965	1970	1980
Austria	0	0.20	0.33	0.44	0.50
Belgium	0	0.18	0.31	0.43	1.70
Canada			2.04	2.65	
Denmark	0	0.05	0.10	0.16	0.60
Finland	0	0.01	0.04	0.11	0.50
France	0.13	0.17	0.64	1.54	
Germany		2.70	3.37	4.11	8.00
Greece	0	0	0.01		
Iceland	0	0	0	0	
Ireland	0	0	0	0	
Italy	0.51	1.17	1.74	3.91	5.15
Japan	0	0	0.18	0.64	
Luxemburg	0	0	0	0.01	
Netherlands	0.21	0.35	0.58	0.98	2.20
Norway	0	0	0.02	0.08	0.30
Portugal	0.01	0.06	0.06	0.07	
Spain	0	0	0	0.27	3.00
Sweden	0	0.07	0.22	0.38	
Switzerland	0	0.02	0.11	0.30	0.70
Turkey	0	0	0	0	
United Kingdom		0.24	0.57	1.00	3.20
United States			31.19	51.04	

Source: Published in *Environmental Implications of Options in Urban Mobility*, OECD, 1973.

Table 8 Total Annual Vehicle Mileage, Passenger Cars + Commercial Vehicles (in million kms)

	1955	1960	1965	1970
Austria	5,237	10,866	10,866	27,714
Belgium		16,000	21,000	17,000
Canada	55,840	68,912	93,278	111,987
Denmark	5,020	9,050	16,810	22,500
Finland		4,548	7,340	9,370
France	44,300	64,400	114,450	169,500
Germany		99,840	190,600	235,900
Italy	36,940	51,528	88,184	127,081
Japan	12,062	28,163	82,155	225,000
Luxemburg				
Netherlands	7,500	12,700	26,500	49,500
Norway	2,556	4,068	7,136	10,836
Portugal	2,810	3,940	6,320	10,150
Spain		8,224	16,662	26,430
Sweden		24,400	32,600	
Switzerland		9,600	17,100	
Turkey		2,064	3,202	4,640
United Kingdom	65,000	98,000	152,000	198,000
United States	974,600	1,156,900	1,420,400	1,708,600

Source: Published in *Environmental Implications of Options in Urban Mobility*, OECD, 1973.

Table 9 Average Annual Mileage Per Vehicle, Passenger Cars (in kms)

	1955	1960	1965	1970
Austria	18,000	18,000	18,000	16,000
Belgium		12,500	11,500	10,200
Canada		14,500	14,500	14,500
Denmark	15,000	14,500	16,100	15,900
Finland			15,800	
France	8,900	8,400	9,700	10,600
Germany		17,800	17,000	16,500
Italy	29,000	19,200	13,000	13,000
Japan	16,400	20,800	19,000	15,300
Netherlands	20,000	18,700	18,000	17,500
Norway	12,000	12,000	12,000	12,000
Portugal	18,400	16,900	17,700	17,100
Spain		15,900	11,700	
Sweden			14,000	15,000
Switzerland		15,000	15,000	15,000
United Kingdom	12,000	13,000	13,000	13,000
United States	15,100	15,200	15,100	15,700

Source: Published in *Environmental Implications of Options in Urban Mobility*, OECD, 1973.

Table 10 Average Annual Mileage Per Vehicle, Commercial Vehicles (in kms)

	1955	1960	1965	1970
Austria	33,000	33,000	33,000	33,000
Belgium		24,700	23,900	20,000
Canada		40,000	40,000	40,000
Denmark	18,500	18,500	22,000	22,000
Finland			31,300	
France	12,900	11,600	14,900	19,500
Germany		25,300	25,000	25,000
Italy	35,700	29,900	26,000	26,000
Japan	10,800	13,700	15,900	18,600
Netherlands	21,500	21,500	21,100	22,000
Norway	14,000	14,000	14,000	14,000
Portugal	24,000	22,000	23,000	22,300
Spain		22,400	18,400	
Sweden		23,000	27,000	
Switzerland		23,000	25,000	27,000
United Kingdom	19,000	20,000	22,000	22,000
United States	17,200	17,000	18,600	18,600

Source: Published in *Environmental Implications of Options in Urban Mobility*, OECD, 1973.

Table 11 Scrappage, Passenger Cars (in thousands)

	1955	1960	1965	1970	1975
Denmark	2	9	14	56	
France		154	427	739	1,189
Germany	105	141	520	771	
Netherlands			50	155	283
Spain			5	21	
Sweden				120	207
United Kingdom	1,300	1,900	1,600	1,800	
United States	3,773	4,200	6,163	6,327	

Source: Published in *Environmental Implications of Options in Urban Mobility*, OECD, 1973.

NOTES TO PART VI

Notes to chapter 26

1. "Urban Air Pollution, With Particular Reference to Motor Vehicles," report of a World Health Organisation Expert Committee, W.H.O. Report No. 410, 1969.

2. "Air Pollution and Countermeasures in Japan," Air Quality Bureau, Environmental Protection Agency, Tokyo, February 1972.

3. "Noise," Final Report of the Wilson Committee on the Problem of Noise, H. M. Stationary Office, 1963; "Urban Traffic Noise: Strategy for an Improved Environment," OECD, 1971; "Report to the President and Congress on Noise," Environmental Protection Agency, December 31, 1971.

4. "La gêne due au bruit de la circulation automobile," C. Lamure and M. Bacelon, Cahier du CSTB No. 88/762, 1967; "Traffic Noise in Residential Areas," National Swedish Institute for Building Research and National Swedish Institute for Public Health, Report No. 36, 1968; "Report to the President and Congress on Noise," Environmental Protection Agency, December 31, 1971.

5. "Etude de la gêne due au trafic automobile urbain," Centre Scientifique et Technique du Bâtiment, 1971.

6. "Passenger Car Noise Survey," California Highway Patrol, January 1970.

7. "Bruit automobile en circulation urbaine" (Motor Vehicle Noise in Urban Traffic), Institut de Recerche des Transports, Paris, 1971.

8. "A Review of Road Traffic Noise," Road Research Laboratory Report LR 347, United Kingdom, 1970.

Notes to chapter 28

1. Urban Planning and Noise from Road Traffic: Draft of general guide to planners, National Swedish Board of Urban Planning, p. 28. In London between 1966–1969 commercial vehicles increased from 260,000 to 266,000 and the biggest proportion increase was in the group weighing over 8 tons. Source: Greater London Development Plan Inquiry Proof E 12/1, paragraph 9.1.9, November 1970.

2. "Madison Mall," Office of Midtown Planning and Development, Office of the Mayor, City of New York, October 1971, p. 12.

3. "Policy Towards the Creation of Vehicle-Free Areas in Cities," OECD Sector Group on the Urban Environment, April 1971.

4. "The Göteborg Traffic Restraint Scheme," Curt M. Elmberg, Department of Planning, Göteborg Transport Autority, May 1971.

5. Memorandum on road pricing by the Department of the Environment to the Greater London Development Plan Inquiry, March 1971.

6. "Study of Air Pollution Aspects of Various Roadway Configurations," Report to the New York City Department of Air Resources, September 1, 1971.

Notes to chapter 29

1. A paper presented in November, 1971 at the U.K. Institution of Mechanical Engineers Symposium: "Air Pollution Control in Transport Engines" claimed that deposits increase with *unleaded* gasoline. A study in the United States, on the other hand, showed an increase in hydrocarbon emissions of 20 per cent in rapid mileage accumulations and 7 per cent in typical customer driving due to the use of *leaded* gasoline. However, other laboratory and field tests undertaken in the U.S. have suggested that the replacement of leaded by unleaded gasoline in 1971 model cars may result in exhaust emissions which increased soiling and decreased light transmission significantly (these effects were strongest when a comparison was made between a leaded, low aromatic and an unleaded, high aromatic gasoline).

2. The power-to-weight ratio of a typical U.S. car is one and a half times greater than that of a typical European car.

3. Fuel consumption would be increased for two reasons: Larger engines require more fuel. Low-compression engines operate at lower temperatures and are therefore less efficient. The cumulative penalty in fuel economy could be as large as 12 per cent.

4. "Automotive Fuels and Air Pollution," Report of the Panel on Automotive Fuels and Air Pollution, U.S. Department of Commerce, 1971.

5. Bonner and Moore, *op. cit.*, pp. 3–12 and 3–13.

6. Bonner and Moore, *op. cit.*, p. 4–3.

7. Bonner and Moore, *op. cit.*, p. 4–3.

8. Bonner and Moore, *op. cit.*, addendum 1, p. 3–8.

9. Press report issued by Bonner and Moore, Oct. 4, 1971.

10. U.S. reply to OECD questionnaire.

11. The variation among estimates of aromatic increase might be attributed to different assumptions of current aromatic content.

12. The cost of such modification has been estimated at $2 per car. Semi-annual Report by the Committee on Motor Vehicle Emissions of the National Academy of Sciences, January 1, 1972, Appendix H.

13. "Inter-Industry Emission Control, A Cooperative Research Program for Automotive Emission Control," SP-361, Society of Automotive Engineers, Inc.